STOCHASTIC PROCESSES IN MAGNETIC RESONANCE

STOCHASTIC PROCESSES IN MAGNETIC RESONANCE

Dan Gamliel & Haim Levanon

The Hebrew University of Jerusalem

World Scientific

Singapore • New Jersey • London • Hong Kong

Published by

World Scientific Publishing Co. Pte. Ltd.

P O Box 128, Farrer Road, Singapore 9128

USA office: Suite 1B, 1060 Main Street, River Edge, NJ 07661

UK office: 57 Shelton Street, Covent Garden, London WC2H 9HE

Library of Congress Cataloging-in-Publication Data

Gamliel, Dan.

Stochastic processes in magnetic resonance / by Dan Gamliel and Haim Levanon.

p. cm.

Includes bibliographical references and index.

ISBN 9810222270

1. Magnetic resonance. 2. Stochastic processes. 3. Simulated annealing (Mathematics) I. Levanon, Haim. II. Title.

QC762.G36 1995

538'.36--dc20 95-16547

CIP

Printed in Singapore.

INTRODUCTION

This book is about the study of stochastic, i.e., random processes in magnetic resonance. In particular, a certain method for calculating the effect of such processes on magnetic resonance line shapes is treated in detail. This method is known as the stochastic Liouville method. Relevant stochastic processes are various diffusion or exchange processes. They are responsible for broadening of spectral lines in some cases, narrowing such lines in other cases, shifting lines, merging and separating lines. It is therefore of great importance to be able to calculate the outcome of an experiment when such random processes are operating. This usually requires some kind of semi-classical equation for the density matrix, in the context of quantum mechanics. The reason for this will now be explained.

A rigorous account of many physical problems, magnetic resonance included, must make use of quantum mechanical concepts. In standard quantum theory one deals with wave functions of given physical systems. However, very often practical situations arise where such a standard treatment is inadequate. For complex systems (e.g. various solids) or macroscopic incoherent assemblies of many small quantum systems (e.g. paramagnetic species in a liquid solution), one needs to describe the experimentally accessible quantum systems by a density matrix rather than by a wave function. This is true even when the relevant system has no interactions with its environment. Moreover, in many physical situations the interesting system has non-negligible interactions with its surroundings. These interactions are often complicated and changing with time. In particular, for many quantum mechanical systems the influence of the environment takes the form of random processes acting on the system. To describe the system properly it is then not sufficient to use the density matrix formalism. This formalism has to be augmented with a semi-classical treatment of the effect of the random processes on the system of interest. In magnetic resonance this is often the case.

There are two main approaches to the calculation of the effect of random processes on magnetic resonance experiments. Both of them lead to a semi-classical equation of motion for the quantum mechanical density matrix. Each of these equations is a mathematical tool for simulating of the behavior of some quantum systems, in which one can describe some features "exactly" - with a quantum mechanical density matrix, and some features approximately with classical stochastic (random) processes. Such a formalism is useful in magnetic resonance, because a quantum description is necessary (and can be given) to the spin system in order to analyze its observed transitions, but there may be many processes affecting the spins indirectly. It is very difficult to describe these processes fully, and it is not needed in practice - one only needs a method to account for the effect of these processes on the spins.

One approach is the Bloch-Redfield treatment, based mainly on perturbation theory, valid for relatively fast random motions with a relatively small effect on the interactions. Such a theory is often useful in NMR (nuclear magnetic resonance), where time scales are relatively long, so the diffusion processes are usually fast on the NMR time scale. The quantum mechanical description of chemical exchange processes, which is not limited to the fast motion regime, is closely related to this approach.

The other approach was initially developed mainly by Kubo, and then further developed and applied to magnetic resonance mainly by Freed. In this method, the emphasis is on a relatively detailed treatment of the stochastic process, without the restriction to fast motions. The resulting equation, known as the stochastic Liouville equation, is applicable to a broad range of problems in magnetic resonance. The "Stochastic Liouville Equation" (SLE) is a stochastic version of the Liouville - von Neumann equation, which is an exact quantum mechanical equation of motion for the density matrix.

This equation is valid even for relatively slow random processes, and is therefore especially suitable for EPR (electron paramagnetic resonance), where the natural time scale is short so that the random processes are not usually fast on this time scale. Many papers on the subject, mainly by Freed's group, have shown that this approach is indeed useful for a very wide variety of experimental situations in magnetic resonance.

The purpose of the book is to present a unified treatment of the different relaxation theories of magnetic resonance, with an emphasis on the stochastic Liouville equation, on which it is difficult to find introductory material at the textbook level. The unified presentation makes it possible to put that equation and the resulting relaxation theory in a proper perspective. As a background to this theoretical discussion, a brief review is given of the theory of the quantum mechanical density matrix. Some introductory mathematical material is reviewed in the Appendices. The general theoretical part of the book is followed by a fairly detailed treatment of some applications of the theory. The final chapter gives a brief introduction to the relevant experimental methods for which the theory has been applied.

An overview of the contents of the book is now given. Chapter I begins with a very brief review of the subject of Liouville's equation in classical mechanics, which is in a sense a precursor of the Liouville-von Neumann equation. The basic formalism of quantum mechanics is then introduced, leading to the main subject of the chapter - the quantum mechanical density matrix and its equation of motion. Examples are given, illustrating the actual application of the formalism in magnetic resonance.

Chapter II starts with the phenomenological Bloch equations, which give a simple classical description of magnetic resonance. This is followed by a quantum mechanical treatment of elementary magnetic resonance, and by some definitions needed for the treatment of stochastic processes. These subjects serve as an introduction to the main part of the chapter, which is the well known relaxation theory in magnetic resonance, due to Bloch and Redfield. This theory, as well as its introductory subjects, have received excellent coverage in several books. They are included here as a background for the following chapters, and in order to make it possible to compare the different relaxation theories described in the present book. Examples are given of stochastic processes which are relevant in this context.

Chapter III deals with the subject of chemical exchange processes, both intra-molecular and inter-molecular, with an emphasis on the former. The main features here are the comparison of the theory with the standard Bloch-Redfield relaxation equations, and the use of various types of symmetry properties to simplify the treatment, both on the theoretical level of deriving selection rules and on the practical level of numerical computations.

Chapter IV develops in a fairly general context the stochastic Liouville equation. This is done first by constructing a relaxation theory in which a relaxation function or operator play the central role, and then by an alternative approach, in which a statistical distribution function plays a central role. Both the classical and quantum mechanical versions of the equation are derived, and their applicability to magnetic resonance is demonstrated. At the same time the relationship between these equations and the formalism developed in the previous two chapters is discussed.

In chapter V several methods for solving the SLE are described. Following the general method of cumulants some numerical methods are presented. The emphasis in this Chapter is on the method of eigenfunctions, which is used in the subsequent formal development in the rest of the book. An application of this method to a problem in nuclear magnetic resonance is included here. The equations result in each case in a set of coupled algebraic equations, which is often of very large dimensions. Therefore a special section is devoted to some numerical techniques which can handle such problems, cutting down the size of numerical computations to a minimum.

Applications of the general theory (with the method of eigenfunctions) to several typical cases in CW EPR are described in chapter VI. The main purpose of this treatment is to develop the relevant equations for these cases, mostly for non-saturated lines. Some typical examples are given for EPR doublets and for photoexcited EPR triplets. A discussion of the special expression for the line shape of transient species is included, which is especially important for photoexcited triplets. The theoretical treatment is followed by some examples of relevant experimental results. Finally, applications to NMR are briefly discussed.

Chapter VII indicates how the theory is extended to more general types of experiments. This includes the study of saturated lines and double resonance methods such as ELDOR and ENDOR, as well as multiple pulse methods. Here, too, the theory is supplemented by some experimental examples.

Chapter VIII is a short introduction to experimental methods in magnetic resonance, for which the theory in this book is relevant. The main emphasis is on methods in EPR, particularly some modern methods which are useful in EPR, where the study of stochastic processes may require a formalism such as that treated in the present book.

The Appendices present some mathematical details which are needed as a background to some of the material in this book. The first four appendices deal with various aspects of angular momentum theory in quantum mechanics. This theory is of fundamental importance in the study of magnetic resonance, including the particular subjects treated here. The fifth one includes some elementary definitions and results in group theory, relevant to the discussion of symmetries in Chapter III.

We would like to emphasize that the book is not intended to give a comprehensive review of the vast existing literature on the subjects treated in this book. Certain aspects of the subject matter have been selected so as to construct a coherent framework, to develop and clarify the central points. Consequently, we have not attempted to refer to all important work on the SLE. Rather, in each chapter only a few references are given, either because they are directly related to the material included in that chapter, or because they can serve as typical examples of concepts or applications discussed in that chapter. However, some of the references are review articles which do have a comprehensive bibliography, so the interested reader may use these in order to find out about additional work that has been done on the study of stochastic processes by means of the SLE.

Finally, it is a pleasure to acknowledge the help of some members of our group. Mr. Kobi Hasharoni, Ms. Tamar Galili, Ms. A. Berman and Dr. Ayelet Regev read part of the manuscript and made some useful comments. A visitor in our group, Dr. Beth

Brauer from Columbus State Community College is thanked for reading part of the manuscript and making detailed comments on both style and contents. We thank Ms. Miriam Achituv for preparation of the Figures. Last but not least, we thank Professor Jack Freed for his encouragement to have this book written.

CONTENTS

CHAPTER I

DENSITY MATRICES IN QUANTUM MECHANICS

I.1 Liouville's Original Equation in Classical Mechanics

The so called Liouville's equation in quantum mechanics, with which we shall be concerned throughout most of this book, does not have a true classical analog. Nevertheless it is similar to an equation developed by Liouville in the context of classical statistical mechanics, and this is the reason that Liouville's name is connected with the quantum mechanical equation. Before dealing with the quantum mechanical Liouville's equation, it is interesting to look first at Liouville's original equation. To do this it is necessary to recall some basic concepts and laws in classical mechanics. At the most elementary level, the motion of material bodies is described by the solution of some equations of motion, usually given as differential equations. For a single body (a point mass, or the center of mass of a rigid body) there is just one equation, Newton's second law, usually written in the form:

$$\boldsymbol{F} = m\boldsymbol{a} = m\frac{d^2\boldsymbol{x}}{dt^2} \tag{I-1}$$

In this equation **F** is the force acting on the body in question, **a** the resulting acceleration and m the mass. Since the acceleration is the second time derivative of the position vector **x**, two integrations of this equation with appropriate initial conditions (and constraints on the motion, if there are any) will yield the time-dependence of the position vector. An equivalent form of this equation in Newtonian mechanics relates the force to the change in linear momentum:

$$\boldsymbol{F} = \frac{d\boldsymbol{p}}{dt} = \frac{d}{dt}m\boldsymbol{v} \tag{I-2}$$

In principle it is the second form of Newton's law (which was actually his original formulation) which is always valid in classical mechanics. In relativistic cases, i.e. at speeds close to the speed of light, these two forms of the law are not equivalent, and then Eq. (I-1) is not valid any more. For our purposes, however, this distinction can be safely ignored.

In many cases the force can be derived from a function of position, called potential energy V(**x**), such that: $\mathbf{F} = -\nabla V(\mathbf{x})$. When such a relation holds, the force is called conservative (or non-dissipative), because then the total energy E = K+V is conserved, where $K=\frac{1}{2}mv^2$ is the kinetic energy (**v** = d**x**/dt is the velocity) and V is the potential energy. This description does not include dissipative phenomena such as friction, where energy is lost from the macroscopic system of interest to microscopic components of a surface, which cannot be probed in an ordinary classical experiment. Such situations, in which energy (of the system under consideration) is not conserved, result from approximate macroscopic descriptions for systems which are too complicated for an exact description. This is where some randomness gets into the problem, even without quantum mechanics.

An alternative formulation of classical mechanics, due to Lagrange, allows one to use a variational principle to derive the equations of motion. The ordinary coordinates $(x_1,x_2,...)$ are replaced by generalized coordinates $(q_1,q_2,...)$ which contain implicitly the constraints on the motion. Using the *scalars* K, V which are functions of $\{q_i, dq_i/dt\}$ one obtains equations of motion for these coordinates. Here the kinetic energy expressed as: $K = \frac{1}{2}m(d\mathbf{q}/dt)^2$.

A third formulation is due to Hamilton. Here: $(\mathbf{q}, d\mathbf{q}/dt)$ are replaced by $(\mathbf{q},\mathbf{p})$ where **p** is momentum, resulting in first order (rather than second order) differential equations of motion, and **q**,**p** appear symmetrically in the equations of motion. This formulation is very convenient from a theoretical point of view, because of its analogy to basic relations in quantum mechanics. Here (still in classical physics) the kinetic energy is written as: $K = \mathbf{p}^2/2m$ $(\mathbf{p}=m\mathbf{v})$. The total energy, which is a number, is equal to Hamilton's function (the Hamiltonian) $H = K + V$. In addition to providing information about positions and momenta, knowledge of these variables also provides information on any other dynamic variables. Thus the state of a single particle at a given moment can be specified by the values of the two three-dimensional vectors **q**, **p** at that moment. Therefore a set of six numbers is mathematically equivalent to a state of the system. For a system with n particles, a microscopic state is equivalent to a point in "phase space". This space is the set of all vectors defined in 6n-dimensional space, in which the abstract coordinates are, using three dimensional physical coordinates and momenta: $\{q_1,q_2,...,q_n;p_1,p_2,...,p_n\}$. If one wishes to use scalar physical coordinates q_i,p_i, the index i runs from 1 to 3n.

For any function U(q,p,t) one can show that:

$$\frac{dU}{dt} = [U,H] + \frac{\partial U}{\partial t} \tag{I-3}$$

where the square brackets are **Poisson's brackets**, defined by

$$[A,B] \equiv \sum_i \left[\frac{\partial A}{\partial q_i}\frac{\partial B}{\partial p_i} - \frac{\partial A}{\partial p_i}\frac{\partial B}{\partial q_i} \right] \tag{I-4}$$

Using these variables it is possible to describe the system by dealing with its representation in an abstract mathematical space.

With this formalism of exact dynamical calculations it is now possible to consider statistical mechanics. In classical statistical mechanics one is concerned with macroscopic states of the system, specified by macroscopic information. Each such state is compatible with many different microscopic states of the system, differing in the positions and momenta of individual particles (e.g. molecules) of which the system is made, but not in the overall description of the system (e.g. center of mass position and total linear momentum). Thus instead of considering an initial state (initial condition) of such a macroscopic system at a given time t_0, one has to consider many possible initial (microscopic) states, with some probability density specifying the probability of being in any of these states (or, in a continuum, in an infinitesimal neighborhood of a state). The possible states are described by points in phase space: $\{q_1^{(i)},q_2^{(i)},\ldots,q_n^{(i)},p_1^{(i)},p_2^{(i)},\ldots,p_n^{(i)};t_0\}$ where the index i is usually continuous. Starting from any of these states the system evolves in time according to Hamilton's equations. However, since one does not know the initial microscopic state of the system, one has to know also the probability density $\rho(q_1,q_2,\ldots,p_1,p_2,\ldots)$ for finding the system in a particular microscopic state. This probability density may be regarded as the density of occupied points in phase space in the infinitesimal neighborhood of $(q_1,q_2,\ldots,p_1,p_2,\ldots)$. In fact, if the system is at thermodynamic equilibrium, the density function is given according to Gibbs by the familiar Boltzmann distribution function:

$$\rho = \exp\left\{-\frac{E(q_1,\ldots,q_n,p_1,\ldots,p_n)}{k_B T}\right\} \tag{I-5}$$

The time dependence of the density function can always be calculated from:

$$\frac{d\rho}{dt} = [\rho,H] + \frac{\partial\rho}{\partial t} \tag{I-6}$$

On the other hand, in statistical mechanics it is shown that:

$$\frac{d\rho}{dt} = 0 \tag{I-7}$$

namely, the density of (occupied) points in phase space is constant in time. An alternative form of this result is **Liouville's theorem**:

$$\frac{\partial\rho}{\partial t} = -[\rho,H] = [H,\rho] \tag{I-8}$$

in which the square brackets are still Poisson's brackets.

The density function in phase space may be utilized in principle in two different ways, depending on whether one is interested in the situation at a certain time or in the dynamics of a system. On the one hand, if one makes a macroscopic measurement on the system, observing "the value" of the variable A, what is actually measured is the average or expectation value:

$$\langle A\rangle = \frac{\int A(q_1,\ldots,q_n,p_1,\ldots,p_n)\,\rho(q_1,\ldots,q_n,p_1,\ldots,p_n)\,d^{3n}q\,d^{3n}p}{\int \rho(q_1,\ldots,q_n,p_1,\ldots,p_n)\,d^{3n}q\,d^{3n}p} \tag{I-9}$$

where $d^{3n}q$ and $d^{3n}p$ are the products of integration elements for the three-dimensional space of all q_i and p_i. On the other hand, if the system is in a time-dependent state, one may use this expression to follow the change with time of $\langle A\rangle$ as the system evolves, e.g., as it relaxes to equilibrium. The time dependence of this expression is due to that of ρ, which can be calculated from Liouville's theorem.

I.2 Wave Functions and Expectation Values in Quantum Mechanics

As an introduction to the quantum mechanical density matrix, a brief review will be given of some basic features of quantum mechanics. Whereas in classical physics one aims at a complete description of position and momentum, and thus of all physical variables, in quantum physics Heisenberg's uncertainty principle makes this goal unattainable. Moreover, in quantum mechanics one does not calculate directly the behavior of the actual physical variables of interest. Rather, anything that can be known about a physical system can only be found by applying some well-defined rules to some auxiliary quantities. In the "wave mechanics" approach, the auxiliary quantity is the **wave function** of the system. Physical properties of the system such as position, momentum, angular momentum (orbital and spin) etc. are represented by mathematical operators named **observables.** All the relevant quantum mechanical information is contained in the results of mathematical operations with these and other operators on appropriate wave functions.

As a result of this different situation, it is impossible to write down an equation of motion for a microscopic physical observable property such as position. Such an equation can only be obtained in quantum mechanics through an averaging process, bringing about a loss of information. What is written instead is an equation of motion for the wave function, from which all relevant information is found. It turns out that the Hamiltonian operator H, analogous to the Hamiltonian of classical physics, plays a central role in the quantum theory. The Hamiltonian operator is again defined as a sum over kinetic energy and potential energy: **H**(q,p) = **K** + **V** as in Hamilton's classical theory, but now the classical variables are replaced by operators. The operator V simply multiplies a given function by the potential energy V(x) (a one dimensional case is assumed here for simplicity), whereas the kinetic energy is represented by a differential operator:

$$p^2 \rightarrow -\hbar^2 \frac{\partial^2}{\partial x^2} \qquad \textit{so that} \qquad K \rightarrow -\frac{\hbar^2}{2m}\frac{\partial^2}{\partial x^2} \tag{I-10}$$

The wave function evolves according to Schrödinger's equation:

$$i\hbar \frac{\partial}{\partial t}\psi = H\psi \tag{I-11}$$

Formally, Schrödinger's equation is similar to the elementary differential equation:

$$i\hbar \frac{d}{dt}x = bx \tag{I-12}$$

where b is a constant, which is solved by

$$i\hbar \int_{x_0}^{x} \frac{dx}{x} = \int_0^t b\,dt \quad \Rightarrow \quad i\hbar \ln\left[\frac{x}{x_0}\right] = bt \tag{I-13}$$

Therefore for the case of the simple equation (I-12):
$x = x_0 \exp\{-ibt/\hbar\} = \exp\{-ibt/\hbar\}\, x_0$, provided b is a constant. In the case of Schrödinger's equation, the constant is replaced by a differential operator, so the solution is not so trivial, and depends on the potential function V(x). In practice it is usually easier to solve the equation by separation of variables, separating the time dependence from the dependence on other variables. If **H** is time independent such a separation is possible. If one assumes the wave function is a product of a position dependent function and a time dependent function:

$$\psi(\mathbf{r},t) = u(\mathbf{r}) f(t) \tag{I-14}$$

where **r** is the three dimensional position vector, then the time dependence is found to be simply:

$$f(t) = \exp(-iEt/\hbar) \tag{I-15}$$

where E is the total energy of the system. The remaining equation is the more difficult one to solve:

$$\mathbf{H} u(\mathbf{r}) = E u(\mathbf{r}) \tag{I-16}$$

In the "matrix mechanics" formalism of quantum mechanics, the wave function is represented by a column vector and an operator like the Hamiltonian is represented by a matrix. Using these forms in Schrödinger's equation one still has an equation of the basic form (I-12), but x is replaced by a vector and b is replaced by a square matrix. If the matrix has constant coefficients, Eq. (I-12) can still be solved as shown above, except that matrix operations do not necessarily commute, so one has to be careful about the correct order of operations. The formal solution involves the exponential of a matrix, defined by its series expansion (convergence is assumed):

$$\frac{d}{dt}\mathbf{v} = \mathbf{A}\mathbf{v} \quad \Rightarrow \quad \mathbf{v} = \exp\{\mathbf{A}t\}\,\mathbf{v}_0 \equiv \left[\mathbf{I} + \mathbf{A}t + \frac{1}{2!}\mathbf{A}^2 t^2 + \ldots\right] \mathbf{v}_0 \tag{I-17}$$

where column vectors appear on both sides of each equation, and **I** is the unit matrix. Note that $v_0 \exp^{At}$ is not defined if v_0 is a column vector and **A** is a matrix. Therefore if **H** is time independent (which is the usual case), the formal solution to Schrödinger's equation can be obtained from the expansion:

$$\psi(t) = \exp\{-i\mathbf{H}t/\hbar\}\,\psi(0) \equiv \left[\mathbf{I} - \frac{it}{\hbar}\mathbf{H} + \frac{1}{2!}\left(\frac{-it}{\hbar}\right)^2 \mathbf{H}^2 + \ldots\right] \psi(0) \tag{I-18}$$

where **I** is the identity operator.

The general solution to the Schrödinger equation is a linear superposition of independent solutions. It is usually most convenient to use as a basis the set of eigenfunctions of the Hamiltonian. Assuming the time dependence was separated as shown above, the general wave function is of the form

$$\psi(r,t) = \sum_n c_n(t)\, u_n(r) \tag{I-19}$$

where all the time dependence is contained in the coefficients $c_n(t)$.

If the system is described by a certain wave function ψ, then what can be observed in practice in a physical measurement of a property A is the expectation value of the corresponding quantum mechanical operator:

$$\langle A \rangle \equiv \int \psi^*(r,t)\, A\, \psi(r,t)\, d^3r \tag{I-20}$$

where $\int d^3\mathbf{r} = \int\int\int dx\, dy\, dz$ refers to integration over ordinary three-dimensional space. When the wave function is given as a superposition, as in Eq. (I-19), the expectation value of any observable operator **A** is:

$$\langle A \rangle \equiv \sum_{m,n} c_m^*(t)\, c_n(t) \int u_m^*(r)\, A\, u_n(r)\, d^3r \tag{I-21}$$

In the matrix mechanics formalism the wave functions $u_n(\mathbf{r})$ of Eqs. (I-19), (I-21) are column vectors and the coefficients $c_n(t)$ are scalars. The operator **A** is represented by a matrix, and the complex conjugate of the column vector corresponding to $u_m(\mathbf{r})$ is replaced by the corresponding row vector.

At this point it is convenient to convert to the Dirac "bracket" formalism, in which the conventional wave function $u_n(r)$ is represented by a "ket" vector $|u\rangle$ in Hilbert space. Then one defines a linear functional ("bra" vector) $\langle u|$ by its operation on arbitrary "ket" vectors $|v\rangle$ in that space:

$$\langle u | v \rangle \equiv \int u^*(r)\, v(r)\, d^3r \qquad or: \quad \langle u | \equiv \int d^3r\, u^*(r) \tag{I-22}$$

Thus the operation of a "bra" on a "ket" gives a scalar product, known (due to its usual notation) as a scalar product bracket. The scalar product defined here is not necessarily an integral over a continuous variable (like **r**). It can also be a sum over a discrete variable, like a spin angular momentum eigenvalue, which is actually the more common case for most of the work in magnetic resonance. There is a very close relationship between the vector space of such functionals and the vector space on which these functionals operate. The two spaces are isomorphic, i.e., they have exactly the same mathematical structure, and the space of functionals is known as the **dual space** to the original one. A specific functional $\langle u|$ is known as the **dual** of the corresponding vector $|u\rangle$.

Using the Dirac notation, Eq. (I-21) can be written compactly as:

$$\langle A \rangle \equiv \langle \psi | A | \psi \rangle = \sum_{m,n} c_m{}^*(t)\, c_n(t) \langle u_m | A | u_n \rangle \tag{I-23}$$

The set of scalar products $\langle u_m|\mathbf{A}|u_n\rangle$ for all m,n can be arranged in a matrix, and this matrix can be used to represent the operator **A** in the given basis. In this formula it is obviously not needed that the basis functions be eigenfunctions of **H**. Using any complete set of wave functions ϕ_m one may represent the operator by a matrix, the elements of which are:

$$A_{m,n} = \int \phi_m{}^*(r)\, A\, \phi_n(r)\, d^3r = \langle m | A | n \rangle \tag{I-24}$$

The expectation value can be calculated from these matrix elements once the coefficients $c_n(t)$ are known. These expressions may be used for any operator on Hilbert space, even if it does not represent a quantity which can be measured directly in an experiment. If **A** is an observable, then the corresponding operator (and its matrix) is hermitian, i.e.:

$$A_{m,n} = A^\dagger{}_{m,n} \equiv (A_{n,m})^* \tag{I-25}$$

This guarantees that the measured quantities $\langle\psi|\mathbf{A}|\psi\rangle$ are always real numbers.

I.3 Level Shift Operators on Hilbert space

Matrix elements of operators appear in a natural way also in a different representation of the operators. It is generally known in linear algebra that the set of all linear operators acting on a given vector space V is itself a vector space, sometimes denoted as L(V,V). We shall use here the standard Dirac notation for these vectors, except that double brackets $\langle\!\langle\rangle\!\rangle$ instead of $\langle\rangle$ will be used, to emphasize that these are not Hilbert space vectors. In this vector space one can define a scalar product using the standard trace operation

$$\langle\!\langle A | B \rangle\!\rangle \equiv Tr\left(A^\dagger B\right) \tag{I-26}$$

Suppose one constructs a basis to the space of operators from the basis of the original (Hilbert) vector space. This can be achieved by defining, for each pair of values of m and n, the operator $|u_m\rangle\langle u_n|$ through its action on an arbitrary vector $|v\rangle$ in Hilbert space:

$$|u_m\rangle\langle u_n| \left(|v\rangle\right) \equiv |u_m\rangle\left(\langle u_n | v \rangle\right) = u_m(r) \int d^3r\, u_n{}^*(r)\, v(r) \tag{I-27}$$

The set of operators $\{|u_m\rangle\langle u_n|\ ;\ m=1,2,...;n=1,2,...\}$ is a basis for the space of linear operators on Hilbert space. These operators are orthogonal to each other and are also normalized, provided the original set $\{|u_m\rangle\}$ was an orthonormal basis of Hilbert space. This can be seen by calculating the scalar product of two such operators:

$$\langle\langle (|i\rangle\langle j|) \,|\, (|k\rangle\langle l|) \rangle\rangle = Tr(|j\rangle\langle i|k\rangle\langle l|) = \sum_m \delta_{i,k}\langle m|j\rangle\langle l|m\rangle = \delta_{i,k}\delta_{j,l} \tag{I-28}$$

This expression is non-zero (and equal to 1) only if $i = k$, $j = l$, so that the two operators are identical.

Since they form a basis, any operator on Hilbert space can be expressed as a linear combination of such operators:

$$A = \sum_{m,n} a_{m,n} |u_m\rangle\langle u_n| \tag{I-29}$$

The coefficients $a_{m,n}$ may be found from:

$$\langle u_i|A|u_j\rangle = \sum_{m,n} a_{m,n} \langle u_i|u_m\rangle\langle u_n|u_j\rangle = a_{i,j} \tag{I-30}$$

using the orthonormality of the basis. It is now clear that the coefficients in the expansion of Eq. (I-29) are simply the matrix elements of **A** in the given basis. In particular, if **A** is one of the $|u_m\rangle\langle u_n|$, say $\mathbf{A} = |u_p\rangle\langle u_q|$, then its matrix in this basis is a very simple one:

$$|u_p\rangle\langle u_q| = \sum_{i,j} \delta_{i,p}\delta_{j,q} |u_i\rangle\langle u_j| \quad \Rightarrow \quad a_{i,j} = \delta_{i,p}\delta_{j,q} \tag{I-31}$$

The matrix has 0's almost everywhere, except for the (p,q) element which is equal to 1.

The operation of these matrices on the basis vectors of Hilbert space is also very simple. In this context it is natural to distinguish between matrices for which $m = n$ and matrices for which $m \neq n$. For any value of m, $|u_m\rangle\langle u_m|$ is a **projection operator** on Hilbert space, projecting out of an arbitrary vector its component which is "parallel" to $|u_m\rangle$. Thus if $|v\rangle = \sum c_k |u_k\rangle$ is an arbitrary Hilbert space vector:

$$|u_m\rangle\langle u_m|(|v\rangle) = \sum_k c_k |u_m\rangle\langle u_m|u_k\rangle = c_m |u_m\rangle \tag{I-32}$$

In particular this implies that operating with $|u_m\rangle\langle u_m|$ twice is the same as operating with it once . This is written as

$$(|u_m\rangle\langle u_m|)^2 = |u_m\rangle\langle u_m| \tag{I-33}$$

which can be verified by operating on any vector with each of the operators appearing on the two sides of this equation. This equality is often taken in algebra as the definition of a projection operator.

For $m \neq n$, $|u_m\rangle\langle u_n|$ takes out of a vector the component "along" $|u_n\rangle$ and "rotates" it to the "direction" of $|u_m\rangle$. When these vectors represent energy eigenfunctions, such a "rotation" is equivalent to a change in the energy level of the system. For this reason these operators are known as **level-shift operators.**

The projection operators defined here are useful also for describing the result of diagonalization of operators. Suppose **A** is some diagonalizable operator, with eigenstates $|u_m\rangle$ and corresponding eigenvalues E_m. Using these eigenstates as a basis for the vector space it is straightforward to show (by transforming to this basis) that

$$A = \sum_m E_m |u_m\rangle\langle u_m| \tag{I-34}$$

The matrices of the level-shift operators (including the projection operators as a special case) are thus very convenient to use as a basis for the space of matrices, representing the observable physical operators. As will be seen later on, the expansion (I-29) is very useful in the density matrix formalism.

Once an operator is described by a matrix, the time dependence of this matrix can be found by following the time dependence of each matrix element. It can be shown from the Schrödinger equation that for any operator A and any states $|m(t)\rangle$, $|n(t)\rangle$:

$$\frac{d}{dt}\langle m(t)|A|n(t)\rangle = \langle m(t)|\frac{\partial A}{\partial t}|n(t)\rangle + \frac{1}{i\hbar}\langle m(t)|(AH - HA)|n(t)\rangle \tag{I-35}$$

This equation is appropriate for the "wave mechanics" formalism, often called the **Schrödinger picture** of quantum mechanics, in which the essential time dependence is ascribed to the wave function. In the alternative formulation of "matrix mechanics", the **Heisenberg picture** of quantum mechanics, all time dependence is relegated to the operators, and wave functions are stationary. This can be achieved by starting from the Schrödinger picture and using the definitions:

$$|\psi\rangle_H(t) \equiv |\psi(t{=}0)\rangle \qquad A_H(t) \equiv \exp\{iHt/\hbar\}\, A \exp\{-iHt/\hbar\} \tag{I-36}$$

with the subscript H for the Heisenberg picture. Then, for any operator **A**(t) and for any "Heisenberg picture" states $|m\rangle$, $|n\rangle$ (from which the subscript H will be omitted):

$$\langle m|\frac{d}{dt}A_H(t)|n\rangle = \langle m|e^{iHt/\hbar}\frac{\partial A}{\partial t}e^{-iHt/\hbar}|n\rangle + \frac{1}{i\hbar}\langle m|(A_H(t)H - HA_H(t))|n\rangle \tag{I-37}$$

This holds for any matrix element (m,n) and therefore also the operator equation holds:

$$\frac{d}{dt}A_H(t) = \left(\frac{\partial A}{\partial t}\right)_H + \frac{1}{i\hbar}[A_H(t), H] \tag{I-38}$$

Here $[\mathbf{A},\mathbf{B}] \equiv \mathbf{AB} - \mathbf{BA}$ (commutator brackets). This last equation closely resembles eq. (I-3) above, which belongs to classical mechanics. This may create the misleading impression that Liouville's equation may also be valid in quantum physics. However, because of Heisenberg's uncertainty principle: $\Delta x \Delta p \geq \hbar$, x and p (or q and p) can not be exactly known simultaneously, so one can not use points in phase space for describing states of the system. One can only determine the probability that the system is at a particular position, or has a particular value of momentum. But one cannot determine both at the same time, so it becomes meaningless to talk about locating the system at a particular point in phase space. Therefore one cannot define the same density function ρ as in classical mechanics.

I.4 A Free Two Level System

(a) Possible results of measurements

The simplest non-trivial quantum system is a two level system which does not interact with its environment. Such a system can be found only in one of two energy eigenstates, denoted as $|\alpha\rangle$ and $|\beta\rangle$, for which Eq. (I-16) is:

$$H|\alpha\rangle = E_\alpha|\alpha\rangle \qquad H|\beta\rangle = E_\beta|\beta\rangle \tag{I-39}$$

The most general wave function for the system is a linear combination:

$$\psi = c_\alpha|\alpha\rangle + c_\beta|\beta\rangle \tag{I-40}$$

so the expectation value of any operator **A** is equal to

$$\langle\psi|A|\psi\rangle = c_\alpha^*(t)c_\alpha(t)\langle\alpha|A|\alpha\rangle + c_\alpha^*(t)c_\beta(t)\langle\alpha|A|\beta\rangle$$

$$+ c_\beta^*(t)c_\alpha(t)\langle\beta|A|\alpha\rangle + c_\beta^*(t)c_\beta(t)\langle\beta|A|\beta\rangle \tag{I-41}$$

From Eq. (I-15) it follows that

$$c_\alpha(t) = \exp(-iE_\alpha t/\hbar)c_\alpha(0) \quad ; \quad c_\beta(t) = \exp(-iE_\beta t/\hbar)c_\beta(0) \tag{I-42}$$

Using the polar decomposition of complex numbers one may write: $c(0)=Ce^{i\phi}$ for the coefficients at time $t = 0$. Since for any phase θ, complex conjugation yields $(\exp(i\theta))^* = \exp(-i\theta)$, one obtains here:

$$\langle\psi|A|\psi\rangle = (C_\alpha)^2\langle\alpha|A|\alpha\rangle + (C_\beta)^2\langle\beta|A|\beta\rangle$$

$$+ 2C_\alpha C_\beta Re\left\{(\exp(i\omega_{\alpha\beta}t + i(\phi_\alpha - \phi_\beta))\langle\alpha|A|\beta\rangle\right\} \tag{I-43}$$

where Re{} denotes the real part of the number, and the frequency $\omega_{\alpha\beta} \equiv (E_\alpha - E_\beta)/\hbar$ is the transition frequency between the two states. The first two terms on the right hand side are expected also on the basis of classical physics, since each of them is the probability of being in a given state multiplied by the expectation value of **A** in that state.

The third term, however, describes a quantum interference effect, which cannot be predicted by classical mechanics. This term is time dependent, oscillating at the transition frequency between the two states. This is also known as the Rabi frequency, because it is the same as the transition frequency of the Rabi experiment. However, in that experiment a time dependent perturbation drives the transition, and the system may be initially in one of its two eigenstates, whereas here only a special initial condition with a constant Hamiltonian is treated. Thus, in principle the result of a measurement on such a system should oscillate with time, if the experimental time resolution is sufficient.

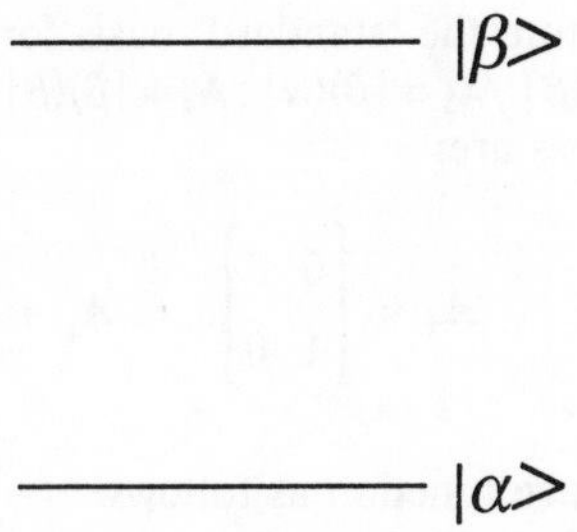

Fig. I.1. *Energy level diagram of a two level system.*

There are two types of measurements in which the oscillating term is not observed in practice. One type is a "slow" measurement, in which the relevant values of t are spread over a time interval which is very large compared with $(\omega_{\alpha\beta})^{-1}$. In such cases what is measured is effectively a time average over Eq. (I-43), and in this average the contribution of the interference term is practically zero. The other type is a (possibly "fast") measurement, which is conducted on a *macroscopic* ensemble of two level systems of this kind. The time dependent factor is common to all systems, but the constant phase factor, which is chosen arbitrarily in quantum mechanics, will vary randomly between different systems. Thus the average over the third term would be zero.

(b) The level-shift operators

Given the basis $\{|\alpha\rangle, |\beta\rangle\}$ for the two-level Hilbert space, any vector may be represented by its expansion coefficients, i.e. ψ of Eq. (I-40) may be represented by the column vector containing its coefficients:

$$\psi \rightarrow \begin{bmatrix} c_\alpha \\ c_\beta \end{bmatrix} \tag{I-44}$$

In particular, the two basis "ket" vectors are, in this representation

$$|\alpha\rangle = \begin{bmatrix} 1 \\ 0 \end{bmatrix} \qquad |\beta\rangle = \begin{bmatrix} 0 \\ 1 \end{bmatrix} \tag{I-45}$$

The corresponding functionals ("bra vectors") are the operations of multiplying from the left with the complex conjugate row vectors, which in this case are simply

$$\langle\alpha| = (1\ \ 0) \qquad \langle\beta| = (0\ \ 1) \tag{I-46}$$

The level-shift operators constitute the "standard" basis for operator space, which is in this case: $\{\mathbf{A_1} \equiv |\alpha\rangle\langle\alpha|\, , \mathbf{A_2} \equiv |\alpha\rangle\langle\beta|\, , \mathbf{A_3} \equiv |\beta\rangle\langle\alpha|\, , \mathbf{A_4} \equiv |\beta\rangle\langle\beta|\}$. The matrices representing these operators in the given basis are:

$$A_1 = \begin{bmatrix} 1 & 0 \\ 0 & 0 \end{bmatrix} \quad A_2 = \begin{bmatrix} 0 & 1 \\ 0 & 0 \end{bmatrix} \quad A_3 = \begin{bmatrix} 0 & 0 \\ 1 & 0 \end{bmatrix} \quad A_4 = \begin{bmatrix} 0 & 0 \\ 0 & 1 \end{bmatrix} \tag{I-47}$$

They operate on the general wave function as follows:

$$A_1 \begin{pmatrix} c_\alpha \\ c_\beta \end{pmatrix} = \begin{pmatrix} c_\alpha \\ 0 \end{pmatrix} \quad A_2 \begin{pmatrix} c_\alpha \\ c_\beta \end{pmatrix} = \begin{pmatrix} c_\beta \\ 0 \end{pmatrix} \quad A_3 \begin{pmatrix} c_\alpha \\ c_\beta \end{pmatrix} = \begin{pmatrix} 0 \\ c_\alpha \end{pmatrix} \quad A_4 \begin{pmatrix} c_\alpha \\ c_\beta \end{pmatrix} = \begin{pmatrix} 0 \\ c_\beta \end{pmatrix} \tag{I-48}$$

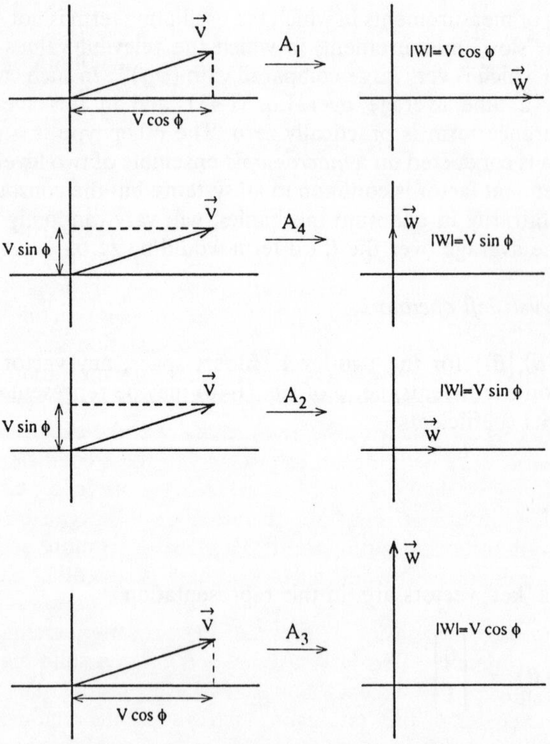

Fig. I.2. *The geometrical operations representing the algebraic operation of the four level-shift operators on a two level system.*

It is easy to see that $\mathbf{A}_1$ is a projection operator on $|\alpha\rangle$, and $\mathbf{A}_4$ is a projection operator on $|\beta\rangle$. It is also obvious that $\mathbf{A}_2$ "rotates" $|\beta\rangle$ to $|\alpha\rangle$, and gives the zero vector when operating on $|\alpha\rangle$. Similarly, $\mathbf{A}_3$ "rotates" $|\alpha\rangle$ to $|\beta\rangle$, and gives the zero vector when operating on $|\beta\rangle$. If this two dimensional vector space is defined just over the field of real numbers, rather than over the field of complex numbers (namely, c_α and c_β are restricted to be real numbers), these operators can be described by a very simple geometrical picture. The general vector can be given a polar representation (namely, using circular coordinates):

$$c_\alpha = V\cos(\phi) \qquad c_\beta = V\sin(\phi) \tag{I-49}$$

where V is its length and ϕ is the angle it makes with the **x** (horizontal) axis. Then the level-shift operators generate the geometrical transformations shown in Fig. I.2.

I.5 The Quantum Mechanical Density Matrix

(a) Motivation for defining a density matrix

It was shown above that if a system is described by a wave function $|\psi\rangle$, which is expanded in a complete basis as: $|\psi\rangle = \sum_k c_k |\phi_k\rangle$ then the expectation value of any observable **A** in the system is:

$$\langle A\rangle = \sum_{k,l} c_k^* c_l \langle\phi_k|A|\phi_l\rangle = \sum_{k,l} c_k^* c_l A_{k,l} \tag{I-23'}$$

Here c_k is directly related to the $|\phi_k\rangle$ component in $|\psi\rangle$, c_l is related to the $|\phi_l\rangle$ component in $|\psi\rangle$, and any phase relationship between them affects the expectation value. For example, if $\{|\phi_k\rangle\}$ are eigenstates of **H**, then $c_k^* c_l$ is proportional to $\exp(iE_k t/\hbar - iE_l t/\hbar) = \exp(i\omega_{kl}t)$. Note that the off-diagonal terms ($k\neq l$) in this equation acquire their time dependence from the fact that each wave function has a dynamic phase factor $\exp(-iE_k/t)$. In these equations it is evident that the expectation value of **A** is determined by the set of numbers $c_k^* c_l$ which multiply its matrix elements. This fact suggests that instead of working directly with the coefficients c_k in the expansion of the wave function, it is possible to work with the set of products of these coefficients, as in Eq. (I-23'). Both sets of numbers, the set $\{c_k\,;\, k=1,...,n\}$ and the set $\{c_k^* c_l\,;\, k,l=1,...,n\}$ contain the full information on the system, and therefore each set can be used to calculate any relevant quantum mechanical quantity. The first set is more economical and thus appears more natural, but the advantages of working with the other alternative will soon become clear.

Now suppose the system being considered is part of a larger complex, conveniently described as: system + environment. This larger complex is often a solid matrix or a liquid solvent in which the "interesting" molecules or atoms are contained. The complex may even include other molecules of the "interesting" type, when their influence on a particular molecule is considered. The measurements refer to variables characterizing the "system" only, e.g. its total energy or total angular momentum. However, the "system" is not closed. It interacts with the environment, and the experimental apparatus which is useful for measuring in the "system" may not be adequate for measuring the same variables in the environment. This is a very important case in practice, and it is therefore very important to find out how the quantum mechanical formalism deals with measurements under such conditions. Since the (smaller) system interacts with the environment, it is impossible to separate system variables from environment variables by expressing the total wave function as a product: $|\psi_{total}\rangle = |\psi_{system}\rangle\,|\psi_{environment}\rangle$. One can only express it as a general linear combination of such products:

$$|\psi_{total}\rangle = \sum_{i,j} c_{i,j}\,|\phi_{i(system)}\rangle\,|\theta_{j(environment)}\rangle \tag{I-50}$$

The expectation value of an operator which operates only on the system (e.g. spin operator for that system) is therefore equal to:

$$\langle A\rangle = \sum_{i,j,k,l} c_{i,j}{}^{*}\,c_{k,l}\,\langle\theta_{j(env)}|\langle\phi_{i(sys)}|A|\phi_{k(sys)}\rangle|\theta_{l(env)}\rangle = \sum_{i,j,k} c_{i,j}{}^{*}\,c_{k,j}\langle\phi_{i(sys)}|A|\phi_{k(sys)}\rangle =$$

$$= \sum_{i,k}\left(\sum_{j} c_{i,j}{}^{*}\,c_{k,j}\right)\langle\phi_i|A|\phi_k\rangle \equiv \sum_{i,k}\rho_{k,i}A_{i,k} \tag{I-51}$$

The second equality on the first line follows from the orthonormality of the wave functions. In the final step in the second line a matrix ρ has been defined (a more formal definition will be given below). This matrix will be called the density matrix. One consequence of its definition is that the density matrix is hermitian. Here ρ_{ki} is not simply related to $|\phi_i\rangle$ or $|\phi_k\rangle$ because it is a sum over products of coefficients. Comparing the end result in Eq. (I-51) to Eq. (I-23') it looks as if some "ensemble average" was taken over $c_k c_i^*$ to get ρ_{ki}, but in fact there is no such ensemble here. In this general case it is impossible to work directly with the wave function, which is only partly known. The only practical alternative is to use the set of numbers that makes up the density matrix, since this set of numbers contains all that may be known about the system. The general formulation given here includes an isolated system as a special case. In that case the total wave function is a product function, and then the basis for the wave functions of the environment may be chosen so that in Eq. (I-51) there is only one single j value. The index j becomes unnecessary, and then $\rho_{ki} = c_i^* c_k$. In this case ρ reduces to the set of coefficients encountered above for the two level case (previous Section).

Returning to the isolated two level system, we find that the diagonal elements of the density matrix are the real numbers $|c_\alpha|^2$, $|c_\beta|^2$ and the off-diagonal elements are $c_\alpha^* c_\beta$ and $c_\beta^* c_\alpha$. The first two numbers are the probabilities of the system being found in states α,β respectively. Thus if one is concerned not with a single microscopic system but with a macroscopic collection of such systems (assumed to be non-interacting) these numbers are the populations of the two states. On the other hand it is clear from Eq. (I-49) that the time development of the system requires knowledge of the off-diagonal elements of the density matrix. Each such element is directly related to a transition between quantum states, since it oscillates at the transition frequency between the corresponding states. As shown in the previous Section, these off-diagonal elements contribute to the expectation value if the wave function is a coherent superposition of states. That is why such elements are named "coherences".

(b) *Formal definition of the density matrix*

From the above discussion it is clear that when interactions between system and environment cannot be neglected, whatever is observed can only be calculated by means of a density matrix. Thus in the most general case a quantum mechanical system will be described by a **density matrix** $\boldsymbol{\rho}$, formally defined by the following conditions:

(i) The expectation value of any operator **A** can be calculated using the density matrix as follows: $\langle \mathbf{A} \rangle = \mathrm{Tr}(\rho \mathbf{A}) = \sum_{i,j} \rho_{ij} \mathbf{A}_{ji}$.
(ii) ρ is hermitian.
(iii) $\mathrm{Tr}(\rho) = 1$ (normalization)
(iv) If w_k are the eigenvalues of ρ (they have to be real due to the hermiticity of ρ) then all w_k are non-negative: $w_k \geq 0$ and: $\sum_k w_k = 1$.

In the special case in which the system is in "a pure state" - i.e., it has a well-defined wave function $|\psi_{(sys.)}\rangle$, one can still define a density matrix by $\rho_{ki} = c_k c_i^*$ where $|\psi_{sys.}\rangle = \sum_j c_j |\phi_j\rangle$ in a given basis. It is straightforward to show that this expression for the density matrix elements results from $\rho = |\psi\rangle\langle\psi|$.

It is clear that the density matrix in quantum mechanics is not the same as the density function in classical phase space, for which Liouville's equation holds. For example, the density matrix does not refer to a fully specified state of the system (position *and* linear momentum given simultaneously). Nevertheless, in terms of calculating average values of observable quantities such as position, linear momentum, angular momentum etc. the density matrix is the analog of the classical density function. This is evident when the definition given above is compared with Eq. (I-9) for the classical function. This similarity is the reason for the name "density matrix". Moreover, it can be shown that at thermal equilibrium the density matrix is equal to

$$\rho_{eq.} = \exp\left[- \frac{H}{k_B T} \right] \tag{I-52}$$

This is the natural generalization of Gibbs' classical expression (Eq. (I-5) above).

Like any hermitian operator, ρ can be diagonalized, and its eigenvectors $|\phi_k\rangle$ form a complete basis. According to Eq. (I-34) it may be expanded in this basis as:

$$\rho = \sum_k w_k |\phi_k\rangle\langle\phi_k| \tag{I-53}$$

where w_k are the eigenvalues. In this basis, a diagonal element of ρ is:

$$\rho_{i,i} = \sum_k w_k \langle i|k\rangle\langle k|i\rangle = w_i \tag{I-54}$$

where $|k\rangle$ is a short notation for $|\phi_k\rangle$. In particular, if the system is in a pure state, so that $\rho = |\psi\rangle\langle\psi|$, $|\psi\rangle$ is an eigenfunction of ρ . The eigenvalue corresponding to $|\psi\rangle$ is equal to 1, and all other eigenvalues are equal to zero. It should be noted that the original choice of basis (prior to diagonalization) is unimportant in this context: a pure state is a single basis vector in one basis, and a superposition of several basis vectors in a different basis. Returning to the general case for ρ and to its diagonalizing basis, the expectation value of any operator **A** is:

$$\langle A\rangle = Tr(\rho A) = \sum_{i,j} \rho_{i,j} A_{j,i} = \sum_{k} \left(w_k \langle i|k\rangle\langle k|j\rangle\right)\langle j|A|i\rangle = \sum_{k} w_k \langle k|A|k\rangle \quad \text{(I-55)}$$

This is a weighted average of the expectation values that **A** would have in each of the states $|k\rangle$. The numbers w_k are the statistical weights used in the averaging. Thus w_k can be regarded as a probability that the system is in state $|k\rangle$. This meaning of w_k is consistent with the conditions: $w_k \geq 0$, $\sum_k w_k = 1$. It is also consistent with the explanation of the diagonal elements of ρ as (normalized) populations.

In a general basis, the diagonal elements of ρ are equal to:

$$\rho_{\alpha,\alpha} = \sum_{k} w_k \langle\alpha|k\rangle\langle k|\alpha\rangle = \sum_{k} w_k |\langle\alpha|k\rangle|^2 \quad \text{(I-56)}$$

Here $|\alpha\rangle$ is a basis state, whereas $|k\rangle$ is an eigenstate of ρ with eigenvalue w_k. The sum is over real positive numbers, so it is always a positive number. Now suppose the system is in state $|k\rangle$, and a measurement is carried out. The numbers $|\langle\alpha|k\rangle|^2$ are the probabilities of finding the system in state $|\alpha\rangle$ (the latter may be, for example, an eigenstate of the Hamiltonian or of an angular momentum operator). Therefore $\rho_{\alpha,\alpha}$ can be regarded as an average of these quantities (i.e., the probabilities) over all the possible initial conditions $|k\rangle$, for which only the relative probabilities are known. Thus $|\langle\alpha|k\rangle|^2$ is indeed a population of the level $|\alpha\rangle$, which is related through Eq. (I-56) to the statistical weights w_k and to the possible results of actual measurements.

As for the off-diagonal elements of ρ, in a general basis they are equal to:

$$\rho_{\alpha,\beta} = \sum_{k} w_k \langle\alpha|k\rangle\langle k|\beta\rangle \quad \text{(I-57)}$$

This is a weighted sum of interference terms, which are complex numbers. The sum may be zero, in which case the averaging cancels the interference effects. If it is non-zero, it means that there is some coherence between the states which avoids the cancellation, and makes possible non-trivial interference in the system.

I.6 The Liouville - von Neumann equation of motion

(a) The equation of motion of the density matrix

Using Eq. (I-53), where $\{|k\rangle\}$ are the eigenvectors of ρ, one can find the equation of motion for the density matrix by applying Schrödinger's equation to these wave functions:

$$i\hbar \frac{d\rho}{dt} = i\hbar \sum_k w_k \left(\left[\frac{d}{dt} |k\rangle \right] \langle k| + |k\rangle \left[\frac{d}{dt} \langle k| \right] \right) =$$

$$= \sum_k w_k \big((H|k\rangle) \langle k| + |k\rangle (\langle k|H) \big) = \sum_k w_k [H, |k\rangle\langle k|] = [H, \rho] \tag{I-58}$$

This result is known as **von Neumann's equation** , usually written as:

$$\frac{d\rho}{dt} = \frac{1}{i\hbar}[H, \rho] \tag{I-59}$$

This looks like Heisenberg's equation of motion for operators (Eq. (I-38) above) except for the sign of the commutator, but it has a very different meaning. Heisenberg's equation of motion refers to operators representing observable quantities, while von Neumann's equation refers to the density operator, which is an auxiliary tool for calculating observable quantities. The opposite sign arises because here the calculation is done in the "Schrödinger picture" where observables are time independent - only the density operator is time-dependent (like wave functions !). This is the opposite of the "Heisenberg picture", where the time dependence is contained in the observables. Due to the similarity between von Neumann's equation and Liouville's equation, and due to the similarity in roles between the quantum mechanical density matrix and the classical density in phase space, von Neumann's equation is usually called the **Liouville-von Neumann equation.**

From the previous derivation of von Neumann's equation it is not clear how to solve the equation directly. Before attempting a direct solution, a solution will be obtained indirectly, through a different derivation. Now the solution (I-18) of Schrödinger's equation for $|k\rangle$, the eigenfunctions of ρ , is used:

$$|k(t)\rangle = \exp(-iHt/\hbar)\, |k(0)\rangle \tag{I-60}$$

Substituting this result into the expression (I-53) for the density matrix:

$$\rho(t) = \sum_k w_k \exp(-iHt/\hbar)\, |k(0)\rangle\langle k(0)|\, \exp(iHt/\hbar) =$$

$$= \exp(-iHt/\hbar)\, \rho(0)\, \exp(iHt/\hbar) \tag{I-61}$$

This gives the time dependence of ρ, which is the solution of the equation of motion. The density matrix at time t is thus related by a unitary ("magnitude preserving") transformation to the density matrix at the initial time (t=0). This is an important characteristic of von Neumann's equation, which does not hold when the equation is modified by including relaxation processes, as will be seen in the following chapters. The equation of motion can be rederived from here by differentiation, keeping the order of possibly non-commuting terms:

$$\frac{d\rho}{dt} = -\frac{i}{\hbar}H\exp(-iHt/\hbar)\rho(0)\exp(iHt/\hbar) + \exp(-iHt/\hbar)\rho(0)\exp(iHt/\hbar)\frac{i}{\hbar}H =$$

$$= \frac{1}{i\hbar}[H,\rho] \tag{I-62}$$

For practical calculations it may be useful to expand each eigenfunction of ρ at time t=0 in the eigenfunctions $|u_n\rangle$ of **H** , for which the corresponding energies are E_n. The expansion is:

$$|k(0)\rangle = \sum_n |u_n\rangle\langle u_n|k(0)\rangle \tag{I-63}$$

Here $\langle u_n|k(0)\rangle$ is the numerical coefficient for the wave function $|u_n\rangle$. Using this expansion in Eq. (I-60) one obtains:

$$|k(t)\rangle = \exp(-iHt/\hbar)\sum_n |u_n\rangle\langle u_n|k(0)\rangle = \sum_n\left(\exp(-iHt/\hbar)|u_n\rangle\right)\langle u_n|k(0)\rangle \tag{I-64}$$

which is convenient for calculations, since (using Eq. (I-17)):

$$\exp(-iHt/\hbar)|u_n\rangle = \left[1 - \frac{it}{\hbar}E_n + \frac{1}{2!}\left(\frac{-it}{\hbar}\right)^2 E_n^{\,2} + \dots\right]|u_n\rangle = \exp(-iE_nt/\hbar)|u_n\rangle \tag{I-65}$$

In other words, in the basis which diagonalizes **H**: $(e^{aH})_{ij} = e^{aE_i}\delta_{ij}$, so that in the same basis also matrix elements of ρ are easy to calculate:

$$\rho(t)_{i,j} = \sum_{k,l}\left(\exp(-iHt/\hbar)\right)_{i,k}\rho(0)_{k,l}\left(\exp(iHt/\hbar)\right)_{l,j} = \exp(-iE_it/\hbar)\rho(0)_{i,j}\exp(iE_jt/\hbar) =$$

$$= \exp(-i\omega_{i,j}t)\rho(0)_{i,j} \tag{I-66}$$

Thus, as long as Eq. (I-61) holds (i.e., in the absence of relaxation processes), if one works in the basis of **H** the time evolution of ρ is trivial to calculate. Each element of ρ evolves separately, uncoupled to any other element of the density matrix, oscillating at the transition frequency of the relevant states.

(b) The Liouville - von Neumann equation in Liouville space

The ordinary form of von Neumann's equation was solved above only indirectly, since the commutator is not simple to handle. However, the equation can be solved directly if it is written in a form for which the solution is known. This can be achieved by regarding ρ as a vector - which is legitimate, because the space of operators (including ρ) which operate on the Hilbert space of wave functions is itself a vector space. This space of operators is called Liouville space, because it arises as a natural vector space for dealing with the Liouville-von Neumann equation. The vectors in this space are called supervectors, to distinguish them from Hilbert space vectors, and the operators on this space are called superoperators. In this space the equation of motion for the density matrix is written in the following way:

$$\frac{d}{dt}\rho_{i,j} = -\frac{i}{\hbar}\sum_{k,l}\left(H_{i,k}\delta_{l,j} - \delta_{i,k}H_{l,j}\right)\rho_{k,l} \equiv -\frac{i}{\hbar}\sum_{k,l} L_{ij,kl}\,\rho_{kl} \tag{I-67}$$

This equation defines the **Liouville superoperator** (or: **Liouvillian**) **L**, which operates on Hilbert space operators (like $\mathbf{S}_z$, or **H**, or ρ). It is represented by a matrix (a "**supermatrix**"), in which each row and each column are labeled by two indices. In the same manner, ρ (or $\mathbf{S}_z$, or **H** etc.) are represented by vectors ("**supervectors**"), in which each element is labeled by two indices. The order in which the elements are arranged in such a supervector is arbitrary, but some ways of ordering the elements are especially convenient for specific purposes, as will be seen in the examples below. In any case, once the order of elements is chosen for supervectors, this imposes a corresponding order for the elements of supermatrices. Unless a specific arrangement of the elements is mentioned, it will be assumed that the supervector is written by copying from the Hilbert space matrix the first column, then the second column, etc.

Supervectors belonging to Liouville space will be denoted by $|A\rangle\rangle$, to distinguish them from Hilbert space vectors $|v\rangle$. The Liouville space vectors $|A\rangle\rangle$ are "superkets". The corresponding "superbras" $\langle\langle A|$ are naturally defined as the functionals which, operating on any given superket $|B\rangle\rangle$, will give as a result the appropriate scalar product in this space. The scalar product is the trace operation, as defined in Eq. (I-26) above.

Working in Liouville space, Eq. (I-67) can be written as:

$$\frac{d}{dt}\rho = -\frac{i}{\hbar}L\rho \tag{I-68}$$

If the Hamiltonian is time independent, **L** is a matrix with constant elements, and then the formal solution is

$$\rho(t) = \exp\left(-iLt/\hbar\right)\rho(0) \tag{I-69}$$

As will be seen in the following chapters, when a stochastic relaxation process is present the supermatrix **L** will have a more general form than that appearing in Eq. (I-67). Nevertheless, Eq. (I-68) will still be valid, and under appropriate conditions the formal solution will still be of the form (I-69), with the appropriate **L**.

(c) The formal structure of the Liouville operator

Understanding the formal structure of **L** is important for studying symmetry properties, and in some cases simplifying equations. The definition of **L** is similar, but not identical, to that of a tensor product (direct product) of operators. A simple example of a tensor product is the product of two ordinary operators, each of them operating in general on a different vector space, like $\mathbf{S}_z\mathbf{I}_z$ when **S** is an electron spin operator and **I** is a nuclear spin operator. For electron wave functions one may use (in the context of magnetic resonance, when only the interactions with magnetic fields are important) the set $\{|m_S\rangle\}$ of eigenfunctions of $\mathbf{S}_z$. In a similar manner, for nuclear wave functions one may use the basis set $\{|m_I\rangle\}$ of eigenfunctions of $\mathbf{I}_z$. To calculate the matrix element of such a product operator one only needs to take the matrix element of each operator in the space on which it operates, and then take the product of the two elements. For this case:

$$\langle n_s n_I | \mathbf{S}_z \mathbf{I}_z | m_s m_I \rangle = \langle n_S | \mathbf{S}_z | m_S \rangle \langle n_I | \mathbf{I}_z | m_I \rangle = (\mathbf{S}_z)_{n_s, m_s} (\mathbf{I}_z)_{n_I, m_I} \tag{I-70}$$

where $|m_S m_I\rangle$ is a short notation for the direct product function $|m_S\rangle \otimes |m_I\rangle$. In this equation the product operator operates formally on a direct product of wave functions. One may therefore define the direct product operator as operating in the space of direct product wave functions (or vectors), where the result of such operations is defined in terms of the operations of the component operators in ordinary Hilbert space. On the basis of this rule one may calculate the result of operating with the tensor product of operators on a general wave function:

$$\mathbf{S}_z \mathbf{I}_z \sum_{m_s, m_I} c(m_s, m_I) \, |m_s m_I\rangle = \sum_{m_s, m_I} c(m_s, m_I) (\mathbf{S}_z |m_S\rangle)(\mathbf{I}_z |m_I\rangle) \tag{I-71}$$

What was demonstrated here for a specific case can be cast in a general form as follows. If **A**,**B** are operators in Hilbert space, and $|i\rangle$, $|j\rangle$ etc. are Hilbert space vectors, then the elements of the tensor product of operators are of the form:

$$\langle ij | \mathbf{A} \otimes \mathbf{B} | kl \rangle = \mathbf{A}_{i,k} \mathbf{B}_{j,l} \tag{I-72}$$

On the other hand, Eq. (I-67) given above for the elements of **L** can be expressed as:

$$\langle\langle (|i\rangle\langle j|) | \mathbf{H} \otimes \mathbf{I} - \mathbf{I} \otimes \mathbf{H} | (|k\rangle\langle l|) \rangle\rangle = Tr\left\{ (|i\rangle\langle j|)^{\dagger} (\mathbf{H} \otimes \mathbf{I} - \mathbf{I} \otimes \mathbf{H}) \, |k\rangle\langle l| \right\} =$$

$$= \mathbf{H}_{i,k} \delta_{l,j} - \delta_{i,k} \mathbf{H}_{l,j} = \mathbf{H}_{i,k} \delta_{j,l} - \delta_{i,k} \mathbf{H}^{t}_{j,l} \tag{I-73}$$

where $\mathbf{H}^t$ is the transpose of **H**. The product defined through this equation, for which the tensor product symbol was used, is *not* a tensor product, but a related operation . A tensor product operates on direct product wave-functions of the form $|i\rangle|j\rangle$, whereas **L** operates on Hilbert space operators, which can be expanded in terms of operators - not wave functions - of the form $|i\rangle\langle j|$. However, the vector space $\mathbf{V_P}$ of direct product wave functions of the form $|i\rangle|j\rangle$ is closely related to the vector space $\mathbf{V_L}$ of operators of the form $|i\rangle\langle j|$.

Three vector spaces are relevant to this relationship. The first is $\mathbf{V_H}$ - ordinary Hilbert space, consisting of vectors (wave functions) $|i\rangle$. On this space the standard scalar product $\langle i|j\rangle$ (quantum mechanical "bracket") is defined. The second is $\mathbf{V_L}$ - Liouville supervector space, which is just the space of linear operators operating on $\mathbf{V_H}$, regarded as supervectors $|A\rangle\rangle$. The Hamiltonian, the density matrix and ordinary spin operators like $\mathbf{S}_z$ belong to $\mathbf{V_L}$. This space is spanned by the level-shift operators $|i\rangle\langle j|$. The scalar product is defined here through the trace operation $\langle\langle A|B\rangle\rangle \equiv Tr(A^{\dagger}B)$. The third vector space is $\mathbf{V_P}$, the space of direct products of wave functions $|ij\rangle \equiv |i\rangle \otimes |j\rangle$. Formally this space may be defined as $\mathbf{V_H} \otimes \mathbf{V_H}$.

The relation between $\mathbf{V_P}$ and $\mathbf{V_L}$ can be defined by mapping each vector $|ij\rangle$ in $\mathbf{V_P}$ to a supervector $|i\rangle\langle j|$ in $\mathbf{V_L}$. This relation is not an isomorphism, because when complex numbers are used the requirements for a linear transformation do not hold in this mapping. Nevertheless, due to this simple relation it is convenient to use the tensor product notation also for operators like the Liouvillian, which operate on $\mathbf{V_L}$.

An interesting problem is calculating the product (or successive application) of two operators, each of which is itself a tensor product, or the related product appearing in the Liouvillian. This problem occurs, for example, when the exponential of such an operator is calculated. For the tensor product, it follows from Eq. (I-72) that:

$$\langle ij|(A \otimes B)(C \otimes D)|kl\rangle = \sum_{m,n} \langle ij|A \otimes B|mn\rangle\langle mn|C \otimes D|kl\rangle =$$

$$= \sum_{m,n} A_{i,m} B_{j,n} C_{m,k} D_{n,l} = (AC)_{i,k}(BD)_{j,l} = \langle ij|(AC) \otimes (BD)|kl\rangle \tag{I-74}$$

This equation is correct for any matrix element of the product, so it can be written as an operator equation:

$$(A \otimes B)(C \otimes D) = (AC) \otimes (BD) \tag{I-75}$$

For the related product in the Liouvillian the situation is slightly different:

$$\langle\langle\, |i\rangle\langle j|(A \otimes B)(C \otimes D)|k\rangle\langle l|\,\rangle\rangle = \sum_{m,n} \langle\langle\, |i\rangle\langle j|A \otimes B|m\rangle\langle n|\,\rangle\rangle\langle\langle\, |m\rangle\langle n|C \otimes D|k\rangle\langle l|\,\rangle\rangle =$$

$$= \sum_{m,n} A_{i,m} B_{n,j} C_{m,k} D_{l,n} = (AC)_{i,k}(DB)_{l,j} \tag{I-76}$$

so that the corresponding operator equation is:

$$(A \otimes B)(C \otimes D) = (AC) \otimes (DB) \tag{I-77}$$

Thus the order of operators **B, D** is reversed relative to the order in a tensor product. However, it should not lead here to any confusion, since tensor products of operators will always appear in this book without the tensor product notation, and their application will be straightforward, as in Eq. (I-70) above. The tensor product notation will be used throughout this book solely for the "pseudo tensor product" relevant for the Liouvillian, as in Eq. (I-73) or Eq. (I-77).

I.7 Density Matrices for Continuously Irradiated S = ½ spins

The meaning of the concepts defined in the previous Sections may be clarified by applying them to a concrete example. A simple example which is relevant to magnetic resonance is that of non-interacting spins (magnetic moments), to which a constant magnetic field is applied. In such a system the spins have energy levels related to their interaction with the external field. Transitions between these levels can be induced by irradiating the spins, i.e. applying a time dependent magnetic field. This situation will be treated in the present Section.

(a) The density matrix of a single free S = ½ spin

Non-interacting spins have, by definition, only interactions with the external magnetic field. Thus their behaviour can be studied by considering each of them as if it were the only spin in the system. A single free spin with S = 1/2 is defined by having the number S = 1/2 in the eigenvalue equation:

$$S^2 |\psi\rangle = S(S+1)\hbar^2 |\psi\rangle \tag{I-78}$$

for any wave function of the system, where **S** is the total spin angular momentum of the system. This is a two level system, in which the states are angular momentum eigenstates. The two wave functions |m = ½> and |m = -½>, defined by

$$S_z |m = ½\rangle = ½\hbar |m = ½\rangle \qquad S_z |m = -½\rangle = -½\hbar |m = -½\rangle \tag{I-79}$$

are often denoted as $|\alpha\rangle$ and $|\beta\rangle$, respectively. If the Hamiltonian is proportional to $\mathbf{S_z}$ they are also energy eigenstates. This is the actual situation when the constant magnetic field is along the **z** axis. The two states can be represented by coefficient vectors as seen in Section I.4 above. A general wave function is a linear combination: $|\psi\rangle = c_+|½\rangle + c_-|-½\rangle$ which can be represented by its coefficient vector as:

$$|\psi\rangle = \begin{bmatrix} c_+ \\ c_- \end{bmatrix} \tag{I-80}$$

If the spin is in a state described by this wave function, it is in a "pure state", and its density matrix is:

$$\rho = |\psi\rangle\langle\psi| =$$

$$= c_+|½\rangle\langle½|c_+^* + c_+|½\rangle\langle-½|c_-^* + c_-|-½\rangle\langle½|c_+^* + c_-|-½\rangle\langle-½|c_-^* \tag{I-81}$$

In this case it is very simple to construct $\boldsymbol{\rho}$ directly, multiplying the bra vector by the ket vector:

$$\rho = |\psi\rangle\langle\psi| = \begin{pmatrix} c_+ \\ c_- \end{pmatrix} \begin{pmatrix} c_+^* & c_-^* \end{pmatrix} = \begin{pmatrix} |c_+|^2 & c_+c_-^* \\ c_-c_+^* & |c_-|^2 \end{pmatrix} \tag{I-82}$$

which is just a different way of writing Eq. (I-77). In order to calculate the eigenvalues of $\boldsymbol{\rho}$ it is convenient to carry out the polar decomposition of the complex coefficients:

$$c_+ = ae^{i\phi} \qquad c_- = be^{i\xi} \tag{I-83}$$

(a, b, ϕ and ξ are real numbers). Then the characteristic equation: $\det\{\rho - \lambda \mathbf{I}\} = 0$ results in

$$\lambda = \frac{a^2+b^2}{2} \pm \left\{ \left(\frac{a^2-b^2}{2} \right)^2 + a^2b^2 \right\} \qquad or:$$

$$\lambda_1 = a^2+b^2 \qquad \lambda_2 = 0 \tag{I-84}$$

and the corresponding eigenvectors are found to be:

$$|v_1\rangle = \begin{pmatrix} c_+ \\ c_- \end{pmatrix} \qquad |v_2\rangle = \begin{pmatrix} c_-^* \\ -c_+^* \end{pmatrix} \tag{I-85}$$

Using Eq. (I-53) the density matrix can be written as:

$$\rho = \lambda_1 |v_1\rangle\langle v_1| + \lambda_2 |v_2\rangle\langle v_2| = (a^2+b^2)\,|v_1\rangle\langle v_1| = (a^2+b^2)\,|\psi\rangle\langle\psi| \tag{I-86}$$

This result corresponds to Eqs. (I-81),(I-82) above, except for the factor (a^2 + b^2). However, if the eigenvectors are normalized, this factor is equal to 1. In fact, since one usually works with normalized wave functions, it can be assumed that $|\psi\rangle$ is normalized, which implies in this case the normalization of the eigenfunctions.

According to Eq. (I-82) above, the elements of $\boldsymbol{\rho}$ in the static situation considered so far are:

$$\rho_{+,+} = a^2 \qquad \rho_{+,-} = ab \cdot \exp(i(\phi - \xi)) \qquad \rho_{-,+} = (\rho_{+,-})^* \qquad \rho_{-,-} = b^2 \tag{I-87}$$

Now assume a constant magnetic field is applied along the **z** axis, so that the Hamiltonian is

$$H = \omega_0 S_z \tag{I-88}$$

Then the two states $|½\rangle$, $|-½\rangle$ are energy eigenstates, with energies $E_+ = ½\hbar\omega_0$ and $E_- = -½\hbar\omega_0$ respectively. The time dependence of the system, starting from the initial condition of Eq. (I-80) (or (I-81)) in the absence of any time dependent external field was calculated in Sec. I.4 above. It was shown there that in a time dependent situation the off-diagonal elements are modified (Eq. (I-43)):

$$\rho_{+,-}(t) = ab \cdot \exp\left(i\omega_{+,-}t + i(\phi - \xi)\right) \qquad \rho_{-,+}(t) = \left(\rho_{+,-}(t)\right)^{*} \tag{I-89}$$

where $\omega_{+,-} = (E_+ - E_-)/\hbar$ is the transition frequency. In this basis **H** is diagonal, so it is easy to show that

$$\frac{1}{i\hbar}[H,\rho] = \begin{pmatrix} 0 & i\omega_{+,-}\rho_{+,-} \\ -i\omega_{+,-}\rho_{-,+} & 0 \end{pmatrix} \tag{I-90}$$

From Eqs. (I-87), (I-89) it follows that exactly the same matrix is obtained for $(d/dt)\rho$, as expected on the basis of the Liouville - von Neumann equation.

In Liouville space ρ has to be written as a vector, and **L** as a matrix. Since the basis used here diagonalizes **H**, the Liouville operator has a very simple form:

$$L_{ij,kl} = H_{i,k}\delta_{l,j} - \delta_{i,k}(H_{j,l})^{t} = E_i\,\delta_{i,k}\,\delta_{j,l} - \delta_{i,k}E_j\delta_{j,l} = \hbar\omega_{i,j}\,\delta_{i,k}\,\delta_{j,l} \tag{I-91}$$

and therefore the Liouville-von Neumann equation becomes:

$$\frac{d}{dt}\rho_{i,j} = -i\sum_{k,l}\omega_{i,j}\,\delta_{i,k}\,\delta_{j,l}\,\rho_{k,l} = -i\omega_{i,j}\,\rho_{i,j} \tag{I-92}$$

where the "diagonal" frequencies $\omega_{i,i}$ are zero. At this point one has to choose a particular order for arranging the elements of the density matrix in a vector. A useful order for magnetic resonance is based on the "quantum order" (or quantum multiplicity) of the elements. Here $\rho_{+,+} = \langle \frac{1}{2}|\rho|\frac{1}{2}\rangle$ and $\rho_{-,-} = \langle -\frac{1}{2}|\rho|-\frac{1}{2}\rangle$ are related to $\Delta m_s = 0$, whereas $\rho_{+,-} = \langle \frac{1}{2}|\rho|-\frac{1}{2}\rangle$ and $\rho_{-,+} = \langle -\frac{1}{2}|\rho|\frac{1}{2}\rangle$ are related to $|\Delta m_s| = 1$. In terms of transition frequencies for the system, the former are zero quantum transition elements, whereas the latter are single-quantum transition elements. Arranging the elements of ρ in the order: $\rho_{+,+}, \rho_{-,-}, \rho_{+,-}, \rho_{-,+}$ (two "zero quantum" elements followed by two "single quantum" elements), the Liouville-von Neumann equation is:

$$\frac{d}{dt}\begin{pmatrix} \rho_{+,+} \\ \rho_{-,-} \\ \rho_{+,-} \\ \rho_{-,+} \end{pmatrix} = \begin{pmatrix} 0 & 0 & 0 & 0 \\ 0 & 0 & 0 & 0 \\ 0 & 0 & -i\omega & 0 \\ 0 & 0 & 0 & i\omega \end{pmatrix}\begin{pmatrix} \rho_{+,+} \\ \rho_{-,-} \\ \rho_{+,-} \\ \rho_{-,+} \end{pmatrix} \tag{I-93}$$

Since the supermatrix is diagonal, the solution of the equation is immediate:

$$\rho_{+,+}(t) = \rho_{+,+}(0) \qquad \rho_{+,-}(t) = \rho_{+,-}(0) \cdot \exp\left(i\omega t\right)$$

$$\rho_{-,+}(t) = \left(\rho_{+,-}(t)\right)^* \qquad \rho_{-,-}(t) = \rho_{-,-}(0) \tag{I-94}$$

In the presence of relaxation processes it is usually convenient to work in this basis, so that the non-trivial structure of the Liouville matrix will arise only from relaxation.

Having found the time dependence of the density matrix one may calculate time dependent expectation values of observables. Thus, for example

$$\langle S_z \rangle(t) = Tr\{S_z \rho\} = Tr\left\{ \tfrac{1}{2} \begin{bmatrix} 1 & 0 \\ 0 & -1 \end{bmatrix} \begin{bmatrix} \rho_{+,+} & \rho_{+,-} \\ \rho_{-,+} & \rho_{-,-} \end{bmatrix} \right\} = \frac{a^2 - b^2}{2} \tag{I-95}$$

and

$$\langle S_y \rangle(t) = Tr\{S_y \rho\} = Tr\left\{ \tfrac{1}{2} \begin{bmatrix} 0 & -i \\ i & 0 \end{bmatrix} \begin{bmatrix} \rho_{+,+} & \rho_{+,-} \\ \rho_{-,+} & \rho_{-,-} \end{bmatrix} \right\} = ab \cdot \sin(\omega_{+,-} t + (\phi - \xi)) \tag{I-96}$$

(b) *Irradiating an ensemble of free spins ($S = ½$)*

If an ensemble of spins $S = ½$, which do not interact among themselves, is placed in a strong magnetic field then each spin has the Hamiltonian of Eq. (I-88). At equilibrium, which can be reached through interactions with the environment (the "lattice"), the system is described with the following density matrix:

$$\rho_{eq.} = \frac{1}{Z} \exp\left[-\frac{H}{k_B T} \right] = \frac{1}{Z} \exp\left[-\frac{\omega_0 S_z}{k_B T} \right] \tag{I-97}$$

In this equation Z is the "partition function" of statistical mechanics, which is equal to

$$Z = Tr\left\{ \exp\left[-\frac{H}{k_B T} \right] \right\} \tag{I-98}$$

The 1/Z factor, equal to ½ in the present case, is needed for normalization, i.e., to ensure that $Tr(\rho) = 1$. In magnetic resonance the resonance energies are much smaller than thermal energy, except in very low temperatures (of the order of 10^{-1} K for NMR or 10^{1} K for EPR). It is therefore common practice to invoke the "high temperature approximation":

$$\rho \approx \frac{1}{Z}\left[I - \frac{\omega_0}{k_B T} S_z \right] = ½\left(I - q S_z\right) \tag{I-99}$$

where **I** is the unit matrix and $q \equiv \omega_0/k_BT$.

Now assume that starting from this equilibrium situation, the system is irradiated with an oscillating magnetic field, having a Hamiltonian of the form: $\mathbf{H_1} = 2\omega_1 S_x \cos(\omega t)$. This time dependent field, the magnitude of which is small compared with that of the constant magnetic field, induces transitions between the energy levels corresponding to the main magnetic field. Transforming to the "rotating frame", which is simply an appropriate interaction representation, the time dependence is eliminated. The transformation will be discussed in more detail in Chapter II. For the present calculation it is only necessary to know that the total Hamiltonian, which is the sum of the main constant term and the irradiation term, now has the form:

$$H = (\omega - \omega_0)\, S_z + \omega_1\, S_x \tag{I-100}$$

where ω is the frequency of irradiation and ω_0 is the Larmor frequency of the spins. The simplest case is that of exact resonance - exact equality between the frequency of irradiation and the Larmor frequency. In that case:

$$H = \omega_1\, S_x \tag{I-101}$$

The time dependence of the density matrix and the magnetization of the system will now be calculated for this special case. The calculation will be done in four different ways, in order to demonstrate on this simple example the general alternatives that exist for doing calculations with the density matrix.

(c) Calculating the density matrix by manipulation of operators

One method of calculation is to work analytically with the quantum mechanical operators, so as to get an analytical result for the density matrix and for the relevant observable expectation values. Only after such a result is obtained will actual matrix representations be used, in order to get a numerical answer. This method has the advantage of providing a physical insight to the meaning of the final answer, but in complicated cases the use of this method may not be feasible. For the simple problem treated here the method can be applied easily, and this will be done now.

During irradiation the density matrix evolves according to Eq. (I-61):

$$\rho(t) = \exp\left(-iHt/\hbar\right)\rho(0)\exp\left(iHt/\hbar\right) \approx \exp\left(-i\omega_1 t S_x/\hbar\right) \tfrac{1}{2}(I - qS_z) \exp\left(i\omega_1 t S_x/\hbar\right)$$

$$= \frac{1}{2}I - \frac{1}{2}q\left(S_z\cos(\omega_1 t) - S_y\sin(\omega_1 t)\right) \qquad \text{(I-102)}$$

using the well known commutation relations of angular momentum operators (see Appendix A). The time is measured from the beginning of irradiation, which for simplicity is assumed to have a constant amplitude (a constant value of ω_1). Now assume the irradiation is applied for a limited time, stopping at time t, so that it is actually a pulse. If $\omega_1 t = \pi/2$ ("a $\pi/2$ pulse"):

$$\rho(t) = \frac{1}{2}I + \frac{1}{2}qS_y \qquad \text{(I-103)}$$

at the end of the pulse. What is interesting in practice is only the relevant (or non-trivial) part of the density operator:

$$\rho_{rel.} \equiv \rho - \frac{1}{Z}I \qquad \left(here:\ \rho_{rel.} = \rho - \frac{1}{2}I\right) \qquad \text{(I-104)}$$

This is because for any operator **A**:

$$\langle A\rangle = Tr(\rho A) = \sum_{i,j}\rho_{i,j}A_{j,i} = \sum_{i,j}\left[\frac{1}{Z}I + \rho_{rel.}\right]_{i,j}A_{j,i} = \sum_{i,j}\left[\frac{1}{Z}\delta_{i,j}A_{j,i} + (\rho_{rel.})_{i,j}A_{j,i}\right] =$$

$$= Tr(A) + Tr(\rho_{rel.}A) \qquad \text{(I-105)}$$

Since the calculations are done here in the "Schrödinger picture", the time dependence is contained in ρ(t) (actually in ρ_r(t)), so Tr(**A**) is a constant, which gives no information on any changes in the system. For example, for the elementary spin operators:

$$Tr(S_x) = Tr(S_y) = Tr(S_z) = 0 \qquad \text{(I-106)}$$

Therefore all dynamic information on the system is contained in that part of the density matrix which is *not* proportional to the unit matrix.

Now suppose that the macroscopic $\mathbf{S}_+$ is measured. In practice this means that $\mathbf{S}_x$ and $\mathbf{S}_y$ are measured (through "quadrature detection"), and the results are combined as $\mathbf{S}_+ = \mathbf{S}_x + i\mathbf{S}_y$. Then before irradiation:

$$Tr(S_+\rho(0)) = -\frac{1}{2}q\,Tr(S_+S_z) = 0 \qquad \text{(I-107)}$$

According to Eq. (I-103) above, a $\pi/2$ pulse along **x** will result in:

$$\langle S_+\rangle(t) = \tfrac{1}{2} q\, Tr\{S_+ S_y\} = Tr\left\{ \tfrac{1}{2} q \cdot \tfrac{1}{2} \hbar^2 \begin{bmatrix} 0 & 1 \\ 0 & 0 \end{bmatrix} \begin{bmatrix} 0 & -i \\ i & 0 \end{bmatrix} \right\} = \frac{iq\hbar^2}{4} \neq 0 \qquad \text{(I-108)}$$

Thus the macroscopic magnetization rotates from the **z** direction to the **x-y** plane, because the relevant part of the density operator is "rotated" in this way. This is the justification for the pictorial description of the spin being rotated from the **z** direction to the **x-y** plane.

(d) Calculation with matrix representations

An alternative method is to substitute matrices for operators at the outset. Here each quantum mechanical operator is represented by its matrix in a given basis of Hilbert space. A natural basis in the present problem is that of the spin functions, i.e., eigenfunctions of $\mathbf{S}^2$ and $\mathbf{S}_z$. In this basis:

$$\rho(0) = \tfrac{1}{2} \left\{ \begin{bmatrix} 1 & 0 \\ 0 & 1 \end{bmatrix} - q' \begin{bmatrix} 1 & 0 \\ 0 & -1 \end{bmatrix} \right\} = \tfrac{1}{2} \begin{bmatrix} 1-q' & 0 \\ 0 & 1+q' \end{bmatrix} \qquad \text{(I-109)}$$

with $q' \equiv \tfrac{1}{2} q\hbar$. To apply Eq. (I-61) one has to calculate the exponential of the Hamiltonian, which requires knowledge of all powers of the Hamiltonian matrix. For the Hamiltonian of Eq. (I-101), the lowest powers are:

$$\frac{1}{\hbar} H = \frac{\omega_1}{2} \begin{bmatrix} 0 & 1 \\ 1 & 0 \end{bmatrix} \qquad \left\{ \frac{1}{\hbar} H \right\}^2 = \left[\frac{\omega_1}{2} \right]^2 \begin{bmatrix} 1 & 0 \\ 0 & 1 \end{bmatrix} = \left[\frac{\omega_1}{2} \right]^2 I$$

$$\left\{ \frac{1}{\hbar} H \right\}^3 = \left[\frac{\omega_1}{2} \right]^3 \begin{bmatrix} 0 & 1 \\ 1 & 0 \end{bmatrix} \qquad \left\{ \frac{1}{\hbar} H \right\}^4 = \left[\frac{\omega_1}{2} \right]^4 I \quad \ldots \qquad \text{(I-110)}$$

Thus the exponential is equal to

$$\exp\{iHt/\hbar\} \equiv I + \frac{it}{\hbar}H + \frac{1}{2!}\left(\frac{it}{\hbar}\right)^2 H^2 + \frac{1}{3!}\left(\frac{it}{\hbar}\right)^3 H^3 + \frac{1}{4!}\left(\frac{it}{\hbar}\right)^4 H^4 + \ldots =$$

$$= \left\{1 + \frac{1}{2!}\left(\frac{i\omega_1 t}{2}\right)^2 + \frac{1}{4!}\left(\frac{i\omega_1 t}{2}\right)^4 + \ldots\right\} I + \left\{\left(\frac{i\omega_1 t}{2}\right) + \frac{1}{3!}\left(\frac{i\omega_1 t}{2}\right)^3 + \ldots\right\}\sigma_x =$$

$$= \left\{\cos\left(\frac{\omega_1 t}{2}\right)\right\} I + i\left\{\sin\left(\frac{\omega_1 t}{2}\right)\right\}\sigma_x = \begin{bmatrix}\cos(\omega_1 t/2) & i\cdot\sin(\omega_1 t/2) \\ i\cdot\sin(\omega_1 t/2) & \cos(\omega_1 t/2)\end{bmatrix} \qquad \text{(I-111)}$$

using the Pauli matrix σ_x (see Appendix A). Calculating the complex conjugate of this exponential is equivalent to inverting the sign of $\omega_1 t$ in the exponent and using the known parity properties of the cosine and the sine:

$$\exp(-iHt/\hbar) = \left\{\cos\left(\frac{\omega_1 t}{2}\right)\right\} I - i\left\{\sin\left(\frac{\omega_1 t}{2}\right)\right\}\sigma_x =$$

$$= \begin{bmatrix}\cos(\omega_1 t/2) & -i\cdot\sin(\omega_1 t/2) \\ -i\cdot\sin(\omega_1 t/2) & \cos(\omega_1 t/2)\end{bmatrix} \qquad \text{(I-112)}$$

All that remains is to multiply out the three matrices of Eqs. (I-112), (I-109) and (I-111) and obtain:

$$\rho(t) = ½\left\{\begin{bmatrix}1 & 0 \\ 0 & 1\end{bmatrix} - q'\begin{bmatrix}1 & 0 \\ 0 & -1\end{bmatrix}\cos(\omega_1 t) + iq'\begin{bmatrix}0 & -1 \\ 1 & 0\end{bmatrix}\sin(\omega_1 t)\right\} =$$

$$= \frac{1}{2}\{I - q\,S_z\cos(\omega_1 t) + q\,S_y\sin(\omega_1 t)\} \qquad \text{(I-113)}$$

which is the same as Eq. (I-102) obtained above analytically. With this final form of the density matrix one may calculate expectation values of operators as above. The use of matrices right from the beginning is admittedly less elegant than the use of operators, and at least in this example results in relatively lengthy calculations. However, in a similar manner one can work with matrices (in Hilbert space) also in more general problems, when non-commuting terms make it difficult to work analytically with operators. Moreover, when the dimensions of a problem are too big for a direct solution one must resort to various numerical methods to solve the problem. A matrix representation of operators is then essential. In cases in which dynamics is involved the dimensions of the calculation are generally large, so it is usually the case that, even if a formal analytical solution may be written down, its practical implications can only be found by such numerical work.

(e) Analytical calculation in Liouville space

Liouville space is defined in such a way that the equation of motion of the density matrix is very simple (Eq. (I-68) above). However, as soon as one attempts to identify the "components" of this equation one has to resort to the use of ordinary Hilbert space, with Eq. (I-73) to connect the two spaces. In the solution, Eq. (I-69), one needs the exponential of the Liouvillian, which is:

$$\exp\left(-iLt/\hbar\right) = \exp\left(-i(\boldsymbol{H}\otimes\boldsymbol{I} - \boldsymbol{I}\otimes\boldsymbol{H})t/\hbar\right) \tag{I-114}$$

This, in turn, requires calculating all powers of the Liouvillian. The second power, for example, is equal (using Eq. (I-77)) to

$$\{\boldsymbol{H}\otimes\boldsymbol{I} - \boldsymbol{I}\otimes\boldsymbol{H}\}^2 = (\boldsymbol{H}\otimes\boldsymbol{I})(\boldsymbol{H}\otimes\boldsymbol{I}) - (\boldsymbol{H}\otimes\boldsymbol{I})(\boldsymbol{I}\otimes\boldsymbol{H}) - (\boldsymbol{I}\otimes\boldsymbol{H})(\boldsymbol{H}\otimes\boldsymbol{I}) + (\boldsymbol{I}\otimes\boldsymbol{H})(\boldsymbol{I}\otimes\boldsymbol{H}) =$$

$$= (\boldsymbol{H}^2\otimes\boldsymbol{I}) - 2(\boldsymbol{H}\otimes\boldsymbol{H}) + (\boldsymbol{I}\otimes\boldsymbol{H}^2) \tag{I-115}$$

Higher powers of the Liouvillian can be calculated in the same manner. Operating with such superoperators on the density matrix (as in Eq. (I-68)) involves calculations of the following form:

$$\sum_{k,l} (\boldsymbol{A}\otimes\boldsymbol{B})_{ij,kl}\, \boldsymbol{C}_{k,l} = \sum_{k,l} \langle\langle\, |i\rangle\langle j|\, \boldsymbol{A}\otimes\boldsymbol{B}\, |k\rangle\langle l|\, \rangle\rangle \langle k|\boldsymbol{C}|l\rangle =$$

$$= \sum_{k,l} \boldsymbol{A}_{i,k}\boldsymbol{B}_{l,j}\boldsymbol{C}_{k,l} = (\boldsymbol{ACB})_{i,j} \tag{I-116}$$

where **A, B, C** are Hilbert space operators ("**C**" in this case is the density matrix ρ). The corresponding operator equation is

$$(\boldsymbol{A}\otimes\boldsymbol{B})\boldsymbol{C} = \boldsymbol{ACB} \tag{I-117}$$

For example, when the exponential superoperator of Eq. (I-114) operates on the density matrix, the first order term is (up to a scalar):

$$\boldsymbol{L}\rho = (\boldsymbol{H}\otimes\boldsymbol{I} - \boldsymbol{I}\otimes\boldsymbol{H})\rho = \boldsymbol{H}\rho - \rho\boldsymbol{H} \tag{I-118}$$

which is just the commutator, which served as a basis for the definition of the Liouvillian. The second order term is proportional to:

$$L^2\rho = (H\otimes I - I\otimes H)L\rho = H^2\rho - 2H\rho H + \rho H^2 \tag{I-119}$$

as follows from Eq. (I-118) or from Eq. (I-115). This is a second order commutator, and it is obvious that higher order commutators will have a more complicated form. In the general case it is impossible to obtain a closed expression for all commutators. However, in the present case the Hamiltonian (of Eq. (I-101)) is simple, and is simply related to the initial density matrix, so the calculations can be carried through. The first step is:

$$\frac{-it}{\hbar}L\rho = \frac{-it}{\hbar}(H\rho - \rho H) = \frac{-i\omega_1 t}{2\hbar}\left\{(S_x I - qS_xS_z) - (IS_x - qS_zS_x)\right\} = \frac{-iq\omega_1 t}{2\hbar}[S_z,S_x] =$$

$$= \frac{q\omega_1 t}{2}S_y \tag{I-120}$$

Thus the next power needed for the exponential is

$$\left[\frac{-it}{\hbar}L\right]^2\rho = \left[\frac{-it}{\hbar}\right]^2(H(L\rho) - (L\rho)H) = \frac{-iq}{2\hbar}(\omega_1 t)^2[S_x,S_y] =$$

$$= \frac{q(\omega_1 t)^2}{2}S_z \tag{I-121}$$

This is equal to the non-trivial part of the zeroth order term ,i.e., of ρ itself (see Eq. (I-99)) multiplied by $-(\omega_1 t)^2$. Therefore operating on the second order term with $(-it/\hbar)$L as in Eq. (I-120) will yield the result of Eq. (I-120) multiplied by the same factor, and so on. The result for the exponential is therefore:

$$\rho(t) = \exp\{-iLt/\hbar\}\rho(0) =$$

$$= \frac{1}{2}I - \frac{q}{2}S_z + \frac{q\omega_1 t}{2}S_y - \frac{1}{2!}\left[\frac{q(\omega_1 t)^2}{2}\right]S_z - \frac{1}{3!}\left[\frac{q(\omega_1 t)^3}{2}\right]S_y + \frac{1}{4!}\left[\frac{q(\omega_1 t)^4}{2}\right]S_z + \dots =$$

$$= \frac{1}{2}I - \frac{q}{2}\cos(\omega_1 t)S_z + \frac{q}{2}\sin(\omega_1 t)S_y \tag{I-122}$$

This is, as expected, the same as the result in Eq. (I-102).

(f) *Matrix representations in Liouville space*

In this method the Liouville supermatrix must be written explicitly from the start. The first step to be done is to choose a particular order for the elements of the density matrix, arranged as a Liouville space vector. In Hilbert space the density matrix is written as

$$\rho = \begin{bmatrix} \rho_{1,1} & \rho_{1,2} \\ \rho_{2,1} & \rho_{2,2} \end{bmatrix} \tag{I-123}$$

In Liouville space, an ordering of the matrix elements useful for magnetic resonance is based on the "quantum order" (or quantum multiplicity) of these elements. Here $\rho_{11} = \langle ½|\rho|½\rangle$ and $\rho_{22} = \langle -½|\rho|-½\rangle$ are related to $\Delta m_s = 0$, whereas $\rho_{12} = \langle ½|\rho|-½\rangle$ and $\rho_{21} = \langle -½|\rho|½\rangle$ are related to $|\Delta m_s| = 1$. In terms of transition frequencies for the system, the former are zero quantum transition elements, whereas the latter are single-quantum transition elements. We therefore put ρ_{11},ρ_{22} first, and then ρ_{12},ρ_{21}. Von Neumann's equation is then:

$$\frac{d}{dt}\begin{pmatrix} \rho_{1,1} \\ \rho_{2,2} \\ \rho_{1,2} \\ \rho_{2,1} \end{pmatrix} = -i \begin{pmatrix} L_{11,11} & L_{11,22} & L_{11,12} & L_{11,21} \\ L_{22,11} & L_{22,22} & L_{22,12} & L_{22,21} \\ L_{12,11} & L_{12,22} & L_{12,12} & L_{12,21} \\ L_{21,11} & L_{21,22} & L_{21,12} & L_{21,21} \end{pmatrix} \begin{pmatrix} \rho_{1,1} \\ \rho_{2,2} \\ \rho_{1,2} \\ \rho_{2,1} \end{pmatrix} \tag{I-124}$$

where $\mathbf{L}_{ij,kl} = \mathbf{H}_{ik}\delta_{lj} - \delta_{ik}\mathbf{H}_{lj} = (\mathbf{L}_{kl,ij})^*$ (**L** is hermitian because **H** is hermitian). In this case

$$H = \frac{\hbar\,\omega_1}{2}\begin{bmatrix} 0 & 1 \\ 1 & 0 \end{bmatrix} \quad \Rightarrow \quad H_{1,1} = H_{2,2} = 0 \qquad H_{1,2} = H_{2,1} = \frac{\hbar\,\omega_1}{2} \tag{I-125}$$

The super-matrix and the super-vector are thus:

$$L = \frac{\hbar\,\omega_1}{2}\begin{pmatrix} 0 & 0 & -1 & 1 \\ 0 & 0 & 1 & -1 \\ -1 & 1 & 0 & 0 \\ 1 & -1 & 0 & 0 \end{pmatrix} \qquad \rho(0) = \frac{1}{2}\begin{pmatrix} 1-q' \\ 1+q' \\ 0 \\ 0 \end{pmatrix} \tag{I-126}$$

and the formal solution of Von Neumann's equation is:

$$\rho(t) = \exp\{-iLt/\hbar\}\,\rho(0) = \cos(Lt/\hbar)\,\rho(0) - i\cdot\sin(Lt/\hbar)\,\rho(0) \tag{I-127}$$

In order to calculate all powers of the Liouvillian supermatrix, it is convenient to define

$$b \equiv \frac{\hbar\,\omega_1}{2} \qquad M \equiv \begin{bmatrix} -1 & 1 \\ 1 & -1 \end{bmatrix} \qquad \Rightarrow \qquad L = b \begin{bmatrix} \mathbf{0} & M \\ M & \mathbf{0} \end{bmatrix} \tag{I-128}$$

Note that **L** is expressed here in terms of sub-matrices - both **M** and the **0** appearing in this form of **L** are 2 by 2 matrices and not scalars. It is easy to verify that

$$M^2 = \begin{bmatrix} 2 & -2 \\ -2 & 2 \end{bmatrix} = -2M \tag{I-129}$$

and thus:

$$L^2 = b^2 \begin{bmatrix} M^2 & \mathbf{0} \\ \mathbf{0} & M^2 \end{bmatrix} = -2b^2 \begin{bmatrix} M & \mathbf{0} \\ \mathbf{0} & M \end{bmatrix}$$

$$L^3 = -2b^3 \begin{bmatrix} \mathbf{0} & M^2 \\ M^2 & \mathbf{0} \end{bmatrix} = 4b^3 \begin{bmatrix} \mathbf{0} & M \\ M & \mathbf{0} \end{bmatrix}$$

$$L^4 = 4b^4 \begin{bmatrix} M^2 & \mathbf{0} \\ \mathbf{0} & M^2 \end{bmatrix} = -8b^4 \begin{bmatrix} M & \mathbf{0} \\ \mathbf{0} & M \end{bmatrix} \qquad etc. \tag{I-130}$$

Therefore

$$\cos(Lt/\hbar) = \begin{bmatrix} I & \mathbf{0} \\ \mathbf{0} & I \end{bmatrix} + \frac{1}{2}\{1 - \cos(\omega_1 t)\} \begin{bmatrix} M & \mathbf{0} \\ \mathbf{0} & M \end{bmatrix} \tag{I-131}$$

$$\sin(Lt/\hbar) = \frac{1}{2}\sin(\omega_1 t) \begin{bmatrix} \mathbf{0} & M \\ M & \mathbf{0} \end{bmatrix} \tag{I-132}$$

When these supermatrices operate on the supervector of $\boldsymbol{\rho}(0)$, the result is:

$$\rho(t) = \frac{1}{2}\begin{pmatrix} 1-q' \\ 1+q' \\ 0 \\ 0 \end{pmatrix} + \frac{1}{2}\{1-\cos(\omega_1 t)\}\begin{pmatrix} q' \\ -q' \\ 0 \\ 0 \end{pmatrix} + \frac{i}{2}\sin(\omega_1 t)\begin{pmatrix} 0 \\ 0 \\ q' \\ -q' \end{pmatrix} =$$

$$= \frac{1}{2}\begin{pmatrix} 1 \\ 1 \\ 0 \\ 0 \end{pmatrix} - \frac{q'}{2}\cos(\omega_1 t))\begin{pmatrix} 1 \\ -1 \\ 0 \\ 0 \end{pmatrix} + \frac{q'}{2}\sin(\omega_1 t)\begin{pmatrix} 0 \\ 0 \\ -i \\ i \end{pmatrix} =$$

$$= \frac{1}{2}I - \frac{q}{2}\cos(\omega_1 t)\, S_z + \frac{q}{2}\sin(\omega_1 t)\, S_y \qquad \text{(I-133)}$$

The last line was obtained here by noting the way that $\mathbf{S}_y$ and $\mathbf{S}_z$ are written as supervectors, using the same ordering of elements as specified above. Again, this final form of the density matrix is identical to that obtained previously.

This form of calculation may look even more cumbersome than the previous one, which involved the ordinary Hilbert space. However, as will be seen in the following chapters, if the appropriate equation of motion for the system is the "stochastic Liouville equation" (which will be defined there) one must work in Liouville space. This fourth method of calculation is then the relevant method. In fact, throughout much of the book the emphasis will be on developing certain operators in Liouville space, analogous to those used in subsection (e) above. However, in most cases it will not be practical to obtain a solution analytically. Therefore the main goal in such situations will be to develop analytically the expressions for writing down a Liouvillian matrix, as done in the present subsection.

Suggested References

* On the classical Liouville equation:

1. F. Reif, *"Fundamentals of statistical and thermal physics"*, (McGraw-Hill, New York, 1965) Appendix A.13.
2. K. Huang, *"Statistical Mechanics"*, 2nd ed. (Wiley, New York, 1987) Ch. 3.
3. L.D. Landau and E.M. Lifshitz, *"Statistical Physics"* (Part I), 3rd ed. (Pergamon Press, Oxford, 1980) Ch. I.

* On the quantum mechanical density matrix:

Ref. 3 above, as well as:

4. U. Fano, *Rev. Mod. Phys.* **29**, 74 (1957).
5. D. ter Haar, *Rept. Progr. Phys.* **24**, 304 (1961).
6. R.P. Feynman, *"Statistical Mechanics"* (Benjamin-Cummings, Reading MA, 1972) Ch. 2 .
7. L.D. Landau and E.M. Lifshitz, *"Quantum Mechanics"*, 3rd ed. (Pergamon Press, Oxford, 1977) Ch. 1.
8. C. Cohen-Tannoudji, B. Diu and F. Laloë, *"Quantum Mechanics"*, Vol. I (Wiley, New York, 1977), Complement E_{III}.

* On the formal structure of the quantum mechanical Liouville operator:

9. P.L. Corio, *"Structure of High Resolution NMR Spectra"* (Academic Press, New York, 1966) Ch. 4.
10. C.N. Banwell and H. Primas, *Mol. Phys.* **6**, 225 (1963).
11. G. Binsch, *J. Am. Chem. Soc.* **91**, 1304 (1969); D.A. Kleier and G. Binsch, *J. Mag. Reson.* **3**, 146 (1970).
12. I. Prigogine, C. George, F. Henin and L. Rosenfeld, *Chem. Scr.* **4**, 5 (1973).
13. J. Jeener, in *Adv. Mag. Res.* vol. 10, edited by J. S. Waugh (Academic Press, New York, 1982), p.1.
14. R.R. Ernst, G. Bodenhausen and A. Wokaun, *"Principles of Nuclear Magnetic Resonance in One and Two Dimensions"* (Clarendon Press, Oxford, 1987) Ch. 2.

CHAPTER II

PERTURBATIVE RELAXATION THEORY

A rigorous treatment of relaxation in magnetic resonance requires using the density matrix formalism, including in it those effects which are responsible for relaxation. The actual theory is quite involved, and therefore simplifications are made where possible. The classical equations which describe magnetic resonance with relaxation will be presented here first, so that some qualitative features of the subject will be clear before the more sophisticated mathematical theory is studied. This is followed by some approximate quantum mechanical treatments of relaxation, as an introduction to the main theories. After this brief review, two important relaxation theories will be presented. One theory, due to Bloch and Redfield, is presented in this Chapter. It relies on an iterative perturbation scheme to derive relaxation equations for the density matrix when the processes causing relaxation are relatively fast. These processes are assumed to be stochastic in nature, so the theory also relies on the randomness of these processes. However, the main emphasis in this method is on the perturbation approach. An alternative theory, in which the randomness plays the central role, will be presented in Chapters IV and V.

II.1 The Phenomenological Bloch Equations for Magnetic Resonance

In the most basic magnetic resonance experiments one employs a strong static magnetic field, and a much weaker oscillating magnetic field perpendicular to it. The sample is assumed to be paramagnetic. i.e., the atomic nuclei (for NMR) or the molecular arrangement of electrons (in EPR) possess a non-zero magnetic moment. This moment is due to the spin angular momentum of the nucleus in the case of NMR. In EPR it is mainly due to the spin of unpaired electrons, although it is also due to the influence of surrounding electrons or ions. For this reason, a nucleus or a molecule which having a non-zero magnetic moment will often be referred to as "a spin". Normally the atoms or molecules are randomly oriented, and thus the sample does not exhibit net paramagnetism. The strong field creates a preference of the paramagnetic moments to align parallel to it due to the energy of interaction between a magnetic moment $\boldsymbol{\mu}$ and an external magnetic field **B**:

$$E_{\mu} = -\boldsymbol{\mu} \cdot \boldsymbol{B} \qquad \text{(II-1)}$$

This alignment effect creates an excess (or deficiency - depending on the sign of the magnetic moment) of spins parallel to the field compared with those anti-parallel to it, and thus net magnetization is induced in the sample. Moreover, the interaction causes the magnetic moment to precess around the direction of the external magnetic field. The time dependent field induces changes in the magnetization of the sample, by changing the instantaneous value of the total magnetic field. If the time dependent field is strong, one may reach partial or full saturation, but in this Chapter only the case of no saturation will be treated.

Much insight can be gained by formulating the main characteristics of the experiment in classical terms. This is done by the classical Bloch equations for the macroscopic magnetization vector of the sample.

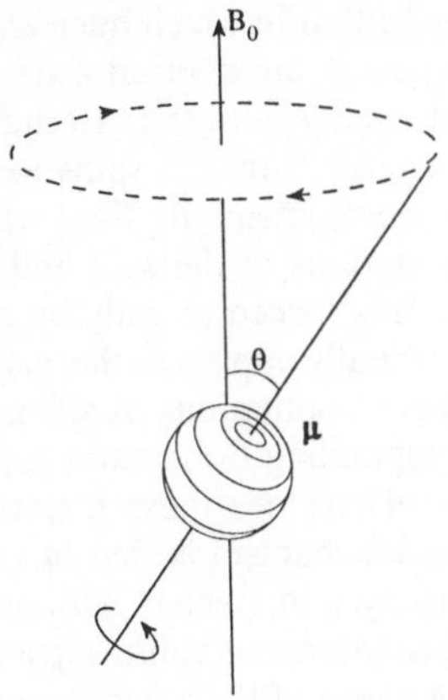

Fig. II.1. *The precession of a magnetic moment in a constant magnetic field.*

The direction of the static field is conventionally defined as the **z** direction, and that of the oscillating field as the **x** direction. If the static field is equal to

$$\boldsymbol{B}_0 = -\frac{\omega_0}{\gamma} z \tag{II-2}$$

where γ is the magnetogyric ratio, then a small transverse field of the form

$$\boldsymbol{B}_1 = -\frac{\omega_1}{\gamma} 2\cos(\omega t)\, x \tag{II-3}$$

will induce changes. Using the electromagnetic equations for precession of a magnetic moment around an external magnetic field, one obtains a set of three equations for the three components of the magnetization vector **M**. The equations are transformed to a frame of reference rotating around $\mathbf{B}_0$ at the irradiation frequency ω. These equations describe precession around the effective field, defined in the rotating frame as

$$\boldsymbol{B}_{eff} = -\frac{\omega_e}{\gamma}\left\{\frac{\omega_0-\omega}{\omega_e} z + \frac{\omega_1}{\omega_e} x\right\} \tag{II-4}$$

with the effective frequency $\omega_e \equiv \{(\omega_0\text{-}\omega)^2 + \omega_1^2\}^{1/2}$. The measured energy absorption is proportional to the magnetization component M_y . The magnetogyric ratio γ which relates frequency (and thus also energy) to the magnetic field is proportional to e/m, where e is the electric charge and m is the mass. This dependence is the source of two important differences between NMR and EPR. First, since the mass of a nucleus is greater by three orders of magnitude than the mass of an electron, typical frequencies in EPR are greater by three orders of magnitude than typical frequencies in NMR. Thus EPR is characterized by time scales which are shorter by $\approx 10^3$ than NMR time scales. The second difference, due to the sign of the electric charge, is that the electronic spins tend to align in a direction opposite to that in which nuclear spins align. In the quantum mechanical treatment this is expressed in an inverted order of the energy levels.

In addition to the transfer of energy from the external oscillating field to the spin system, there is also some loss of energy from the spins to their environment. This is partly due to the fluctuating local electromagnetic field which is experienced by each magnetic moment, due to random motions of the spin and of other molecules. These interactions ensure that if the system is forced initially by external perturbations to be in a non-equilibrium state, it will gradually approach thermal equilibrium . Equilibrium is characterized by having the relative populations of spins parallel and antiparallel to the static field related by the corresponding Boltzmann factors $\exp(-E/k_BT)$, where E is the energy of the relevant state. Therefore these interactions serve as a relaxation mechanism fo the system. Another mechanism is that of various interactions between each spin and other parts of the system, which will not directly influence these populations, but will influence the instantaneous value of transverse magnetization. The first type of mechanism will cause a decay of the **z** component of magnetization back to its equilibrium value, with a characteristic time constant T_1. The second type of mechanism will cause a decay of the transverse components of magnetization back to zero, which is their equilibrium value, at a typical rate of T_2 .

The combination of Larmor precession and a simultaneous decay of magnetization leads to the phenomenological description of relaxation in the Bloch equations:

$$\frac{d}{dt}M_x = -\frac{1}{T_2}M_x + \Delta\omega M_y \qquad \text{(II-5a)}$$

$$\frac{d}{dt}M_y = -\Delta\omega M_x - \frac{1}{T_2}M_y - \omega_1 M_z \qquad \text{(II-5b)}$$

$$\frac{d}{dt}M_z = \omega_1 M_y - \frac{1}{T_1}(M_z - M_0) \qquad \text{(II-5c)}$$

where $\Delta\omega$ is the difference between the irradiation frequency ω and the Larmor frequency ω_0: $\Delta\omega=\omega-\omega_0$. Intra-molecular interactions are not included in these equations; they are included, however, in the general quantum mechanical treatment, to be presented later on in this Chapter. From the solution of these phenomenological equations it is found that the energy absorbed in the system depends on $\Delta\omega$. The form of the dependence is that of a resonance phenomenon, i.e. , the energy absorption is significant only if $\Delta\omega$ is very small compared to the irradiation strength ω_1.

By solving the equations one can predict the time evolution of the magnetization, which is generally damped oscillations as a function of time. The oscillation frequency is the distance (in frequency) from resonance, and the decay rate is related to the time constants T_2 and T_1. A Fourier transform (with respect to time) of the time dependent magnetization yields the frequency domain spectrum, a broadened Lorentzian line in the present case. If the magnetization has a time evolution of the form

$$M(t) = \exp\left(i\omega_a t - bt\right) \tag{II-6}$$

then the corresponding frequency domain spectrum is:

$$\overline{M}(\omega) = \int_0^\infty \exp\left(-i\omega t\right)\exp\left(i\omega_a t - bt\right)dt = \frac{1}{b + i\left(\omega_a - \omega\right)} \tag{II-7}$$

The real part of this expression is $b\{b^2+(\omega_a-\omega)^2\}^{-1}$, representing a Lorentzian line. Its half-width is equal to b. In practice the role of the parameter "b" is taken by $1/T_2$, as will be seen in the next section for FID following a single pulse. The position (in frequency domain) of the center of the line is determined by the resonance frequency ω_a, which is the Larmor frequency modified by various interaction terms.

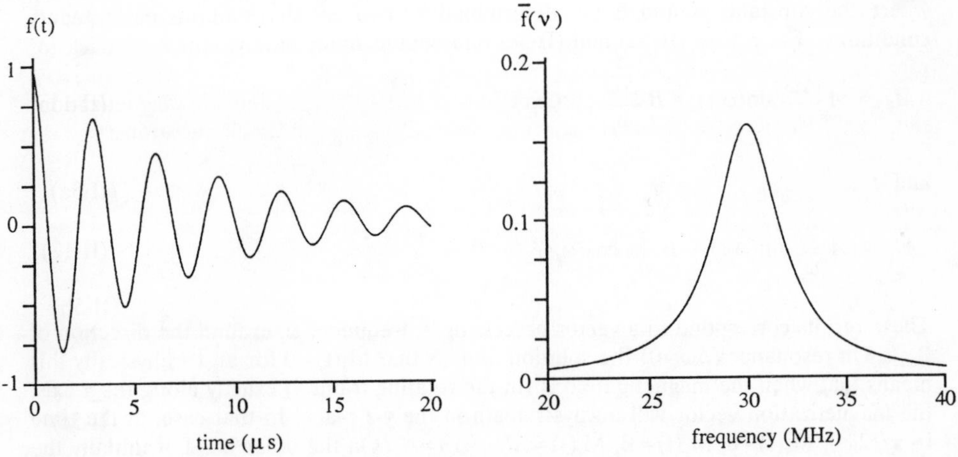

Fig. II.2. *A decaying signal f(t) = cos(ωt)*exp(-bt) and its Fourier transform.*

II.2 Applying the Bloch Equations

In order to illustrate the most prominent features of the Bloch equations they will be applied here to three simple cases.

(a) No relaxation ($T_2, T_1 \to \infty$)

In this case all interactions with the environment are taken to be so weak that the decay is infinitely slow, and therefore negligible. Taking the time derivative of Eq. (II-5b) and using Eqs. (II-5a), (II-5c):

$$\frac{d^2}{dt^2}M_y = -\Delta\omega \frac{d}{dt}M_x - \omega_1 \frac{d}{dt}M_z = -(\Delta\omega)^2 M_y - (\omega_1)^2 M_y \tag{II-8}$$

Defining the effective frequency as $\omega_e^2 \equiv (\Delta\omega)^2 + \omega_1^2$, this equation can be written in the form:

$$\frac{d^2}{dt^2}M_y + (\omega_e)^2 M_y = 0 \tag{II-9}$$

which is the same as the standard equation of motion for harmonic oscillations. The well-known solution is

$$M_y = A\cos(\omega_e t) + B\sin(\omega_e t) \tag{II-10}$$

where the constants A and B are determined by two initial conditions or boundary conditions. From Eqs. (II-5a) and (II-5c) one obtains immediately:

$$M_x = A\frac{\Delta\omega}{\omega_e}\sin(\omega_e t) - B\frac{\Delta\omega}{\omega_e}\cos(\omega_e t) \tag{II-11}$$

and

$$M_z = A\frac{\omega_1}{\omega_e}\sin(\omega_e t) - B\frac{\omega_1}{\omega_e}\cos(\omega_e t) \tag{II-12}$$

These results correspond to a vector precessing at frequency ω_e around the direction of $\mathbf{B}_{eff}$. On resonance ($\Delta\omega=0$) the solution implies that $M_x(t)=0$ for all t. Physically this means that when the magnetic field is (in the rotating frame !) exactly along the **x** axis, the magnetization vector will always remain in the **y-z** plane. In that case, at the time $t=\pi/(2\omega_e)$: $M_x(t)=0$, $M_y(t)=B$, $M_z(t)=A(\omega_1/\omega_e)=A$. On the other hand, if initially the system was at equilibrium , $M_x(0)=0$, $M_y(0)=0$, $M_z(t)=M_0$ and then $A=0$. Therefore in this special case $M_z(t)=0$, $M_y(t)=-M_z(0)$ i.e., the magnetization has been rotated from the **z** direction to the **y** direction after the time interval $t=\pi/(2\omega_e)$. This is a $\pi/2$ rotation, so the irradiation in this time interval is called a $\pi/2$ pulse (assuming irradiation does not continue).

If the system is initially at equilibrium, so that only $M_z\neq 0$, and the irradiation is off resonance, one must have $A=B=0$ (from $M_x(0)=M_y(0)=0$) but then also $M_z(0)=0$. This means the equations (with a non-zero $\Delta\omega$) are inconsistent, because they do not allow a non-zero initial M_z if both M_x and M_y are initially zero. The reason is that by omitting the relaxation term of Eq. (II-5c) the longitudinal magnetization component is forced to relax to zero, like the transverse components. This is physically incorrect, because the longitudinal component should relax to a non-zero equilibrium value. This inconsistency shows that the longitudinal relaxation term is essential for a correct description of evolution from or to equilibrium. However, when this term is included the solution is not so simple any more. A partial solution to the problem is to subtract from the left hand side of Eq. (II-5c) the time derivative of M_0, which is equal to zero. Then the initial zero value of M_x, M_y will only imply an initial zero value of M_z-M_0, which is physically correct. Nevertheless, these initial values will remain unchanged for all t. A related case is treated in the following subsection.

(b) No irradiation ($\omega_1=0$)

In this case the spins are only under the influence of the static magnetic field. In such a case it is meaningless to define a rotating frame, so the calculations are done in the laboratory frame. The Bloch equations are therefore modified by $\Delta\omega \rightarrow -\omega_0$, and are thus given by

$$\frac{d}{dt}M_x = -\frac{1}{T_2}M_x - \omega_0 M_y \tag{II-13a}$$

$$\frac{d}{dt}M_y = \omega_0 M_x - \frac{1}{T_2}M_y \tag{II-13b}$$

$$\frac{d}{dt}M_z = -\frac{1}{T_1}(M_z - M_0) \tag{II-13c}$$

The third equation is trivial to solve, because it is not coupled to the other two equations. Solving first for M_z-M_0, which has the same time derivative as M_z, and then adding M_0:

$$M_z(t) = M_0 + \{M_z(0) - M_0\}\exp\left[-\frac{t}{T_1}\right] \tag{II-14}$$

This solution corresponds to a gradual decay (or growth) from an arbitrary initial value $M_z(0)$ to the equilibrium value M_0, with a time constant T_1. For the transverse components one could try to differentiate the first two equations and substitute (II-13a) into the differentiated (II-13b), and vice versa. This would give for M_y

$$\frac{d^2}{dt^2}M_y + \frac{1}{T_2}\frac{d}{dt}M_y + \frac{\omega_0}{T_2}M_x + (\omega_0)^2 M_y = 0 \tag{II-15}$$

and a similar equation for M_x. These are basically the equations for damped harmonic oscillators which are coupled to one another, and the coupling complicates the problem. It is therefore preferable to solve directly by regarding M_x, M_y as the elements of a two dimensional vector. Define $v_1 \equiv M_x$, $v_2 \equiv M_y$, $a \equiv \omega_0$ and $b \equiv 1/T_2$. Then Eqs. (II-13a), (II-13b) can be combined to the vector equation

$$\frac{d}{dt} v = C v \tag{II-16}$$

with the definitions

$$v = \begin{bmatrix} v_1 \\ v_2 \end{bmatrix} \qquad C = \begin{bmatrix} -b & -a \\ a & -b \end{bmatrix} \tag{II-17}$$

The formal solution is given by (cf. Eq. (I-17)):

$$v(t) = \exp\{Ct\}\, v(0) = \left[I + Ct + \frac{1}{2!} C^2 t^2 + \ldots \right] v(0) \tag{II-18}$$

which requires knowledge of all powers of the matrix **C**. This may be impossible for a general matrix, but it is very simple for a diagonal matrix, which is exponentiated simply by exponentiating separately each of its diagonal elements. If **C** is diagonalized by

$$T^{-1} C T = \Lambda \tag{II-19}$$

then

$$\exp\{Ct\} = \left[I + t\left(T\Lambda T^{-1}\right) + \frac{1}{2!} t^2 \left(T\Lambda T^{-1}\right)\left(T\Lambda T^{-1}\right) + \ldots \right] =$$

$$= T \left[I + t\Lambda + \frac{1}{2!} t^2 \Lambda^2 + \ldots \right] T^{-1} = T \exp(\Lambda t)\, T^{-1} \tag{II-20}$$

In terms of matrix elements the equation is:

$$\left(\exp(Ct)\right)_{i,j} = \sum_{k,l} T_{i,k} \left(\exp(\Lambda t)\right)_{k,l} \left(T^{-1}\right)_{l,j} = \sum_{k,l} T_{i,k} \exp\left(\Lambda_{k,k} t\right) \delta_{k,l} \left(T^{-1}\right)_{l,j} =$$

$$= \sum_k T_{i,k} \exp(\lambda_k t) \left(T^{-1}\right)_{k,j} \tag{II-21}$$

where $\lambda_k \equiv \Lambda_{k,k}$ is an eigenvalue of **C**. In the present case, the eigenvalues of **C** are:

$$\lambda_1 = -b + ia \qquad \lambda_2 = -b - ia \tag{II-22}$$

and the corresponding eigenvectors are

$$c_1 = \frac{1}{\sqrt{2}}\begin{bmatrix}1\\-i\end{bmatrix} \quad and \quad c_2 = \frac{1}{\sqrt{2}}\begin{bmatrix}1\\i\end{bmatrix} \tag{II-23}$$

These eigenvectors are the columns of the transformation matrix **T**. The matrix **C** is a "normal matrix", i.e., it commutes with its hermitian adjoint: $\mathbf{CC}^\dagger = \mathbf{C}^\dagger\mathbf{C}$. Therefore it is diagonalized by a unitary matrix, namely: $\mathbf{T}^{-1} = \mathbf{T}^\dagger$. Therefore

$$\exp(Ct) = \frac{1}{2}\begin{bmatrix}1 & 1\\-i & i\end{bmatrix}\begin{bmatrix}\mu_1 & 0\\0 & \mu_2\end{bmatrix}\begin{bmatrix}1 & i\\1 & -i\end{bmatrix} = \frac{1}{2}\begin{bmatrix}\mu_1+\mu_2 & i(\mu_1-\mu_2)\\-i(\mu_1-\mu_2) & \mu_1+\mu_2\end{bmatrix} \tag{II-24}$$

with the definition $\mu_k \equiv \exp\{\lambda_k t\}$ (k=1,2). Thus, using the explicit values of μ_1, μ_2, the general solution is

$$\begin{bmatrix}M_x(t)\\M_y(t)\end{bmatrix} = \exp(-t/T_2)\begin{bmatrix}\cos(\omega_0 t) & -\sin(\omega_0 t)\\\sin(\omega_0 t) & \cos(\omega_0 t)\end{bmatrix}\begin{bmatrix}M_x(0)\\M_y(0)\end{bmatrix} \tag{II-25}$$

which corresponds to rotation in the **x-y** plane with a gradual decay to zero, which is the equilibrium value of the transverse components. Combining this with the result for the longitudinal component, there is a precession of the magnetization vector about the static magnetic field, and a simultaneous decay to equilibrium. In the special case in which there is no static field (e.g., the static field was turned off together with the oscillating field) the transformation matrix in Eq. (II-25) is just the unit matrix, and the magnetization decays without any precession.

An important practical application of this calculation is to the FID (Free Induction Decay) experiment, where a strong irradiation pulse is followed by free evolution . Typically the pulse is a $\pi/2$ pulse, defined by $\omega_1\tau=\pi/2$, where τ is the duration of the pulse. Normally the system is initially at thermal equilibrium, so the magnetization is along **z**. The pulse itself is short, so relaxation during the pulse may be neglected. Assuming for simplicity that the irradiation is on resonance, it follows from the calculation done above for case (i) that the pulse will exactly rotate the magnetization by an angle of $\pi/2$, from the **z** to the **y** axis. Then immediately after the pulse the initial condition for FID is: $M_x(0)=0$, $M_y(0)=-M_0$ and $M_z(0)=0$. Eqs. (II-25) and (II-14) then lead to the solution:

$$M_x(t) = M_0\sin(\omega_0 t)\exp(-t/T_2) \tag{II-26a}$$

$$M_y(t) = -M_0\cos(\omega_0 t)\exp(-t/T_2) \tag{II-26b}$$

$$M_z(t) = M_0\{1-\exp(-t/T_1)\} \tag{II-26c}$$

where t is measured from the end of the pulse.

(c) *Steady state CW irradiation*

Another case of practical importance is that of continuous irradiation, in which transient effects due to the early stage of operating the irradiation have decayed away. In this case one needs the Bloch equations with both irradiation and relaxation, but the magnetization is assumed to be constant in the rotating frame, so the three time derivatives are set equal to zero. The solution in the rotating frame is then

$$M_x(t) = \frac{M_0}{F}\left\{\Delta\omega \cdot \omega_1 \cdot (T_2)^2\right\} \tag{II-27a}$$

$$M_y(t) = \frac{M_0}{F}\{\omega_1 \cdot T_2\} \tag{II-27b}$$

$$M_z(t) = \frac{M_0}{F}\left\{1 + (\Delta\omega \cdot T_2)^2\right\} \tag{II-27c}$$

where $F \equiv \{1 + (\Delta\omega\, T_2)^2 + \omega_1^2\, T_1\, T_2\}$. In the laboratory frame the solution is:

$$M_x^{(L)}(t) = M_x \cos(\omega_0 t) - M_y \sin(\omega_0 t) \tag{II-28a}$$

$$M_y^{(L)}(t) = M_x \sin(\omega_0 t) + M_y \cos(\omega_0 t) \tag{II-28b}$$

$$M_z^{(L)}(t) = M_z \tag{II-28c}$$

In conclusion, the phenomenological equations of Bloch are useful for many practical situations, giving the main features of magnetic resonance experiments. Their success implies that any sophisticated description of these experiments must include both coherent effects (due to irradiation) and incoherent effects (causing relaxation), as in the Bloch equations. A correct description of a general spin system with all its transitions has to be quantum mechanical. Due to the macroscopic size of the sample in magnetic resonance, what is measured in practice is only an ensemble average over many microscopic systems. Different spin systems have wave functions which differ at least in their overall phases, and these random differences require the use of a density matrix rather than a wave function. Accounting for random influences of environment leading to relaxation will require a stochastic version of the Liouville - von Neumann equation. This will be developed in the following sections.

II.3 Quantum Mechanical Treatment of Elementary Magnetic Resonance

In the basic applications of magnetic resonance one deals with magnetic dipole moments of nuclei or electrons, due to their spin angular momentum, without any intra-molecular interactions. The quantum mechanical system is thus defined as a "spin" in an external magnetic field. The static magnetic field is assumed to be in the **z** direction, and there may also be a weaker oscillating field along **x**, regarded as a perturbation. The spin operator **S** and its component $\mathbf{S_z}$ have simultaneous eigenfunctions $|s,m\rangle$ defined by:

$$S^2\,|s,m\rangle \;=\; s\,(s+1)\,\hbar^2\,|s,m\rangle \qquad\qquad S_z\,|s,m\rangle \;=\; m\,\hbar\,|s,m\rangle \tag{II-29}$$

where s is a non-negative integer or half-integer, called "the spin" of the system, and the possible values of m are: m=-s, -s+1,..., s-1,s. In many practical cases s=½ or s=1. Since the only interactions are between the spin and the external field, the Hamiltonian is obtained by taking a scalar product between the magnetic field, which is a vector in ordinary space, and the magnetic moment, which is a quantum mechanical operator and a vector in spin space (cf. Eq. (II-1)). In the present case:

$$H \;=\; \omega_0\,S_z \;+\; \omega_1\cos(\omega t)\,S_x \;\equiv\; H_0 \;+\; H_1(t) \tag{II-30}$$

The main part of the Hamiltonian is proportional to the constant magnetic field, and the smaller part to the time dependent field. The eigenstates of $\mathbf{S_z}$ are also eigenstates of $\mathbf{H_0}$, so in many applications it is convenient to work in the spin basis, i.e., in the $\{|s,m\rangle\}$ basis in calculations for magnetic resonance. In this basis:

$$H_0\,|s,m\rangle \;=\; \hbar\,\omega_0 m\,|s,m\rangle \;\equiv\; E_m\,|u_m\rangle \tag{II-31}$$

The result of a measurement is an expectation value of one of the two transverse spin operators, $\mathbf{S_x}$ and $\mathbf{S_y}$, or (in quadrature detection) a combination of these expectation values. Calculating them requires knowledge of a wave function or a density matrix for the system, and therefore the purpose of the quantum mechanical calculations is to solve equations for such a time dependent wave function or density matrix. Once this has been accomplished, the calculation of the time dependent magnetization is a simple expectation value calculation. Note that $\mathbf{S_x}$ and $\mathbf{S_y}$ have non-zero elements only between spin functions differing in their "m" quantum number by 1. Therefore, if the expectation value is calculated with a density matrix (as in Sec I.7, for example), only those elements of the density matrix connecting states differing in their "m" value by 1 will appear in the sum. Thus an ordinary (e.g. CW or FID) magnetic resonance experiment can only probe these elements of the density matrix. Accessing the other elements can only be done indirectly, through their involvement in the evolution of the density matrix in a multiple pulse experiment.

(a) Approximate transition probabilities

If a spin is initially in an eigenstate of $\mathbf{H_0}$, the operation of the time dependent field will make this state a linear superposition of eigenstates of the new Hamiltonian, the full **H**. If the time dependent field is relatively weak, these new eigenstates are not very different from the original ones. There is thus only a small probability of the system being finally in a different $\mathbf{H_0}$ eigenstate than the original one, and this probability may be approximately calculated from standard perturbation theory. Assuming $\omega_1 \ll \omega_0$, also $\|\mathbf{H_1}\| \ll \|\mathbf{H_0}\|$, which is the condition for using perturbation theory. If $|u_n\rangle$ are the eigenstates of $\mathbf{H_0}$, the Schrödinger equation with the full **H** is solved by

$$\psi(t) = \sum_n a_n(t) \exp(-iE_n t/\hbar) |u_n\rangle \tag{II-32}$$

from which the following exact equation is obtained for the coefficients:

$$\frac{d}{dt} a_n(t) = \frac{1}{i\hbar} \sum_{m \neq n} a_m(t) (H_1)_{n,m} \exp(i\omega_{n,m} t) \tag{II-33}$$

The set of these equations for all n values is exactly equivalent to the original Schrödinger equation. Now assume the system is originally in the state $|u_k\rangle$, so the initial condition is: $a_n(0) = \delta_{n,k}$. For a weak perturbation it may be assumed that

$$a_k(t) \approx 1 \qquad |a_m(t)| \ll 1 \quad (m \neq k) \tag{II-34}$$

and therefore

$$\frac{d}{dt} a_n(t) \approx \frac{1}{i\hbar} (H_1)_{n,k} \exp(i\omega_{n,k} t) \Rightarrow a_n(t) \approx \frac{1}{i\hbar} \int_0^t (H_1)_{n,k} \exp(i\omega_{n,k} t') dt' \tag{II-35}$$

The transition probability $P_{n,k}$ from $|u_k\rangle$ to $|u_n\rangle$ is defined as the probability that a transition has taken place at some time up to time t. It is equal to the probability that at time t the spin is found in state $|u_n\rangle$, so that $P_{n,k} = |a_n(t)|^2$. This is true as long as $P_{n,k}\, t \ll 1$. Once this criterion is not satisfied, the effect of the perturbation becomes too big for being correctly described by the perturbative calculation.

For example, suppose that the system is an s=½ system, and the spin is initially in the $|½,½\rangle$ state. The coefficient for being finally in the $|½,-½\rangle$ state is calculated from Eq. (II-35):

$$a_{-1/2}(t) \approx -i\omega_1 \int_0^t \cos(\omega t') \exp(-i\omega_0 t') dt' =$$

$$= \frac{\omega_1}{2} \left\{ \frac{\exp(-i(\omega+\omega_0)t) - 1}{\omega + \omega_0} - \frac{\exp(i(\omega-\omega_0)t) - 1}{\omega - \omega_0} \right\} \tag{II-36}$$

This expression has a resonance at $\omega = \omega_0$ and at $\omega = -\omega_0$. One of the two positions is a true resonance, corresponding to the component of the oscillating field which rotates in the "right" sense for inducing a transition. The other position corresponds to the counter-rotating field, which does not induce a transition.

(b) An exact quantum mechanical description

The system is the same as defined in Eq. (II-30), but for simplicity only one rotating component of the oscillating field will be included. The Schrödinger equation is thus

$$i\frac{\partial\psi}{\partial t} = \left\{\omega_0 S_z + \omega_1\left(\cos(\omega t)\,S_x + \sin(\omega t)\,S_y\right)\right\}\psi = \left\{\omega_0 S_z + \frac{\omega_1}{2}\left(e^{-i\omega t}S_+ + e^{i\omega t}S_-\right)\right\}\psi \tag{II-37}$$

There are two different ways to deal with the time dependence of the Hamiltonian. One is to solve the equation directly with the time dependence. This requires using a "time ordering operator" **T**, which takes care of the fact that in general, the Hamiltonian at different times does not commute with itself, i.e., for $t_1 \neq t_2$: $[\mathbf{H}(t_1)\,,\,\mathbf{H}(t_2)] \neq 0$. The formal solution with this operator is written as

$$\psi(t) = T\left\{\exp\left[-i\int_0^t H(t)\,dt\right]\right\}\psi(0) \tag{II-38}$$

The time ordering is defined in practice by expanding the exponential in a series, writing each term with the correct time order. However, it is very difficult to calculate this solution. The second and much easier way is to transform the equation to another frame of reference, in which the Hamiltonian is time independent. In general, a transformation defined by

$$\psi \rightarrow \psi' = U\psi \qquad\qquad H \rightarrow H' = U H U^{-1} \tag{II-39}$$

leads to the following Schrödinger equation:

$$i\frac{\partial\psi'}{\partial t} = i\left\{\frac{\partial U}{\partial t}\psi + U\frac{\partial\psi}{\partial t}\right\} = i\frac{\partial U}{\partial t}\psi + U\left(H\psi\right) =$$

$$= i\frac{\partial U}{\partial t}\psi + \left(U H U^{-1}\right)\left(U\psi\right) = i\frac{\partial U}{\partial t}U^{-1}\psi' + H'\psi' \equiv H_{eff}\psi' \tag{II-40}$$

The transformed wave function ψ' satisfies an equation which has the same form as the original Schrödinger equation, but the effective Hamiltonian $\mathbf{H}_{\mathbf{eff}}$ in this new equation is *not* the transformed Hamiltonian **H'**. It includes, in addition to **H'**, also a term due to the possible time dependence of the transformation. In our case such a time dependence is essential for removing the time dependence of the Hamiltonian. Thus Eq. (II-40) is an "effective Schrödinger equation" for the rotating frame.

The transformation to the rotating frame is defined quantum mechanically by

$$U = \exp\left(i\omega S_z t\right) \tag{II-41}$$

Using the well-known transformation properties of angular momentum operators (see Appendix A), this leads from the Hamiltonian of Eq. (II-30) to

$$H' = \omega_0 S_z + \omega_1 S_x \qquad \Rightarrow \qquad H_{eff} = (\omega_0 - \omega) S_z + \omega_1 S_x \tag{II-42}$$

One usually defines $\Delta\omega \equiv (\omega_0 - \omega)$ for the off-resonance term. The effective Hamiltonian is indeed time independent, and therefore Eq. (II-40) can be solved directly, with no need for time ordering:

$$\psi'(t) = \exp\left[-i\int_0^t H_{eff}dt\right]\psi'(0) = \exp(-iH_{eff}t)\,\psi'(0) \tag{II-43}$$

This solution can be calculated in the standard spin basis, i.e., in the basis in which $\mathbf{S}^2$ and $\mathbf{S}_z$ are diagonal. It is easier to handle the solution in the basis which diagonalizes $\mathbf{H}_{eff}$. However, this is not the standard spin basis, beacuse $\mathbf{S}_x$ which contributes to $\mathbf{H}_{eff}$ is not diagonal in the standard basis.

A density matrix $\boldsymbol{\rho}$ can be defined for the spin system in the usual manner. Working in the rotating frame of reference and calculating in the basis of $\mathbf{H}_{eff}$, the matrix elements of $\boldsymbol{\rho}$ evolve as:

$$\rho_{m,n}(t) = \exp(-i\omega_{m,n}t)\,\rho_{m,n}(0) \tag{II-44}$$

(as in Eq. (I-66)). On the other hand, if one prefers calculating in the standard spin basis, the density matrix in the rotating frame will be calculated as:

$$\rho(t) = \exp(-iH_{eff}t)\,\rho(0)\exp(iH_{eff}t) \tag{II-45}$$

(as in Eq. (I-61)).

If the spin system has a well defined wave function then it is in a specific quantum mechanical state - a pure state. The density matrix in the rotating frame is then related to the wave function by

$$\rho(t) = |\psi'(t)\rangle\langle\psi'(t)| \tag{II-46}$$

Example II.1: The s = ½ case

The simplest spin system has s = ½ , with only two quantum levels, as mentioned in Chapter I. For such a system, the matrix of the effective Hamiltonian of Eq. (II-42) in the standard spin basis is equal to

$$H_{eff} = \frac{\Delta\omega}{2}\begin{bmatrix}1 & 0\\ 0 & -1\end{bmatrix} + \frac{\omega_1}{2}\begin{bmatrix}0 & 1\\ 1 & 0\end{bmatrix} \equiv \frac{\omega_e}{2}\begin{bmatrix}a & b\\ b & -a\end{bmatrix} \tag{E-1}$$

with the effective frequency ω_e defined in Eq. (II-4), so that $a^2+b^2=1$. The eigenvalues of $\mathbf{H}_{eff}$ are then

$$\lambda_1 = \frac{\omega_e}{2}\left(a^2+b^2\right)^{1/2} = \frac{\omega_e}{2} \qquad \lambda_2 = -\frac{\omega_e}{2}\left(a^2+b^2\right)^{1/2} = -\frac{\omega_e}{2} \tag{E-2}$$

and the corresponding eigenvectors are:

$$v_1 = \frac{1}{\sqrt{2}}\begin{bmatrix}(1+a)^{1/2}\\(1-a)^{1/2}\end{bmatrix} \qquad v_2 = \frac{1}{\sqrt{2}}\begin{bmatrix}(1-a)^{1/2}\\-(1+a)^{1/2}\end{bmatrix} \tag{E-3}$$

In order to calculate the exponential appearing in Eq. (II-44) (or (II-45)) it is convenient to employ Eq. (II-20). Using the notation $x \equiv (1+a)^{1/2}$, $y \equiv (1-a)^{1/2}$, and $\mu_k = \exp(i\lambda_k\alpha)$ for any scalar α (in particular for $\alpha = t$):

$$\exp\left(i\alpha H_{eff}\right) = \frac{1}{2}\begin{bmatrix}x & y\\y & -x\end{bmatrix}\begin{bmatrix}\mu_1 & 0\\0 & \mu_2\end{bmatrix}\begin{bmatrix}x & y\\y & -x\end{bmatrix} =$$

$$= \frac{1}{2}\begin{bmatrix}(\mu_1+\mu_2)+a(\mu_1-\mu_2) & b(\mu_1-\mu_2)\\ b(\mu_1-\mu_2) & (\mu_1+\mu_2)-a(\mu_1-\mu_2)\end{bmatrix} =$$

$$= \begin{bmatrix}\cos\left[\frac{\omega_e\alpha}{2}\right] + i\,a\sin\left[\frac{\omega_e\alpha}{2}\right] & i\,b\sin\left[\frac{\omega_e\alpha}{2}\right]\\ i\,b\sin\left[\frac{\omega_e\alpha}{2}\right] & \cos\left[\frac{\omega_e\alpha}{2}\right] - i\,a\sin\left[\frac{\omega_e\alpha}{2}\right]\end{bmatrix} \tag{E-4}$$

****

II.4 Basic Concepts in Stochastic Processes

The behaviour of the magnetization of a sample depends on the microscopic magnetic dipole moments of nuclei or electrons. These magnetic moments interact not only with the external magnetic fields applied in the laboratory, but also with electromagnetic multipoles of the close environment. Important multipoles are the electric dipoles and quadrupoles, and magnetic dipoles. The latter two kinds of mutipoles interact with the spins, i.e., with the magnetic dipole moments. In a liquid and in a gas the incessant molecular motions change the relative distances and orientations of such multipoles in a practically random manner. Thus molecular motion in a liquid or a gas causes each spin to "see" fluctuating magnetic fields and electric field gradients. These fluctuations partly average out the interactions and cause relaxation. In solids, on the other hand, molecular motion is restricted, and more importantly there are cooperative effects, so the situation is more complex. In this Chapter the situation in liquids and gases will be discussed, and the behaviour of the magnetization will be calculated accordingly. As an introduction to this treatment, some basic concepts and facts concerning random processes will be briefly reviewed in this section. A full treatment of the subject is available in the sources referenced at the end of this Chapter.

A function $y(t)$ is called a **random function** if for each t it has a random value, subject only to the probability distribution $p(y,t)$. This distribution refers to the a priori probability, namely the probability that the function has a particular value y at time t, provided that one has no information on the values of the function at earlier times, except possibly for the definition of the function. If y stands for a physical variable, the variable is called a **random** or **stochastic process**, which is to be distinguished from a "deterministic" process. In ordinary physics, both classical and quantum mechanical, the evolution of a physical variable is determined by the solution of some deterministic differential equations. In classical physics these equations refer directly to the interesting variables, and in quantum mechanics to the wave function. For a stochastic variable, however, the value of the variable at any given moment is in principle completely random, and cannot be predicted in a deterministic way. Even in quantum mechanics, where the use of the wave function imposes a degree of indeterminism, a stochastic process introduces an additional source of indeterminism.

Some macroscopic processes are inherently unpredictable, due to their quantum mechanical origin. This is the case in radioactive decay, for example. As long as a radioactive nucleus has not decayed, the probability per unit time that it will decay remains a constant. Thus the probability that a particular radioactive nucleus will decay cannot be predicted by knowledge of its prior state. In such cases, therefore, the stochasticity is inherent in the process.

In other cases, the process is microscopically deterministic, but the degree of its predictability depends on the level at which it is described. Some processes are completely unpredictable, because of our ignorance of microscopic details. An example is the process of throwing a coin, and observing whether it falls "face up" (assigned the arbitrary value $y=1$) or "face down" (assigned the arbitrary value $y=-1$). If the results of the throws at different times t_1, t_2,... are arranged in a sequence, the result at t_n is in principle completely independent of all earlier results. It is true that the throw of the coin can be described in principle by deterministic classical equations. However, it is impossible in practice to specify all the initial conditions of a throw, and the result is very sensitive to these conditions. At the level of treating the throw as a "black box", the result is completely random.

In many practical cases one has partial knowledge about the system, so it is not treated as a black box, but there is not sufficient information for a deterministic description. Diffusion is an example of such a process, which on the macroscopic level is conveniently defined as stochastic. Nevertheless, in the treatment of diffusion one normally has some information on the initial conditions at a given moment, so the relevant physical variables do not seem to change in a haphazard manner. If, for example, the composition of a gas mixture at a particular volume element is known at a given instant, the composition of the mixture in the same volume element will not be very different after a very short time. Thus the random process of diffusion does have some dependence on its history.

The probability distribution function of a random function is usually normalized so that the total probability of having any value of y is equal to 1:

$$\int p(y,t)\,dy = 1 \tag{II-47}$$

If the possible values of y form a discrete set, the integral $\int \ldots dy$ is replaced by the summation $\sum_y \ldots$, in Eq. (II-47) and in all subsequent equations. The average or expectation value of any function of y at time t is defined as:

$$\langle f(y(t)) \rangle = \overline{f(y(t))} = \int f(y(t))\,p(y,t)\,dy \tag{II-48}$$

In particular, the average value of y is:

$$\langle y(t) \rangle = \overline{y(t)} = \int y(t)\,p(y,t)\,dy \tag{II-49}$$

One also defines the **joint probability** $p(y_1,t_1;y_2,t_2)$ that the random function has the value y_1 at time t_1 and the value y_2 at time t_2. This is related to the **conditional probability** $P(y_2,t_2\,|y_1,t_1)$, that the function has the value y_2 at time t_2 provided it is already known that it has the value y_1 at time t_1. The relation is:

$$p(y_1,t_1;y_2,t_2) = p(y_1,t_1) \cdot P(y_2,t_2\,|y_1,t_1) \tag{II-50}$$

An important class of random functions is that of stationary functions. The random function y(t) is **stationary** if $p(y,t)=p(y)$, namely the probability distribution is independent of time. The function y itself can still assume a different value for each t. For example, in the case of throwing a coin the probability of obtaining $y=1$ (see above) or $y=-1$ is equal to ½ at every throw, but the actual value of y may change from one throw to the next. Stochastic processes which are microscopically deterministic do have some "memory" of their history, but the a priori probability $p(y,t)$ can still be constant. In a stationary process there is no value of time which has any special status, and therefore the conditional probability can only depend on the time difference and not on the specific values of the two times:

$$P(y_2,t_2\,|y_1,t_1) = P(y_1,y_2,t_2-t_1) \tag{II-51}$$

Another important class of random functions, partly overlapping with that of stationary functions, is that of Markovian functions. A random function is **Markovian** if it has only a "one step memory". For a discrete process this means that $y(t_n)$ will only depend on $y(t_{n-1})$ and not on earlier values. For a continuous process, y(t) will only depend on y(t-dt) and not on earlier values of y. Markovian functions are characterized by the Chapman-Kolmogorov equation, expressing the conditional probability of going from a specific initial state to a specific final state by means of intermediate states:

$$P(y_2,t_2\,|y_1,t_1) = \int P(y_2,t_2\,|y',t')\,P(y',t'\,|y_1,t_1)\,dy' \tag{II-52}$$

where t' is a fixed time, and the condition: $t_1 < t' < t_2$ must be satisfied. If the variable is discrete, the equation is

$$P(y_2,t_2\,|\,y_1,t_1) = \sum_m P(y_2,t_2\,|\,y_m,t_m)\,P(y_m,t_m\,|\,y_1,t_1) \tag{II-53}$$

with the same condition on t_m. From this equality in one of its two versions it is possible to derive a very useful equation, which makes it possible to deal with experimentally accessible quantities. This is the **master equation**, specifying the "flow of probability" to and from a specific value of the random function. For a continuous variable, the equation is:

$$\frac{\partial}{\partial t}p(y,t) = \int \{W(y\,|\,y')p(y',t) - W(y'\,|\,y)p(y,t)\}\,dy' \tag{II-54}$$

In this equation $W(y_2|y_1)$ is the transition probability per unit time from y_1 to y_2. The first term on the right hand side of the master equation is a "source" term, summing over the contributions to p(y,t) of transitions from other states, or values, of *y*. The second term is a "sink" term, summing over the losses of probability due to transitions to other values of *y*. The occupation probabilities p(y,t) determine the outcome of an experiment in terms of the probability to find the system in a particular state. The transition probability determines the relative probability to make a particular transition, and are thus also experimentally measurable. If the set of possible values of *y* is a discrete set, the master equation is:

$$\frac{d}{dt}p_k(t) = \sum_m \{W_{k,m}p_m(t) - W_{m,k}p_k(t)\} \tag{II-55}$$

with $p_k(t) \equiv p(y_k,t)$ for any value of k. An important example of a Markovian process is Brownian diffusion. One type of this process, Brownian rotational molecular diffusion, is a very common model for molecular motion in liquids, and is therefore very interesting for magnetic resonance applications. The discrete version of the master equation is often useful also for non-Markovian processes.

The **correlation function** of two functions of random variables, $f_1(y_1(t))$ and $f_2(y_2(t))$ can be defined as:

$$\langle f_1(y_1(t_1))f_2(y_2(t_2))\rangle = \overline{f_1(y_1(t_1))f_2^*(y_2(t_2))} = \int\!\!\int f_1(y_1(t_1))f_2^*(y_2(t_2))\,p(y_1,t_1;y_2,t_2)\,dy_1\,dy_2$$

$$= \int\!\!\int f_1(y_1(t_1))f_2^*(y_2(t_2))\,p(y_1,t_1)\,P(y_2,t_2\,|\,y_1,t_1)\,dy_1\,dy_2 \tag{II-56}$$

Two functions are said to be uncorrelated if

$$\langle f_1(y_1(t_1))f_2(y_2(t_2))\rangle = \langle f_1(y_1(t_1))\rangle\langle f_2^*(y_2(t_2))\rangle \tag{II-57}$$

Namely, for uncorrelated functions the integrations over the random variables may be separated, averaging over each function separately. A special case of a correlation function is the **auto-correlation** of a function $f(y(t))$, which is defined as:

$$G(t_1,t_2) = \langle f(y_1(t_1)) f(y_2(t_2)) \rangle \tag{II-58}$$

If the function is stationary, its auto-correlation is only a function of the time difference, which is a single time variable:

$$G(\tau) = \int\int f(y_1(t)) f^*(y_2(t+\tau)) p(y_1) P(y_1,y_2,\tau) \, dy_1 \, dy_2 \tag{II-59}$$

By definition: $P(y_2,y_1,-\tau) = P(y_1,y_2,\tau)$. It follows that

$$G(-\tau) = G^*(\tau) \tag{II-60}$$

If there is a time symmetry such that also $P(y_1, y_2, -\tau) = P(y_1, y_2, \tau)$, then $G(\tau) = G(-\tau)$ and also $G(\tau) = G^*(\tau)$. The auto-correlation function is then real, and an even function of τ. The value of a function at time t has maximal correlation with its own value at the same time, so the maximum auto-correlation is

$$G(0) = \int dy_1 f(y_1(t)) p(y_1) \int dy_2 f^*(y_2(t)) P(y_1,y_2,0) =$$

$$= \int dy_1 f(y_1(t)) p(y_1) \int dy_2 f^*(y_2(t)) \, \delta(y_2 - y_1) = \int dy \, p(y) \, |f(y)|^2 \tag{II-61}$$

The Fourier transform of an auto-correlation function is called a **spectral density**:

$$J(\omega) = \int_{-\infty}^{\infty} G(\tau) e^{-i\omega\tau} d\tau \quad \Rightarrow \quad G(\tau) = \frac{1}{2\pi} \int_{-\infty}^{\infty} J(\omega) e^{i\omega\tau} d\omega \tag{II-62}$$

The argument ω of $J(\omega)$ has physical dimensions of frequency, and $J(\omega)$ is thus the frequency spectrum of $G(\tau)$. From the inverse transform one finds that the auto-correlation for $\tau=0$ is just the integral over the corresponding spectral density:

$$G(0) = \frac{1}{2\pi} \int_{-\infty}^{\infty} J(\omega) \, d\omega \tag{II-63}$$

Example II.2: Dipole-dipole interaction

One of the most important interactions in magnetic resonance is that between two magnetic dipole moments. It depends on the distance between the dipoles and on the orientation of the axis joining them relative to the external magnetic field. If the two dipoles are in the same molecule, the distance between them is normally fixed, but the orientation chages rapidly due to random molecular rotational motion. Thus one may choose as a random variable for this problem the orientation $y(t) = \theta(t)$ of the dipole-dipole axis with respect to the direction of the external field. In the presence of a strong external magnetic field, the main term in the interaction is of the form:

$$f(y(t)) = A\{1 - 3\cos^2(\theta(t))\} \qquad \text{(E-1)}$$

The constant A is a function of the dipole-dipole distance, which is fixed in this problem. The a priori probability for finding the dipole pair at a given orientation is isotropic, which means that the probability distribution $p(\theta)$ is a constant: $p(\theta) = p_0$. The value of this constant is determined by the normalization condition:

$$1 = \int_0^\pi d\theta \sin(\theta)\, p(\theta) = p_0 \int_0^\pi d\theta \sin(\theta) = 2p_0 \quad \Rightarrow \quad p_0 = \frac{1}{2} \qquad \text{(E-2)}$$

The average value of the interaction vanishes, because

$$\langle f(y)\rangle = A\int_0^\pi \{1 - 3\cos^2(\theta(t))\}\, p_0 \sin(\theta)\, d\theta = \frac{A}{2}\int_{-1}^{1} \{1 - 3x^2\}\, dx = 0 \qquad \text{(E-3)}$$

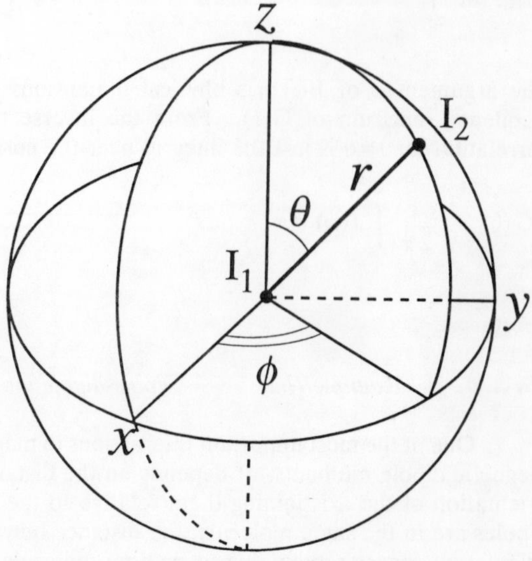

Fig. II.3. *Axis system and polar coordinates for the dipolar coupling of two protons.*

The auto-correlation function depends on the nature of the time dependence of $\theta(t)$. If, for example, $f(t+\tau)$ is uncorrelated with $f(t)$, then Eqs. (II-57) and (E-3) imply that also the auto-correlation $G(\tau)$ would vanish. However, at least at $\tau=0$ there must be a non-zero correlation for any time dependence of $\theta(t)$:

$$G(0) = A^2 \int_0^{\pi} \{1 - 3\cos^2(\theta(t))\}^2 p_0 \sin(\theta)\, d\theta = \frac{A^2}{2} \int_{-1}^{1} \{1 - 3x^2\}^2 dx = \frac{4}{5} A^2 \quad \text{(E-4)}$$

On the other hand, for very long time differences τ one expects zero correlation, since the relevant processes are random. Therefore for any random physical process it is expected that $G(\tau)$ gradually decays to zero from its initial value. The decay may be monotonous, i.e., always decreasing, but is not always of this type. For example, the decay may involve weak oscillations combined with a steady decrease in $G(\tau)$. The ratio

$$g(\tau) = \frac{G(\tau)}{G(0)} \quad \text{(E-5)}$$

measures the decay of the auto-correlation function. For many random processes occurring in physical situations, this decay is exponential:

$$g(\tau) = \exp\left[-\frac{\tau}{\tau_C}\right] \quad \text{(E-6)}$$

The time constant τ_C is known as the **correlation time** of the process.

****

II.5 Approximate Treatments of Relaxation in Simple Cases

(a) Autocorrelation functions in time dependent perturbation theory

Assume: $\mathbf{H} = \mathbf{H_0} + \mathbf{H_1}(t)$ where $\| \mathbf{H_1} \| << \| \mathbf{H_0} \|$, and $\mathbf{H_1}$ represents randomly changing interactions of the spin with its environment. One may then apply the perturbation calculation of Sec. II.3.(a). The difference is that now $\mathbf{H_1}(t)$ is a random function of time, rather than a Hamiltonian with a known time dependence as in that Section. If the system is initially in the state $|u_k\rangle$, which is an eigenstate of $\mathbf{H_0}$, the coefficient of $|u_n\rangle$ in the wave function at time t is given by Eq. (II-35):

$$a_n(t) \approx \frac{1}{i\hbar} \int_0^t (H_1)_{n,k} \exp\left(i\omega_{n,k} t'\right) dt' \quad \text{(II-64)}$$

The total probability for a transition from $|u_k\rangle$ to $|u_n\rangle$ is $P_{n,k}(t) = |a_n(t)|^2$, and the transition probability per unit time is:

$$W_{n,k} = \frac{d}{dt}\left\{a_n(t)\,a_n^*(t)\right\} = \left\{\frac{d}{dt}a_n(t)\right\}a_n^*(t) + a_n(t)\left\{\frac{d}{dt}a_n^*(t)\right\} \qquad \text{(II-65)}$$

The second term on the right-hand side is the complex conjugate of the first term. Thus

$$W_{n,k} = \frac{1}{\hbar^2}(H_1)_{n,k}(t)\exp\left(i\omega_{n,k}t\right)\int_0^t (H_1)_{n,k}^*(t')\exp\left(-i\omega_{n,k}t'\right)dt' + (c.c.) =$$

$$= \frac{1}{\hbar^2}\int_0^t (H_1)_{n,k}(t)\,(H_1)_{n,k}^*(t')\exp\left(i\omega_{n,k}(t-t')\right)dt' + (c.c.) \qquad \text{(II-66)}$$

Due to the random changes in the time dependent interactions, it is impossible to write down a completely definite expression for $\mathbf{H}_1(t)$. One can only calculate stochastic expectation values over the interaction. Averaging over the product of matrix elements in Eq. (II-66) results in the autocorrelation function

$$G_{n,k}(\tau) = \frac{1}{\hbar^2}\langle\,(H_1)_{n,k}(t)\,(H_1)_{n,k}^*(t+\tau)\,\rangle \qquad \text{(II-67)}$$

if the process is assumed to be stationary, and τ=t'-t is the time difference. The stochastic equivalent of Eq. (II-66) is therefore, using Eq. (II-60):

$$\langle W_{n,k}\rangle = \int_{-t}^{0} G_{n,k}(\tau)\exp(-i\omega_{n,k}\tau)\,d\tau + \int_{-t}^{0} G_{n,k}^*(\tau)\exp(i\omega_{n,k}\tau)\,d\tau =$$

$$= \int_{-t}^{0} G_{n,k}(\tau)\exp(-i\omega_{n,k}\tau)\,d\tau - \int_{t}^{0} G_{n,k}^*(\tau')\exp(-i\omega_{n,k}\tau')\,d\tau' =$$

$$= \int_{-t}^{t} G_{n,k}(\tau)\exp(-i\omega_{n,k}\tau)\,d\tau \approx \int_{-\infty}^{\infty} G_{n,k}(\tau)\exp(-i\omega_{n,k}\tau)\,d\tau = J_{n,k}(\omega_{n,k}) \qquad \text{(II-68)}$$

The approximate equality is based on the assumption that the relevant times t are long relative to characteristic times of the random processes. Notice that the indices (n,k) of the spectral density function are the same as the indices of the average transition probability per unit time. However, the argument $\omega_{n,k}$ of the spectral density could have different indices, as will be seen in the following section.

Example II.3: *A one-term stochastic operator*

Suppose the stochastic Hamiltonian consists of a single term, which is a constant spin operator **A** multiplied by some random function of time F(t). For the calculation of a matrix element only the spin operator is relevant:

$$\left(H_1(t)\right)_{n,k} = A_{n,k}F(t) \tag{E-1}$$

The corresponding autocorrelation function is then calculated by averaging only over the function, the spin operator being irrelevant:

$$G_{n,k}(\tau) = A_{n,k}A_{n,k}^{*}\,\langle F(t)\,F^{*}(t+\tau)\rangle \equiv A_{n,k}A_{n,k}^{*}\,g(\tau) \tag{E-2}$$

If the function is normalized in the usual manner, so that g(0) = 1, then

$$G_{n,k}(0) = |A_{n,k}|^2 \quad and \quad G_{n,k}(\tau) = G_{n,k}(0)\,g(\tau) \tag{E-3}$$

The autocorrelation is thus a product of two different terms. One is time independent and depends only on the spin operator, and the other depends only on the stochastic process, describing the relaxation through the decay of the autocorrelation. Since all the dependence on the stochastic process is contained in F(t), it is natural to define a spectral density which is related only to this function, so that Eq. (II-62) becomes:

$$J(\omega) = \int_{-\infty}^{\infty} g(\tau)\,e^{-i\omega\tau}\,d\tau \tag{E-4}$$

Following the perturbation calculation shown above, the transition probability per unit time is

$$\langle W_{n,k}\rangle = |A_{n,k}|^2\,J(\omega_{n,k}) \tag{E-5}$$

****

(b) The master equation for populations

It was shown above that with an initial condition of $a_n(0) = \delta_{n,k}$ and a weak time dependent field, time dependent perturbation theory yields the following transition probability:

$$P_{n,k}(t) = |a_n(t)|^2 \quad \Rightarrow \quad W_{n,k}(t) = \frac{d}{dt}P_{n,k}(t) \tag{II-69}$$

More generally, this result implies for the population Π_n of state n:

$$\Pi_n(t) = \int_0^t W_{n,k}(t)\,|a_k(0)|^2\,dt \approx \int_0^t W_{n,k}(t)\,|a_k(t)|^2\,dt \qquad \text{(II-70)}$$

If the initial condition is more general, with non-zero coefficients for several states, the population $\Pi_m(t) = |a_m(t)|^2$ grows due to transitions from other states $|u_k\rangle$, and decreases due to transitions from $|u_m>$ to other states $|u_k>$. This leads to a master equation:

$$\frac{d}{dt}\Pi_m(t) = \sum_k \{W_{m,k}\,\Pi_k(t) - W_{k,m}\,\Pi_m(t)\} \qquad \text{(II-71)}$$

The first term on the right hand side is the total rate of incoming population, and the second term is the total rate of outgoing population. The probability per unit time for transitions in both directions may be assumed to be equal, in which case

$$\frac{d}{dt}\Pi_m(t) = \sum_k W_{m,k}\,\{\Pi_k(t) - \Pi_m(t)\} \qquad \text{(II-72)}$$

If the system is in a steady state of "dynamic equilibrium", all the time derivatives vanish, which implies (through Eq. (II-72)) that $\Pi_k = \Pi_m$ for all m,k. This is a state of saturation, in which all levels are equally populated. Transitions due to irradiation will occur with equal intensity in both directions (in the frequency range relevant for magnetic resonance, spontaneous emission is negligible), so there will be no net absorption of energy. If the populations are normalized so that

$$\sum_m \Pi_m(t) = \sum_m |a_m(t)|^2 = 1 \qquad \text{(II-73)}$$

and the number of levels is N, then the populations in the saturated system are

$$\Pi_m = \frac{1}{N} \qquad \text{(II-74)}$$

for all levels. The system can then be said to have an "infinite temperature", because the Boltzmann factor characterizing the relative level population will tend to 1 only at an infinite temperature:

$$T \rightarrow \infty \quad \Rightarrow \quad \exp\left[-\frac{E}{k_B T}\right] \rightarrow \exp(0) = 1 \qquad \text{(II-75)}$$

Here E is the energy difference between two levels. Thus, if the system is at equilibrium, only an infinite temperature will equalize the populations of all levels.

This unreasonable result implies that the above treatment needs a modification. The physical justification for such a modification is based on the original assumption of a weak perturbation, which precludes the possibility of saturating the system by irradiation. This means that once the time is sufficiently long that Π_m becomes appreciable, the correct equations cannot be approximated any more by the equations written above. However, a full treatment of the problem can be avoided by the following ad hoc solution. One assumes that the equations can be used for $\Pi_m = \Phi_m \exp(E_m/kT)$ where now Φ_m are the populations, so that the steady state result $\Pi_m = 1/N$ is equivalent to: $\Phi_m = \exp(-E_m/kT)/N$ as required. Using the "high temperature approximation", which is valid in practically all cases in magnetic resonance, one has

$$\Pi_k - \Pi_m = \Phi_k \exp\left[\frac{E_k}{k_B T}\right] - \Phi_m \exp\left[\frac{E_m}{k_B T}\right] \approx$$

$$\approx \Phi_k \left[1 + \frac{E_k}{k_B T}\right] - \Phi_m \left[1 + \frac{E_m}{k_B T}\right] = \Phi_k - \Phi_m + \frac{\Phi_k E_k - \Phi_m E_m}{k_B T} \tag{II-76}$$

On the other hand, the same approximation leads to

$$\Phi_k^{(equi.)} - \Phi_m^{(equi.)} \approx \frac{1}{N}\left[1 - \frac{E_k}{k_B T}\right] - \frac{1}{N}\left[1 - \frac{E_m}{k_B T}\right] = -\frac{\Phi_k E_k - \Phi_m E_m}{k_B T} \tag{II-77}$$

for the equilibrium populations. Therefore

$$\Pi_k - \Pi_m \approx \left(\Phi_k - \Phi_k^{(equi.)}\right) - \left(\Phi_m - \Phi_m^{(equi.)}\right) \tag{II-78}$$

from which the master equation can be written for the true populations as

$$\frac{d}{dt}\Phi_m(t) = \sum_k W_{m,k}\left\{\left(\Phi_k - \Phi_k^{(equi.)}\right) - \left(\Phi_m - \Phi_m^{(equi.)}\right)\right\} \tag{II-79}$$

in the high temperature approximation. This is effectively the same as Eq. (II-72), except that now the variables are the deviations from the equilibrium populations rather than the actual populations. The new form of the equation describes correctly relaxation to thermal equilibrium at a finite temperature. Nevertheless, this is not a complete description of the spin system, because it is based on the assumption that level populations are sufficient to describe the evolution of the system. In the language of the density matrix this means that there are no off-diagonal elements, and as a result the expectation value of the transverse magnetization components is zero.

II.6 The Density Matrix Relaxation Theory

(a) A semi-classical derivation of a master equation for the density matrix

The Bloch-Redfield approach uses a systematic perturbation expansion in order to evaluate the effect of random processes on the system, and then averages over the random variables in order to obtain a manageable relaxation equation. As in the previous section, it is assumed that the Hamiltonian is a sum of two terms,

$$H = H_0 + H_1(t) \tag{II-80}$$

The first part is large and constant, and the second is small and changes rapidly in a random manner. Unlike in the previous treatment, based on wave functions and first order transition probabilities, the present approach uses the density matrix, which is the natural choice for a system in which one deals with a macroscopic ensemble. Using the density matrix allows one to calculate coherences and not only populations, and in general to follow the behaviour of the spin system throughout an experiment. The starting point is the Liouville-von Neumann equation

$$\frac{d}{dt}\rho = \frac{1}{i\hbar}[H_0 + H_1(t), \rho] \tag{II-81}$$

In order to focus on the effect of the time-dependent interactions one transforms to the interaction representation (similar to the "rotating frame" transformation discussed earlier). Using the superscript (I) to denote the interaction representation, one has

$$H_0 \rightarrow H_0^{(I)} \equiv \exp\left(iH_0t/\hbar\right)H_0\exp\left(-iH_0t/\hbar\right) = H_0 \tag{II-82a}$$

$$H_1(t) \rightarrow H_1^{(I)}(t) \equiv \exp\left(iH_0t/\hbar\right)H_1(t)\exp\left(-iH_0t/\hbar\right) \tag{II-82b}$$

$$\rho \rightarrow \sigma \equiv \rho^{(I)} \equiv \exp\left(iH_0t/\hbar\right)\rho\,\exp\left(-iH_0t/\hbar\right) \tag{II-82c}$$

Thus the equation of motion for the transformed density matrix is

$$\frac{d}{dt}\sigma = \frac{1}{i\hbar}[H_0 + H_1^{(I)}(t), \sigma] + \frac{i}{\hbar}[H_0, \sigma] = \frac{1}{i\hbar}[H_1^{(I)}(t), \sigma] \tag{II-83}$$

In other words, in the transformed system the evolution of the density matrix is determined only by the time-dependent processes. This prevents the much larger constant interactions from "masking" the time-dependent effects, which are of interest here.

The main difficulty with the equation, however, is the time dependence of the Hamiltonian, which has not been removed. Because of this time dependence it is impossible to integrate the equation directly. A systematic approximation scheme is to iterate the equation repeatedly, which is similar to performing a Taylor expansion with time ordering, on the basis of the integral equation

$$\sigma(t) = \sigma(0) + \frac{1}{i\hbar}\int_0^t [H_1^{(I)}(t'), \sigma(t')]\,dt' \tag{II-84}$$

A direct solution would require knowing the exact time dependence of $\sigma(t)$ in order to perform the integration, which is needed in order to find σ(t). Since σ(t) is not known in advance, the iterative procedure approximates it in stages. The zeroth order approximation, ignoring the integral, reduces to the initial condition, which contains no information on the evolution of the system:

$$\sigma^{(0)}(t) = \sigma(0) \qquad \text{(II-85)}$$

The first order approximation adds the integral of Eq. (II-84), but replaces σ(t') in the integrand by $\sigma^{(0)}$(t'):

$$\sigma^{(1)}(t) = \sigma(0) + \frac{1}{i\hbar}\int_0^t [H_1^{(I)}(t'), \sigma^{(0)}(t')]\, dt' =$$

$$= \sigma(0) + \frac{1}{i\hbar}\int_0^t [H_1^{(I)}(t'), \sigma(0)]\, dt' \qquad \text{(II-86)}$$

Here the integrand does not contain the variable σ(t) which we are trying to evaluate, so the integration can be carried out directly..In the second order approximation one modifies Eq. (II-84) by replacing σ(t') in the integrand by $\sigma^{(1)}$(t'):

$$\sigma^{(2)}(t) = \sigma(0) + \frac{1}{i\hbar}\int_0^t [H_1^{(I)}(t'), \sigma^{(1)}(t')]\, dt' = \sigma(0) + \frac{1}{i\hbar}\int_0^t [H_1^{(I)}(t'), \sigma(0)]\, dt'$$

$$+ \left(\frac{1}{i\hbar}\right)^2 \int_0^t dt' \int_0^{t'} dt'' \, [H_1^{(I)}(t'), [H_1^{(I)}(t''), \sigma(0)]] \qquad \text{(II-87)}$$

In principle this procedure can be continued to any desired order. In practice it is assumed that second order is sufficient. This assumption is valid provided the correlation time τ_C of the stochastic process is much shorter than T_2 and T_1. It is convenient to work with the equivalent differential equation:

$$\frac{d}{dt}\sigma(t) = \frac{1}{i\hbar}[H_1^{(I)}(t), \sigma(0)] - \frac{1}{(\hbar)^2}\int_0^t dt'' \, [H_1^{(I)}(t), [H_1^{(I)}(t''), \sigma(0)]] =$$

$$= \frac{1}{i\hbar}[H_1^{(I)}(t), \sigma(0)] - \frac{1}{(\hbar)^2}\int_0^t d\tau \, [H_1^{(I)}(t), [H_1^{(I)}(t-\tau), \sigma(0)]] \qquad \text{(II-88)}$$

The randomness of H_1(t) implies that also σ(t) is random, so the observable behaviour of the system is described by its expectation value. At this stage it is necessary to get some information on the time-dependent Hamiltonian. First we shall make the assumption that the average of the stochastic Hamiltonian over its (as yet unknown) probability distribution is zero:

$$\langle H_1^{(I)}(t) \rangle = 0 \tag{II-89}$$

This implies that any non-zero average should be included in $\mathbf{H_0}$, or that if the average is non-zero it is time independent. A process with a time dependent average is "coherent" and not stochastic, and should be treated in a different way, as will be shown later on. The following assumptions will now be made:

(i) The stochastic Hamiltonian at time t, $\mathbf{H_1}(t)$ is uncorrelated with the initial density matrix $\sigma(0)$. Thus their averages can be performed separately (see Eq. (II-57)). This is justified if t is much longer than the decay time of the integrand, which is of the order of the correlation time τ_C.

(ii) $\sigma(0)$ in the integrand can be replaced by $\sigma(t)$, which is justified if: $\tau_C \ll t \ll (T_2)^2/\tau_C$.

(iii) The upper integration limit can be extended to infinity. This requires that the decay time of the integrand is of the order of the correlation time τ_C.

These assumptions imply that the system is characterized by several different time scales, which are clearly separated. The stochastic process is very fast with respect to the experimentally relevant time scale, and the decay of the magnetization is very slow on that time scale. Thus one can average over the stochastic behaviour without losing any essential characteristic of the time dependence of the magnetization. The equation of motion is thus approximated by the Bloch-Redfield relaxation equation, which is the central result of this chapter:

$$\frac{d}{dt}\langle \sigma(t) \rangle = -\frac{1}{(\hbar)^2}\int_0^\infty d\tau \, \langle [H_1^{(I)}(t), [H_1^{(I)}(t-\tau), \sigma(t)]] \rangle \tag{II-90}$$

Matrix elements of this equation can be calculated in the basis of $\mathbf{H_0}$ in the following way. A general matrix element of the double commutator is equal to

$$[H_1^{(I)}(t), [H_1^{(I)}(t-\tau), \sigma(t)]]_{i,j} = \sum_{m,n} H_1^{(I)}(t)_{i,m} H_1^{(I)}(t-\tau)_{m,n} \sigma(t)_{n,j}$$

$$- \sum_{m,n} H_1^{(I)}(t)_{i,m} \sigma(t)_{m,n} H_1^{(I)}(t-\tau)_{n,j} - \sum_{m,n} H_1^{(I)}(t-\tau)_{i,m} \sigma(t)_{m,n} H_1^{(I)}(t)_{n,j}$$

$$+ \sum_{m,n} \sigma(t)_{i,m} H_1^{(I)}(t-\tau)_{m,n} H_1^{(I)}(t)_{n,j} \equiv \sum_{m,n} Q_{ij,mn} \sigma(t)_{m,n} \tag{II-91}$$

where the element of the superoperator **Q** is given by

$$Q_{ij,mn} = \delta_{n,j} \sum_p H_1^{(I)}(t)_{i,p} H_1^{(I)}(t-\tau)_{p,m} - H_1^{(I)}(t)_{i,m} H_1^{(I)}(t-\tau)_{n,j} - H_1^{(I)}(t-\tau)_{i,m} H_1^{(I)}(t)_{n,j}$$

$$+ \delta_{i,m} \sum_p H_1^{(I)}(t)_{n,p} H_1^{(I)}(t-\tau)_{p,j} = \exp\left(i\left(\omega_{i,j} - \omega_{m,n}\right)t\right) \times$$

$$\times \left\{ \delta_{n,j} \sum_p \exp\left(i\omega_{m,p}\tau\right) H_1^{(I)}(t)_{i,p} H_1^{(I)}(t-\tau)_{p,m} - \exp\left(i\omega_{j,n}\tau\right) H_1^{(I)}(t)_{i,m} H_1^{(I)}(t-\tau)_{n,j} \right.$$

$$\left. - \exp\left(i\omega_{m,i}\tau\right) H_1^{(I)}(t-\tau)_{i,m} H_1^{(I)}(t)_{n,j} + \delta_{i,m} \sum_p \exp\left(i\omega_{p,n}\tau\right) H_1^{(I)}(t)_{n,p} H_1^{(I)}(t-\tau)_{p,j} \right\} \tag{II-92}$$

with the notation $\omega_{ab} = (E_a - E_b)/\hbar$, where the energies E_a, E_b are eigenvalues of $\mathbf{H_0}$. Averaging the products of matrix elements of $\mathbf{H_1}$ and integrating over τ one may write the Bloch-Redfield equation as:

$$\frac{d}{dt} \langle \sigma(t) \rangle_{i,j} = \sum_{m,n} \exp\left(i\left(\omega_{i,j} - \omega_{m,n}\right)t\right) R_{ij,mn} \langle \sigma(t) \rangle_{m,n} \tag{II-93}$$

Here **R** is a superoperator, belonging to Liouville space, like **L** or **Q**, and its elements are defined as:

$$R_{ij,mn} = -\frac{1}{(\hbar)^2}\left\{ \delta_{n,j}\sum_p \int_0^\infty \exp(i\omega_{m,p}\tau)\langle H_1^{(I)}(t)_{i,p}H_1^{(I)}(t-\tau)_{p,m}\rangle\, d\tau \right.$$

$$- \int_0^\infty \exp(i\omega_{j,n}\tau)\langle H_1^{(I)}(t)_{i,m}H_1^{(I)}(t-\tau)_{n,j}\rangle\, d\tau - \int_0^\infty \exp(i\omega_{m,i}\tau)\langle H_1^{(I)}(t-\tau)_{i,m}H_1^{(I)}(t)_{n,j}\rangle\, d\tau$$

$$\left. + \delta_{i,m}\sum_p \int_0^\infty \exp(i\omega_{p,n}\tau)\langle H_1^{(I)}(t-\tau)_{n,p}H_1^{(I)}(t)_{p,j}\rangle\, d\tau \right\} \qquad \text{(II-94)}$$

Using the hermiticity of $\mathbf{H}_1$ one may write $\mathbf{H}_1(t)_{ab} = \mathbf{H}_1(t)_{ba}^*$, so the averages in the integrands are in the standard form of auto-correlation functions. If $\mathbf{H}_1(t)$ is a stationary random operator, its auto-correlation depends only on the time difference τ (and not on t), and this dependence is finally integrated over. Thus the elements of **R** are constant in time if $\mathbf{H}_1(t)$ is a stationary stochastic operator.

According to Eq. (II-90) the variations in time of $\sigma(t)$ are slow, because $\mathbf{H}_1(t)$ is small. But the exponent in Eq. (II-93) oscillates very fast unless $\omega_{ij}-\omega_{mn}=0$, because its frequencies are related to $\mathbf{H}_0$, which is large. The implication is that all terms with a rapidly oscillating time dependence are multiplied by very small coefficients, and only those terms with no such oscillations are significant. Therefore all terms in which $\omega_{ij}\neq\omega_{mn}$ can be neglected, to obtain a simpler form of Eq. (II-93):

$$\frac{d}{dt}\langle\sigma(t)\rangle_{i,j} = \sum_{m,n}{}' R_{ij,mn}\langle\sigma(t)\rangle_{m,n} \qquad \text{(II-95)}$$

where the "prime" on the summation indicates the restriction to those states $|u_m\rangle$, $|u_n\rangle$ which satisfy the condition $\omega_{ij}=\omega_{mn}$. This is the usual form of Redfield's equation, which can be regarded as a generalized master equation (compare Eq. (II-55)). It is very simple, because all the coefficients in this differential equation are constants. It is still necessary to construct a model for $\mathbf{H}_1$ and to determine its probability distribution, so as to calculate the coefficients $\mathbf{R}_{ij,mn}$.

So far the calculation was done in the interaction representation. Transforming back to the laboratory frame, one only needs

$$\langle\rho(t)\rangle = \exp(-iH_0t/\hbar)\langle\sigma(t)\rangle\exp(iH_0t/\hbar) \qquad \text{(II-96)}$$

In the basis of $\mathbf{H}_0$ it is easy to relate the matrix elements of $\langle\rho\rangle$ to those of $\langle\sigma\rangle$:

$$\langle\rho(t)\rangle_{m,n} = \exp(-iE_mt/\hbar)\langle\sigma(t)\rangle_{m,n}\exp(iE_nt/\hbar) = \exp(-i\omega_{m,n}t)\langle\sigma(t)\rangle_{m,n} \qquad \text{(II-97)}$$

so their time dependence is also simply related:

$$\frac{d}{dt}\langle \rho(t) \rangle_{i,j} = -i\,\omega_{i,j}\langle \rho(t) \rangle_{i,j} + \exp(-i\omega_{i,j}t)\frac{d}{dt}\langle \sigma(t) \rangle_{i,j} =$$

$$= \frac{1}{i\hbar}[H_0, \langle \rho(t) \rangle]_{i,j} + \sum_{m,n}{}' \exp(-i\omega_{i,j}t)\, R_{ij,mn}\langle \sigma(t) \rangle_{m,n} \tag{II-98}$$

In the exponent one may substitute $\omega_{m,n}$ for $\omega_{i,j}$, so the equation may be summarized in operator form as follows:

$$\frac{d}{dt}\rho = \frac{1}{i\hbar}[H_0, \rho] + R\rho = \left\{\frac{1}{i\hbar}L_0 + R\right\}\rho \tag{II-99}$$

Here the averaging brackets were omitted, but the average over ρ is implied here and in consequent formulas. $\mathbf{L_0}$ is the Liouvillian superoperator corresponding to $\mathbf{H_0}$. This form of Redfield's equation is a generalized Liouville-von Neumann equation, in which the first term on the right hand side is the standard term that would occur in the absence of stochastic processes (compare Eq. (I-59)), and the second term expresses the relaxation effect due to the stochastic processes. This form of the equation is the general form of the Liouville-von Neumann equation with relaxation, which is useful also when the processes leading to relaxation are described in a different manner. This will be seen in the following chapters, where different treatments of relaxation lead to the same kind of equation, except that the relaxation operator **R** has a different meaning in each case.

(b) Generalizations of the equation

The form of Redfield's equation derived here describes a gradual decay to an "infinite temperature" equilibrium, namely it leads to an equilibrium in which all levels are equally populated. A more rigorous quantum mechanical description of the relaxation processes (which will not be shown here, but is developed in the references given at the end of this Chapter) shows that the correct form of the equation for a finite temperature is

$$\frac{d}{dt}\rho = \frac{1}{i\hbar}L_0\rho + R(\rho - \rho_0) \tag{II-100}$$

where ρ_0 is the steady state solution, or equilibrium density matrix:

$$\rho_0 = \frac{\exp(-H_0/k_BT)}{Tr\left(\exp(-H_0/k_BT)\right)} \tag{II-101}$$

The modification of the Bloch-Redfield equation needed to ensure the correct equilibrium is thus analogous to the modification of the simple master equation (II-72) for finite temperatures, Eq. (II-79).

This equation for the density matrix can be used to calculate equations of motion for expectation values of quantum mechanical operators. These equations describe the evolution of the corresponding macroscopic variables, in particular the magnetization vector of the sample. Suppose $\mathbf{m}(t) \equiv \langle \mathbf{M} \rangle (t)$ is such a macroscopic average. Its time dependence is given by

$$\frac{d}{dt} m = Tr\left\{ M \frac{d}{dt} \rho(t) \right\} = \frac{1}{i\hbar} Tr\left\{ M\left(L_0 \rho\right)\right\} + Tr\left\{ M\left(R\left(\rho - \rho_0\right)\right)\right\} =$$

$$= \frac{1}{i\hbar} \sum_{i,j} M_{i,j} \left(L_0 \rho\right)_{j,i} + \sum_{i,j} M_{i,j} \left(R\left(\rho - \rho_0\right)\right)_{j,i} \tag{II-102}$$

Using the values of the relaxation matrix elements given in Eq. (II-94) it is possible to get explicit expressions for the coefficients in the present equation. In this manner one may set equations for the macroscopic magnetization, for example.

Finally, let us return to a point which was hinted at earlier, the possible existence of a time dependent part in the Hamiltonian which is not stochastic in nature. This is in fact an essential component of CW magnetic resonance experiments, in which the irradiation is applied throughout the experiment. Suppose there is a coherent time dependent part $\mathbf{H}_2$ in the Hamiltonian (e.g. $\mathbf{H}_2 = 2\omega_1 \cos(\omega t) \mathbf{S}_x$) then:

$$H = H_0 + H_1(t) + H_2(t) \tag{II-103}$$

where $\mathbf{H}_1(t)$ changes in time randomly, and $\mathbf{H}_2(t)$ changes in time coherently. Calculating in the interaction representation as above, Eq. (II-88) would change to:

$$\frac{d}{dt} \sigma(t) = \frac{1}{i\hbar} [H_1^{(I)}(t) + H_2^{(I)}(t), \sigma(0)]$$

$$- \frac{1}{(\hbar)^2} \int_0^t d\tau \, [H_1^{(I)}(t) + H_2^{(I)}(t), [H_1^{(I)}(t-\tau) + H_2^{(I)}(t-\tau), \sigma(0)]] \tag{II-104}$$

If $\| \mathbf{H}_2 \|$ is small enough one may neglect the effect of $\mathbf{H}_2$ on the integral, so averaging under the same conditions as above, the analog of Eq. (II-90) would be:

$$\frac{d}{dt} \langle \sigma(t) \rangle = \frac{1}{i\hbar} [H_2^{(I)}(t), \langle \sigma(0) \rangle] - \frac{1}{(\hbar)^2} \times$$

$$\times \int_0^\infty d\tau \, \langle [H_1^{(I)}(t), [H_1^{(I)}(t-\tau), \sigma(t)]] \rangle \tag{II-105}$$

Transforming back to the laboratory system, one obtains the analog of Eq. (II-99):

$$\frac{d}{dt}\rho = \frac{1}{i\hbar}[H_0 + H_2(t), \rho] + R\rho \tag{II-106}$$

Once the elements of the relaxation superoperator **R** are calculated, this equation can be solved by going to the rotating frame, because the relaxation superoperator is not affected by this transformation. The correction for a finite temperature equilibrium may be done as in Eq. (II-100).

In fact, the requirement that $\| \mathbf{H}_2 \|$ is small may be eliminated if one performs first the rotating frame transformation, and only at a second stage defines an interaction representation. The rotating frame transformation would lead to:

$$H = H_0 + H_1(t) + 2\omega_1 \cos(\omega t) S_x \rightarrow$$

$$H' = H_0 - \omega S_z + H_1'(t) + \omega_1 S_x = H_0' + H_1'(t) \tag{II-107}$$

One thus has a constant Hamiltonian and a randomly changing Hamiltonian, so the original derivation of the Bloch-Redfield equation can be repeated for them, leading to an equation of the same form as (II-99). The equation would refer to the rotating frame, and $\mathbf{H}_0$ in it would be replaced by $\mathbf{H}_0'$. In this case the relaxation operator should be calculated, in principle, in the rotating frame.

II.7 An Operator Form for the Master Equation

In the derivation presented above no specific details were given concerning the stochastic Hamiltonian. If one constructs the stochastic Hamiltonian in terms of known spin operators, the Redfield relaxation operator may be expressed in terms of known quantities. The stochastic Hamiltonian can be written as a sum:

$$H_1(t) = \sum_q F^{(q)}(t)\, A^{(q)} \tag{II-108}$$

in which $\mathbf{A}^{(q)}$ are constant spin operators and $\mathrm{F}^{(q)}$ are random functions of time, representing the influence of the "lattice" (= environment of spins). The random functions are assumed to be stationary. The $\mathbf{A}^{(q)}$ are normally chosen as components of irreducible spherical tensors constructed of spin operators (see Appendix D), and since the Hamiltonian is a scalar, also the $\mathrm{F}^{(q)}$ are components of irreducible spherical tensors. Due to the Wigner-Eckart theorem this form of the Hamiltonian is completely general. In most practical cases only a few terms are included in this summation. Transforming to the interaction representation:

$$H_1(t) = \sum_q F^{(q)}(t) \exp(iH_0 t/\hbar)\, A^{(q)} \exp(-iH_0 t/\hbar) = \sum_q F^{(q)}(t)\, A^{(q)}{}_p \exp\left(i\omega^{(q)}{}_p t\right) \tag{II-109}$$

The operators $\mathbf{A}^{(q)}{}_p$ appearing in this formula are the Fourier components of the transformed $\mathbf{A}^{(q)}$, which is a coherently time dependent operator. Thus the Bloch-

Redfield equation becomes:

$$\frac{d}{dt}\langle \sigma(t)\rangle = -\frac{1}{(\hbar)^2}\sum_{p,q,r,s} \exp\left(i(\omega^{(q)}_p + \omega^{(r)}_s)t\right)[A^{(r)}_s,[A^{(q)}_p,\sigma(t)]] \times$$

$$\times \int_0^\infty d\tau \exp\left(-i\omega^{(q)}_p\tau\right)\langle F^{(q)}(t)\, F^{(r)}(t+\tau)\rangle \qquad \text{(II-110)}$$

where stationarity of the process is assumed. The spin operators $\mathbf{A}^{(q)}$ are in general non-hermitian, and the hermiticity of $\mathbf{H}_1$ requires that for each term $F^{(q)}(t)\,\mathbf{A}^{(q)}$ there will also be its hermitian conjugate, $F^{(q)*}(t)\,\mathbf{A}^{(q)\dagger}$. In spherical irreducible tensors, for each component with index (q) there is also a corresponding term with index (-q). It is thus natural to choose $F^{(-q)}=F^{(q)*}$, $A^{(-q)}=A^{(q)\dagger}$. These requirements are satisfied, for example, by the spherical harmonic functions, which are components of irreducible spherical tensors. For each $Y^l_m(\theta,\phi)$ there is also a function $Y^l_{-m}(\theta,\phi)$ which satisfies $Y^l_{-m}(\theta,\phi)=Y^{l\,*}_m(\theta,\phi)$. Under such conditions also $\omega^{(-q)}_p=-\omega^{(q)}_p$. Then

$$\langle F^{(q)}(t)\, F^{(r)}(t+\tau)\rangle = \langle F^{(q)}(t)\, F^{(-s)*}(t+\tau)\rangle \equiv G_{q,-s}(\tau) = \delta_{q,-s}\, G_q(\tau) \qquad \text{(II-111)}$$

The last equality on the right-hand side is not completely general, but it is satisfied by the auto-correlation function defined here for many stochastic processes occurring in magnetic resonance. An example is given below. It will be assumed here that this auto-correlation is a real and even function of τ (see discussion of Eq. (II-60)). In such cases it can be shown that the integration results in

$$\int_0^\infty d\tau \exp\left(-i\omega^{(q)}_p\tau\right) G_{q,-s}(\tau) \approx \frac{1}{2}\int_{-\infty}^{\infty} d\tau \exp\left(-i\omega^{(q)}_p\tau\right) G_{q,-s}(\tau) = \frac{1}{2} J_q\left(\omega^{(q)}_p\right) \qquad \text{(II-112)}$$

Therefore, neglecting rapidly oscillating terms as in Eq. (II-95):

$$\frac{d}{dt}\langle \sigma(t)\rangle = -\frac{1}{2(\hbar)^2}\sum_{p,q} J_q\left(\omega^{(q)}_p\right)[A^{(-q)}_p,[A^{(q)}_p,\sigma(t)]] \qquad \text{(II-113)}$$

If the correlation time is relatively short, the spectral densities for all frequencies are not very different from the spectral densities for zero frequency:

$$J_q\left((\omega^{(q)}_p\right) \approx J_q(0) \qquad \text{(II-114)}$$

so the elements of the Redfield operator **R** of Eqs. (II-94), (II-95) above may be calculated from (II-113) using just the spin operators $\mathbf{A}^{(q)}$. This is the "motional narrowing limit". From Eq. (II-94):

$$R_{ij,mn} = -\frac{1}{2(\hbar)^2}\sum_{p,q} J_q(0)\left\{\delta_{j,n}\sum_r \left(A^{(-q)}{}_p\right)_{i,r}\left(A^{(q)}{}_p\right)_{r,m} - \left(A^{(-q)}{}_p\right)_{i,m}\left(A^{(q)}{}_p\right)_{n,j}\right.$$

$$\left. - \left(A^{(q)}{}_p\right)_{i,m}\left(A^{(-q)}{}_p\right)_{n,j} + \delta_{i,m}\sum_r \left(A^{(q)}{}_p\right)_{m,r}\left(A^{(-q)}{}_p\right)_{r,j}\right\} \qquad \text{(II-115)}$$

The elements are indeed time independent, as they have to be for a stationary stochastic process.

More generally, the elements of **R** can be calculated without special assumptions, if the decomposition of $\mathbf{H_1}$ is redefined as follows. Instead of using irreducible tensors, the matrix of $\mathbf{H_1}$ is written in the basis of $\mathbf{H_0}$, and is then decomposed in terms of the level-shift operators (see Eqs. (I-29),(I-30)):

$$H_1(t) = \sum_{i,j} F^{(i,j)}(t)\, a_{i,j}\, |i\rangle\langle j| \equiv \sum_{i,j} F^{(i,j)}(t)\, A^{(i,j)} = \sum_{i,j} F^{(i,j)}(t)\, \left(A^{(i,j)}\right)_{i,j} |i\rangle\langle j| \qquad \text{(II-116)}$$

The typical component $\mathbf{A}^{(i,j)}$ is proportional to $(\mathbf{H_1}(t))_{ij}|i\rangle\langle j|$, for which the only non-zero element is the (i,j) element. Since these elements refer to the basis of $\mathbf{H_0}$, the transformation to the interaction representation results in:

$$H_1(t) = \sum_{i,j} F^{(i,j)}(t)\, A^{(i,j)} \exp\left(i\omega_{i,j}t\right) \qquad \text{(II-117)}$$

Here the frequency is simply defined as $\omega_{ij}=(E_i-E_j)/\hbar$, referring to the energy eigenvalues of the main Hamiltonian. There is no need to sum separately over frequencies. The equivalent of Eq. (II-110) would be:

$$\frac{d}{dt}\langle\sigma(t)\rangle = -\frac{1}{(\hbar)^2}\sum_{i,j,m,n} \exp\left(i(\omega_{i,j}+\omega_{m,n})t\right)\,[A^{(m,n)},[A^{(i,j)},\sigma(t)]] \times$$

$$\times \int_0^\infty d\tau\, \exp\left(-i\omega_{i,j}\tau\right)\langle F^{(i,j)}(t)\, F^{(m,n)}(t+\tau)\rangle \qquad \text{(II-118)}$$

Now define

$$I_{ij,mn} \equiv \int_0^\infty d\tau\, \exp\left(-i\omega_{i,j}\tau\right)\langle F^{(i,j)}(t)\, F^{(m,n)}(t+\tau)\rangle \qquad \text{(II-119)}$$

and then

$$R_{ij,mn} = -\frac{1}{2(\hbar)^2}\sum_p \left\{ \delta_{j,n}\sum_r I_{rm,ir}\left(A^{(r,m)}\right)_{r,m}\left(A^{(i,r)}\right)_{i,r} - I_{nj,im}\left(A^{(n,j)}\right)_{n,j}\left(A^{(i,m)}\right)_{i,m} \right.$$

$$\left. - I_{im,nj}\left(A^{(i,m)}\right)_{i,m}\left(A^{(n,j)}\right)_{n,j} + \delta_{i,m}\sum_r I_{nr,rj}\left(A^{(n,r)}\right)_{n,r}\left(A^{(r,j)}\right)_{r,j} \right\} \quad \text{(II-120)}$$

If one defines the following spectral densities:

$$J_{ij,mn}(\omega) \equiv \int_{-\infty}^{\infty} d\tau \exp(-i\omega\tau) \langle \left(H_1(t)\right)_{i,j} \left(H_1(t+\tau)\right)_{mn}{}^{*} \rangle =$$

$$= \int_{-\infty}^{\infty} d\tau \exp(-i\omega\tau) \langle \left(H_1(t)\right)_{i,j} \left(H_1(t+\tau)\right)_{nm} \rangle \quad \text{(II-121)}$$

it follows from the stationarity of the stochastic process that they obey the symmetry relations:

$$J_{ij,mn}(\omega) = J_{mn,ij}(\omega) \qquad \textit{and} \qquad J_{ji,nm}(\omega) = J_{ij,mn}(-\omega) \quad \text{(II-122)}$$

If the auto-correlation is a real and even function of τ, these spectral densities also satisfy

$$J_{ij,mn}(-\omega) = J_{ij,mn}(\omega) \qquad \Rightarrow \qquad J_{ji,nm}(\omega) = J_{ij,mn}(\omega) \quad \text{(II-123)}$$

Using the approximate relation in Eq. (II-112) and the symmetry relations of these spectral densities, the elements of the Redfield superoperator **R** can be expressed as:

$$R_{ij,mn} = \frac{1}{2(\hbar)^2}\left\{J_{nj,mi}(\omega_{n,j}) + J_{im,jn}(\omega_{i,m}) - \delta_{j,n}\sum_r J_{rm,ri}(\omega_{r,m}) - \delta_{i,m}\sum_r J_{nr,jr}(\omega_{nr})\right\} =$$

$$= \frac{1}{2(\hbar)^2}\left\{J_{im,jn}(\omega_{j,n}) + J_{im,jn}(\omega_{i,m}) - \delta_{j,n}\sum_r J_{ir,mr}(\omega_{r,m}) - \delta_{i,m}\sum_r J_{nr,jr}(\omega_{nr})\right\} \quad \text{(II-124)}$$

Example II.4: *Relaxation by dipolar coupling*

An important case of relaxation is that caused by dipolar coupling, i.e. the interaction between two magnetic dipole moments (or magnetic moments, as we have called them until now). This occurs whenever there is stochastic relative motion between the magnetic moments of two different spins. The change can be a change of distance or a change of relative orientation if the spins belong to different molecules. If the spins belong to the same molecule, such changes will not normally occur, but random molecular motion will still change the orientation of the line connecting the two dipoles with respect to the orientation of the external magnetic field. As far as the dipole-dipole interaction is concerned, one may regard the relative motion of the spins as a tumbling of the dipole-dipole relative position vector in the coordinate system dictated by the extrenal magnetic field. The position vector is conveniently described with spherical coordinates (r,θ,ϕ). For the case of a constant distance r the motion is on the surface of a sphere, as shown in Fig. II.4.

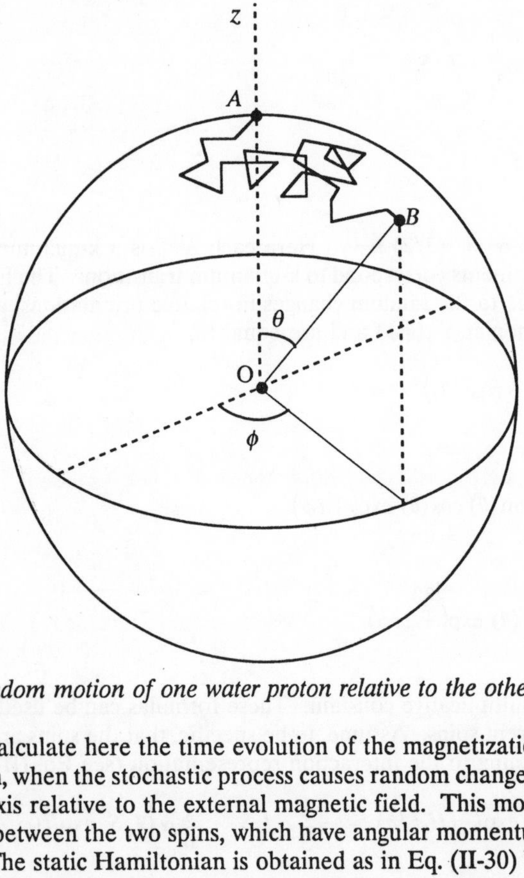

Fig. II.4. *Random motion of one water proton relative to the other.*

We shall calculate here the time evolution of the magnetization from the Bloch-Redfield equation, when the stochastic process causes random changes in the orientation of the spin-spin axis relative to the external magnetic field. This motion modulates the dipolar coupling between the two spins, which have angular momentum operators **I** and **S**, respectively. The static Hamiltonian is obtained as in Eq. (II-30) by a scalar product between the magnetic field and the magnetic moment. Assuming as usual the field is in the **z** direction,

$$H_0 = \omega_I I_z + \omega_S S_z \tag{E-1}$$

The dipolar coupling is a scalar, formed from the products of components of two spherical tensors of rank 2:

$$H_1(t) = \sum_k F^{(2,k)}(t)\, A^{(2,k)} \tag{E-2}$$

in which the spin operators are

$$A^{(2,0)} = \alpha\left\{-\frac{2}{3} I_z S_z + \frac{1}{6}\left(I_+ S_- + I_- S_+\right)\right\} \tag{E-3a}$$

$$A^{(2,\pm1)} = \mp\alpha\left\{I_z S_\pm + I_\pm S_z\right\} \tag{E-3b}$$

$$A^{(2,\pm2)} = \frac{1}{2}\alpha I_\pm S_\pm \tag{E-3c}$$

with the definition $\alpha \equiv -(3/2)\,\hbar\gamma_I\gamma_S$. Here each $A^{(2,k)}$ is a k-quantum operator, i.e., its non-zero matrix elements correspond to k-quantum transitions. The functions, which are time dependent due to the random changes in relative orientations, are proportional to the spherical harmonics $Y^2_k(\theta,\phi)$ and are equal to:

$$F^{(2,0)} = \frac{1}{r^3}\left(1 - 3\cos^2(\theta)\right) \tag{E-4a}$$

$$F^{(2,\pm1)} = \mp\frac{1}{r^3}\sin(\theta)\cos(\theta)\exp(\mp i\phi) \tag{E-4b}$$

$$F^{(2,\pm2)} = \frac{1}{r^3}\sin^2(\theta)\exp(\mp 2i\phi) \tag{E-4c}$$

up to an overall multiplicative constant. These formulas can be used both for identical spins and for different spins. Assume, to be specific, that the spins are different, so that $\omega_I \neq \omega_S$. Transforming to the interaction representation (see Eq. (II-109)):

$$\exp(iH_0 t/\hbar)\, A^{(2,0)} \exp(-iH_0 t/\hbar) = -\frac{2}{3}\alpha I_z S_z + \frac{1}{6}\alpha\left\{I_+ S_- \exp(i(\omega_I - \omega_S)t) + I_- S_+ \times\right.$$

$$\left.\times \exp(-i(\omega_I - \omega_S)t)\right\} \equiv A^{(0)}{}_0 \exp(i\cdot 0\cdot t) + A^{(0)}{}_1 \exp\left(i\omega^{(0)}{}_1 t\right) + A^{(0)}{}_{-1}\exp\left(i\omega^{(0)}{}_{-1} t\right) \tag{E-5}$$

with $\omega^{(0)}{}_1 = \omega_I - \omega_S$ and $\omega^{(0)}{}_{-1} = -\omega^{(0)}{}_1$, and similar formulas apply to the other components of the stochastic Hamiltonian. These results may be substituted into Eq. (II-110), and then equations of the type (II-102) can be set up for the components of the magnetization vector.

In order to find explicitly all coefficients in the resulting equations, it is needed to calculate correlation functions of the type appearing in Eq. (II-110) or (II-121). This requires a model for the time modulation of the orientation. Assume this is described as Brownian rotational diffusion, where the length of the relative position vector is constant but its orientation is time dependent. This is effectively random motion on the surface of a sphere, as shown in Fig. II.4 above. The random variable is $y \equiv \Omega=(\theta,\phi)$, the a priori probability for having any particular orientation is $p(y,t)=1/4\pi$ (isotropic distribution, normalized) and the conditional probability for having at time t a particular orientation Ω, provided at time $t=0$ it was Ω_0, is $P(y_0,y,t)=P(\Omega_0,\Omega,t)$ abbreviated as $P(\Omega,t)$. The relevant auto-correlation functions (see Eq. (II-59)) are of the form

$$G(\tau) = \int\int f(\Omega_0)f^*(\Omega)p(\Omega_0)P(\Omega_0,\Omega,\tau)\,d\Omega_0 d\Omega =$$

$$= \frac{1}{4\pi}\int\int f(\Omega_0)f^*(\Omega)P(\Omega,\tau)\,d\Omega_0 d\Omega \qquad \text{(E-6)}$$

For $P(\Omega,t)$ one assumes, following Debye, the Brownian rotational diffusion equation:

$$\frac{\partial}{\partial t}P(\Omega,t) = R\,\nabla^2 P(\Omega,t) \qquad \text{(E-7)}$$

where R is a constant number - the rate constant, which determines the speed of motion. This equation is solved subject to the initial condition $P(\Omega,0)=\delta(\Omega-\Omega_0)$. The equation can be solved by expanding the conditional probability function in terms of the spherical harmonic functions, which are eigenfunctions of the Laplacian operator appearing in Eq. (E-7). The calculation will not be done here (it can be found, for example, in Ref. [3] below). The result is that the expansion coefficients decay with time exponentially, where the exponent depends on the value of (l) in Y^l_m. In the case of dipolar coupling and in most practical cases, $l=2$ is the relevant order, and the corresponding decay time is the correlation time τ_C of the auto-correlation functions.

****

Suggested References

* On the classical Bloch equations and on quantum mechanical treatment of magnetic resonance:

1. A. Carrington and A.D. McLachlan, "*Introduction to Magnetic Resonance*" (Harper and Row, 1967), Chs. 1,2.
2. C.P. Slichter, "*Principles of Magnetic Resonance*", 3rd ed. (Springer, Berlin, 1990), Ch. 2.
3. A. Abragam, "*Principles of Nuclear Magnetism*" (Oxford, Oxford, 1961) Chs. II, III.

* On stochastic processes in general:

4. N.G. Van Kampen, *"Stochastic Processes in Physics and Chemistry"* (North Holland, Amsterdam, 1981) Chs. I - V .

* On relaxation theory for magnetic resonance:

5. N. Bloembergen, E.M. Purcell and R.V. Pound, *Phys. Rev.* **73**, 679 (1948) ("BPP")
6. R.K. Wangsness and F. Bloch, *Phys. Rev.* **89**, 728 (1953); F. Bloch, *Phys. Rev.* **102**, 104 (1956); **105**, 1206 (1957).
7. A.G. Redfield in *Adv. Mag. Res.*, vol. 1, edited by J.S. Waugh (Academic, New York, 1965), p. 1.
8. Ref. 2, Ch. 5.
9. Ref. 3, Ch. 8.

CHAPTER III

CHEMICAL EXCHANGE

III.1 The Classical Rate Equations

In the beginning of the previous Chapter the concept of a Lorentzian line was briefly discussed. Actual spectra in magnetic resonance usually consist of several or even many spectral lines, each of which has a Lorentzian form. The complete spectrum is known as a line shape, and it is characteristic of the physical system in which it was obtained. In some systems a special type of temperature dependence is observed in the magnetic resonance line shapes. In the simplest cases of this type (see Fig. III.1), one observes at low temperatures two narrow spectral lines. When the temperature is gradually raised, these lines gradually broaden, and at the same time they move closer to each other. At a certain temperature they coalesce into one broad line, and when the temperature is raised further this line gradually narrows down. The final narrow line observed for high temperatures appears at a spectral position (frequency) intermediate between the positions of the original lines observed at low temperatures.

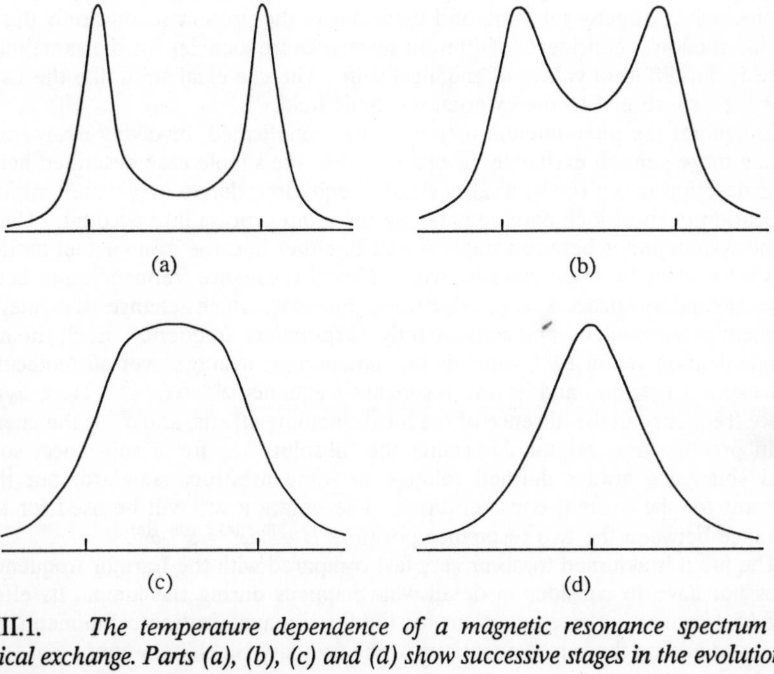

Fig. III.1. *The temperature dependence of a magnetic resonance spectrum due to chemical exchange. Parts (a), (b), (c) and (d) show successive stages in the evolution of the spectrum when the temperature increases.*

This phenomenon is qualitatively explained by a jump mechanism. At very low temperaures, the system can be in one of two alternative states, differing by their resonance frequency but having the same chemical identity. These two states are often equally populated, but may have different populations. Each state gives rise to a separate spectral line. When the temperature is raised, energy is transferred into a mechanism of jumps between these states. The jumps tend to average the resonance position, and that is what is observed at high temperatures. All the other observations reflect stages in a continuous process of merging two narrow lines into one intermediate narrow line.

The processes described here may be observed, in principle, both in NMR and in EPR. However, in practice their rate is usually too slow to be important in EPR, so most of the work on this subject belongs to NMR. For this reason our treatment of the subject will be given in the context of NMR, although it should be relevant in principle also to EPR, wherever such problems occur there. In NMR each resonance frequency is determined by a sum of two terms. The main term is the Larmor frequency characteristic of the nucleus, due to the external magnetic field. For the strength of magnetic fields employed normally in magnetic resonance, this term is larger by several orders of magnitude than the additional term. The second term is thus a small correction term, resulting from local interactions. One of these is a shielding interaction, which screens the nucleus to some degree from the external field. This results in a frequency term called the chemical shift, due to the shielding effects of the local environment of the nuclear spin. For problems involving chemical exchange it is often this term which is directly relevant, and therefore in the present section only this kind of local interaction is considered. Different resonance frequencies for the same nucleus therefore mean different values of chemical shift. The chemical shift, like the Larmor frequency, is proprtional to the external magnetic field.

In general the phenomenon may be more complicated, involving many spectral lines and a more general exchange mechanism. For the simple case described here, an adequate description is given by a set of classical equations due to McConnell, which are a modification of the Bloch equations taking the jump process into account. The idea is that the system jumps between states A and B, either because of an actual motion of the relevant nucleus between two sites with different resonance frequencies, or because of some secondary process e.g., electronic motions which change the magnetic environment of the nucleus, and consequently its resonance frequency. Each site has its own magnetization vector $\mathbf{M}^{(\alpha)}$, which is the macroscopic average over all molecules in which the spin is in site α, and its own resonance frequency $\omega^{(\alpha)}=\omega_0+\delta^{(\alpha)}$. Here ω_0 is the resonance frequency in the absence of the local shielding effects, and $\delta^{(\alpha)}$ is the chemical shift. In practice one cannot determine the "absolute" ω_0 for a substance, so that chemical shifts are always defined relative to some arbitrary standard, but this is unimportant for the present considerations. The notation $\Delta^{(\alpha)}$ will be used for ω-$\omega^{(\alpha)}$. The distance between the two resonance positions is $\omega^{(B)}$-$\omega^{(A)}=\delta^{(B)}$-$\delta^{(A)}$.

The jump is assumed to occur very fast compared with the Larmor frequency, so one does not have to consider in detail what happens during the jump. Its effect is included by constructing rate equations for the three magnetization components in the two sites, and adding these rate equations to the ordinary Bloch equations. For state A (or site A) the equations are:

$$\frac{d}{dt}M_x^{(A)} = -\frac{1}{T_2}M_x^{(A)} + \Delta^{(A)}M_y^{(A)} \qquad - k^{(AB)}M_x^{(A)} + k^{(BA)}M_x^{(B)} \quad \text{(III-1a)}$$

$$\frac{d}{dt}M_y^{(A)} = -\Delta^{(A)}M_x^{(A)} - \frac{1}{T_2}M_y^{(A)} - \omega_1 M_z^{(A)} \qquad - k^{(AB)}M_y^{(A)} + k^{(BA)}M_y^{(B)} \quad \text{(III-1b)}$$

$$\frac{d}{dt}M_z^{(A)} = \qquad \omega_1 M_y^{(A)} - \frac{1}{T_1}\left(M_z^{(A)} - M_0^{(A)}\right) - k^{(AB)}M_z^{(A)} + k^{(BA)}M_z^{(B)} \quad \text{(III-1c)}$$

and for site B the equations are similar, with the obvious interchange of A and B indices. In each of these equations, the exchange part consists of two rate terms, appearing on the extreme right, and the rest of the equation is the same as in the ordinary Bloch equations. In fact, the rate terms are simply a "sink" term and a "source" term, as in the master equation (Eq. (II-55)) discussed in the previous chapter. Here $k^{(\alpha\beta)}$ is the rate of jumps from site α to site β, which is equal to the inverse average lifetime of state α:

$$k^{(\alpha\beta)} = \frac{1}{\tau_\alpha} \quad \text{(III-2)}$$

In a steady state the relative populations of the two sites are determined by the flow into each of them. This can be seen by ignoring the Bloch part of the equation and setting the time derivatives equal to zero. The equilibrium magnetizations are seen to obey:

$$\frac{M_0^{(A)}}{M_0^{(B)}} = \frac{k^{(BA)}}{k^{(AB)}} \quad \text{(III-3)}$$

and since the magnetization is proportional to the population, the population ratio is

$$\frac{P^{(A)}}{P^{(B)}} = \frac{k^{(BA)}}{k^{(AB)}} = \frac{\tau_A}{\tau_B} \quad \text{(III-4)}$$

In many systems the two sites have equal lifetimes and are therefore equally populated. In such cases $k^{(AB)} = k^{(BA)} = k$. For the sake of simplicity only such cases will be treated here, because the main conclusions to be derived below are valid also when the populations are not equal.

As found experimentally, the spectrum changes as a function of temperature, and the fundamental assumption of the rate equations is that the jump rate is temperature dependent, causing changes in the spectrum. The main parameter which determines the effect of the jumps on the line shape is the magnitude of the rate constant k compared with the frequency difference between different spectral lines. In the present problem there is only one rate constant and one frequency difference, so one distinguishes between three regimes, each with its characteristic behaviour:

(*a*) $\dfrac{k}{\delta^{(A)} - \delta^{(B)}} \ll 1$ (*slow jumps*)

(*b*) $\dfrac{k}{\delta^{(A)} - \delta^{(B)}} \approx 1$ (*intermediate rate jumps*)

(c) $\dfrac{k}{\delta^{(A)} - \delta^{(B)}} \gg 1$ (*fast jumps*)

The equations (III-1) will now be solved without any assumptions on the relative magnitude of the various constants, and then the implications of the solution will be examined for each of these regimes. Define $M_\pm^{(\alpha)} = M_x^{(\alpha)} \pm iM_y^{(\alpha)}$ (α = A,B). The first two real equations in (III-1) can be combined to give the complex equations for $M_\pm$, but only the equation for M_+ is needed, because that of M_- is its complex conjugate. The resulting set of equations is:

$$\frac{d}{dt}M_+^{(A)} = -\left[\frac{1}{T_2} + k^{(AB)} + i\Delta^{(A)}\right] M_+^{(A)} + k^{(BA)}M_+^{(B)} - i\omega_1 M_z^{(A)} \tag{III-5a}$$

$$\frac{d}{dt}M_+^{(B)} = -\left[\frac{1}{T_2} + k^{(BA)} + i\Delta^{(B)}\right] M_+^{(B)} + k^{(AB)}M_+^{(A)} - i\omega_1 M_z^{(B)} \tag{III-5b}$$

$$\frac{d}{dt}M_z^{(A)} = -\frac{i\omega_1}{2}\left(M_+^{(A)} - M_-^{(A)}\right) - \frac{1}{T_1}\left(M_z^{(A)} - M_0^{(A)}\right) - k^{(AB)}M_z^{(A)} + k^{(BA)}M_z^{(B)} \tag{III-5c}$$

$$\frac{d}{dt}M_z^{(B)} = -\frac{i\omega_1}{2}\left(M_+^{(B)} - M_-^{(B)}\right) - \frac{1}{T_1}\left(M_z^{(B)} - M_0^{(B)}\right) - k^{(BA)}M_z^{(B)} + k^{(AB)}M_z^{(A)} \tag{III-5d}$$

The signal is determined by the transverse magnetization M_+, but its time-dependent value is coupled to the longitudinal magnetization M_z. At this point the discussion will be restricted to the case of a CW experiment, in which the system is in a steady state, and the two jump rates are equal. The steady state assumption implies the four time derivatives are equal to zero, so that the solution for one instant remains correct for all times. The second assumption implies $M_0^{(A)} = M_0^{(B)} = \frac{1}{2}M_0$ where M_0 is the total equilibrium magnetization. The irradiation is further assumed to be weak (no saturation), so that its effect on the spins is also weak, and therefore at all times the **z** magnetization is almost equal to its equilibrium value:

$$M_z^{(A)} \approx M_0^{(A)} = \frac{M_0}{2} \quad and \quad M_z^{(B)} \approx M_0^{(B)} = \frac{M_0}{2} \tag{III-6}$$

This is an important result, because it means that in order to calculate the transverse magnetization one does not need (with the current approximations) to know the exact value of M_z in order to solve for M_+. The equations (III-5a), (III-5b) for the transverse magnetization components are thus effectively uncoupled from the equations (III-5c), (III-5d) for the longitudinal magnetization. Under these conditions, Eqs. (III-5a), (III-5b) become:

$$-C^{(A)} M_+^{(A)} + k\ M_+^{(B)} = \frac{i\omega_1}{2} M_0 \tag{III-7a}$$

$$k\ M_+^{(A)} - C^{(B)} M_+^{(B)} = \frac{i\omega_1}{2} M_0 \tag{III-7b}$$

with $C^{(\alpha)} = 1/T_2 + k + i\Delta^{(\alpha)}$. Therefore

$$M_+^{(A)} = -\frac{i\omega_1}{2} M_0 \left[\frac{k + C^{(B)}}{C^{(A)} C^{(B)} - k^2} \right] =$$

$$= -\frac{i\omega_1}{2} M_0 \left[\frac{\left[\frac{1}{T_2} + 2k\right] + i\Delta^{(B)}}{\frac{1}{T_2}\left[\frac{1}{T_2} + 2k\right] - \Delta^{(A)}\Delta^{(B)} + i\left(\Delta^{(A)} + \Delta^{(B)}\right)\left[\frac{1}{T_2} + k\right]} \right] \tag{III-8}$$

and a similar equation for $M_+^{(B)}$, with interchange of the indices (A), (B).These results depend on the irradiation frequency ω through $\Delta^{(\alpha)}$. It is convenient to assume irradiating at the central frequency $\omega = \omega_0 + \frac{1}{2}(\delta^{(A)} + \delta^{(B)})$, because then $\Delta^{(A)} = \frac{1}{2}(\delta^{(B)} - \delta^{(A)}) = -\Delta^{(B)}$ so that

$$M_+^{(A)} = -\frac{i\omega_1}{2} M_0 \left[\frac{\left[\frac{1}{T_2} + 2k\right] - i\Delta^{(A)}}{\frac{1}{T_2}\left[\frac{1}{T_2} + 2k\right] + (\Delta^{(A)})^2} \right] \tag{III-9}$$

This formula can now be used for each of the three regimes mentioned above. In the rigid limit ,i.e., in the absence of exchange (k = 0) this result is just the Bloch solution for these conditions. The position of the resonance (distance from the irradiation frequency) is specified by $\Delta^{(A)}$ and its width by $1/T_2$. For slow exchange, the denominator is slightly increased so that the effective absolute value of $\Delta^{(A)}$ is reduced. Thus the two resonance positions are slightly shifted towards each other. At the same time the width parameter is changed to $2k+1/T_2$, divided by approximately the same denominator as before. Since the original linewidth must be small relative to the frequency differences if the spectrum is sharp, the effect on the linewidth can be significant, whereas the frequency shift is of secondary importance. At the other extreme, for fast exchange, the linewidth is much greater than the deviation of the line from the center, since $2k+1/T_2$ is much greater than $\Delta^{(A)}$. Thus the two lines have completely merged. Nevertheless, the combined line is narrow, because the form of the magnetization is approximately the same as for no exchange. For intermediate exchange rates it is clear that there is some broadening and frequency shift, since the dependence on all parameters here is continuous.

The Bloch-McConnell equations thus give a correct description of the main features of simple exchange phenomena. From this description it follows that even when the two sites are at a "dynamic equilibrium", having equal populations, the exchange process has a non-trivial effect, broadening the spectrum in the frequency domain. This is equivalent to faster decay of the signal in the time domain as shown in Chapter II. The result (III-9) obtained here is thus in the form of a Lorentzian line, and therefore its inverse Fourier transform to the time domain is a decaying function, in which the exchange constant k contributes to the decay rate.

III.2 The Quantum Mechanical Density Matrix Formalism

(a) *The equation of motion for the density matrix*

In the Bloch-McConnell equations there are two magnetic sites, which may be occupied simultaneously by two spins, but no interactions between the spins are included. It is easy to generalize the equations for a larger number of non-interacting spins. However, when interactions within the observed spin system are present, one has to deal with the two-spin (or multi-spin) system as a whole, and not with each spin separately. The chemical shift is effectively an interaction between each spin and the magnetic field, but other interactions, e.g. the scalar (J) interaction or the dipolar interactions, involve directly the magnetizations of two spins. When a molecule has several spins involved in an exchange process, and at the same time strongly interacting with one another, the need for a more general theory is much more acute. Strong inter-spin interactions occur especially in liquid crystalline solutions, where dipolar interactions are not averaged out to zero, but the problem is relevant also to ordinary liquid solutions. The approach of Kaplan and Alexander takes interactions within a spin system into account by working with the density matrix for the whole spin system, and using the density matrix to calculate expectation values of magnetization.

As in the classical treatment, it is assumed that the jumps are relatively fast, so one only has to consider the situation before and after a jump. The effect of a jump is to transform wave functions $\psi(t)$ and operators **A** (including the density matrix $\rho(t)$) as

$$\psi(t) \rightarrow \psi'(t) = X\psi(t) \qquad\qquad A \rightarrow A' = XAX^{-1} \qquad\qquad \text{(III-10)}$$

The transformation **X** is effectively a permutation of spin indices in the wave functions and operators. In subsection III.3.(a) below the standard matrix form of **X** will be presented for a particular case, an exchanging two-spin system. In the present Section only the permutation approach will be used, since it gives a clear picture of the exchange process without unnecessary mathematical details. In the simplest relevant case only two spins are involved, with a "secular" interaction i.e., an interaction which has a diagonal matrix in the basis of spin product functions. Thus a general wave function in the spin product basis is

$$|\psi\rangle = |I,m_1\rangle^{(A)}\,|I,m_2\rangle^{(B)} \tag{III-11}$$

I rather than S is used here for the total spin angular momentum, in accordance with the common notation in nuclear magnetic resonance (in electronic paramagnetic resonance, S is normally used). The original Hamiltonian (ignoring exchange) is

$$H = \Delta^{(A)} I_z^{(A)} + \Delta^{(B)} I_z^{(B)} + J I_z^{(A)} I_z^{(B)} \tag{III-12}$$

This corresponds to having a scalar interaction $\mathbf{J}\mathbf{I}^{(A)}\cdot\mathbf{I}^{(B)}$ where J is small compared with the relevant frequency differences, and then relying on first-order perturbation theory to neglect those terms in the interaction which are not diagonal in $\mathbf{I}_z^{(A)}$ and in $\mathbf{I}_z^{(B)}$. The result of applying the transformation is

$$|\psi'\rangle = |I,m_2\rangle^{(A)}\,|I,m_1\rangle^{(B)} \tag{III-13}$$

and

$$H' = \Delta^{(B)} I_z^{(A)} + \Delta^{(A)} I_z^{(B)} + J I_z^{(B)} I_z^{(A)} \tag{III-14}$$

In other words, the two spins exchange their magnetic properties, which is formally equivalent to having the two nuclei exchange positions in a rigid molecular structure. Then also the density matrix is transformed, as in Eq. (III-9). It is now assumed that, as in the Bloch-Redfield relaxation theory, the rate of change of the density matrix is a sum of two contributions. One is the ordinary von Neumann term, and the other describes the effect of the random process. In the present case the random process is the jumps with an average rate of $1/\tau$. The system is assumed to be in "dynamic equilibrium" in terms of the chemical exchange process, although the magnetization is not necessarily in its equilibrium state. In fact, the magnetization must be in a non-equilibrium state at least during part of the experiment, or else there will be no time variations to observe. The jumps may be either intra-molecular or inter-molecular. Alexander reserves the name "chemical exchange" for the latter case, but we shall use this term indiscriminately for both types of exchange processes.

Relaxation effects due to other mechanisms may be added through a simplified Redfield operator, with the following definition:

$$R_{ij,mn} = \delta_{i,m}\,\delta_{j,n}\left[\delta_{i,j}\frac{1}{T_1} + \left(1-\delta_{i,j}\right)\frac{1}{T_2}\right] \tag{III-15}$$

This definition is chosen so that for $i \neq j$:

$$(R\,\rho)_{i,j} = \sum_{m,n} R_{ij,mn}\,\rho_{m,n} = R_{ij,ij}\,\rho_{i,j} = \frac{1}{T_2}\,\rho_{i,j} \qquad \text{(III-16)}$$

whereas for i = j:

$$(R\,\rho)_{i,i} = \sum_{m,n} R_{ii,mn}\,\rho_{m,n} = R_{ii,ii}\,\rho_{i,i} = \frac{1}{T_1}\,\rho_{i,i} \qquad \text{(III-17)}$$

Thus for off-diagonal elements of the density matrix the corresponding diagonal element of **R** is equal to $1/T_2$, and for diagonal elements of the density matrix the corresponding diagonal element of **R** is equal to $1/T_1$. The combined result is a semi-classical equation of motion for the density matrix:

$$\frac{d}{dt}\rho = \frac{1}{i\hbar}[H,\rho] + \frac{1}{\tau}\left(X\rho X^{-1} - \rho\right) - \frac{1}{T_2}\,\rho_{off\text{-}diag.} - \frac{1}{T_1}\left(\rho_{diag.} - \rho_0\right) \qquad \text{(III-18)}$$

The first term on the right hand side of the equation is the usual Liouville - von Neumann term. The second term describes chemical exchange in this model, and the last two terms represent relaxation due to other processes via the simplified Redfield superoperator. Using the direct product notation, justified by the close relation of superoperators here to operator direct products (see Ch. I, subsection I.6.(c)), the equation may be written in Liouville space as:

$$\frac{d}{dt}\rho = \frac{1}{i\hbar}L\rho + \frac{1}{\tau}\left(X\otimes X^{-1} - I\otimes I\right)\rho + R^{(phen.)}\rho \qquad \text{(III-19)}$$

where the phenomenological Redfield term $\mathbf{R}^{(phen.)}$ is only included for completeness. In practice one may do all calculations without it, adding the line width effect in the final stage of calculating a spectrum. Note that this equation is in the same general form as Redfield's equation (II-99), but the main relaxation operator here, namely the exchange superoperator, is very different from that of Redfield. In the Bloch-Redfield formalism the randomness appears in a certain part of the Hamiltonian, and this is responsible for the relaxation supermatrix, which can be calculated by doing some suitable averaging over products of matrix elements of the Hamiltonian. In the chemical exchange formalism the randomness appears directly in the equation of motion for density matrix, and the consideration of the different states does not require an average, but a source-sink balance. The main component here is the chemical exchange term, calculated directly with the transformation operator **X**, which is the main feature of the theory. The additional part due to other processes requires only choosing values for the parameters T_1 and T_2. In principle one could write the equation separately for "site A" and for "site B" as in the classical equations, but the use of the transformation operator **X** makes this unnecessary.

There is also another difference between the chemical exchange equations and the Bloch-Redfield equation. The latter is valid only for fast random processes, whereas the former is valid for any rate of chemical exchange. However, within the chemical exchange formalism one may formulate an approximation for fast exchange. Such an approximation, which is essentially the Bloch-Redfield limit of chemical exchange, is developed in the following subsection.

The equation written above applies to various types of intra-molecular and inter-molecular exchange processes, but at this stage only its most elementary application is considered. The exchange is assumed to be intra-molecular, with only two possible sites for the molecule. The molecule thus undergoes interconversions between two states continuously. Since two exchanges bring the molecule back to its original state, the transformation operator satisfies $\mathbf{X}^2 = \mathbf{I}$ (the unit operator), or $\mathbf{X}^{-1} = \mathbf{X}$.

Example III.1: Ring inversion of s-trioxane

In a molecule of s-trioxane (Fig. III.2) three hydrogen nuclei are in an "axial" position (nuclei 1,2,3 in the left-hand part of the diagram) and three are in an "equatorial" position (nuclei 4,5,6 there). The ring inversion process inverts the positions of these nuclei. In addition to interchanging the chemical shifts $\delta^{(j)}$ (j = 1, ..., 6), it also interchanges spin-spin interaction constants. We shall consider the case of s-trioxane dissolved in a liquid crystal, where strong dipolar interactions are present and therefore a quantum treatment is needed. A typical wave function is (without exchange)

$$|\psi\rangle = |\phi_1(1)\rangle|\phi_2(2)\rangle|\phi_3(3)\rangle|\phi_4(4)\rangle|\phi_5(5)\rangle|\phi_6(6)\rangle \tag{E-1}$$

where the subscript in each component refers to the "site" or to its magnetic characteristics, whereas the argument in parentheses is the identity of the particular nucleus. Originally the Hamiltonian is

$$H = \delta^{(a)}\left(I_z^{(1)} + I_z^{(2)} + I_z^{(3)}\right) + \delta^{(e)}\left(I_z^{(4)} + I_z^{(5)} + I_z^{(6)}\right) + \left(J^{(a,a)} + D^{(a,a)}\right)\left(I_z^{(1)} I_z^{(2)} + \dots\right)$$

$$+ \left(J^{(e,e)} + D^{(e,e)}\right)\left(I_z^{(4)} I_z^{(5)} + \dots\right) + \left(J^{(a,e)} + D^{(a,e)}\right)\left(I_z^{(1)} I_z^{(4)} + \dots\right) \tag{E-2}$$

where J stands for the scalar coupling, present also in isotropic liquids, and D for the much stronger dipolar coupling. The superscripts a,e refer to axial and equatorial hydrogens, respectively. The interchange process transforms the wave function into

$$|\psi\rangle = |\phi_4(1)\rangle|\phi_5(2)\rangle|\phi_6(3)\rangle|\phi_1(4)\rangle|\phi_2(5)\rangle|\phi_3(6)\rangle \tag{E-3}$$

and the Hamiltonian into

$$H = \delta^{(e)}\left(I_z^{(1)} + I_z^{(2)} + I_z^{(3)}\right) + \delta^{(a)}\left(I_z^{(4)} + I_z^{(5)} + I_z^{(6)}\right) + \left(J^{(e,e)} + D^{(e,e)}\right)\left(I_z^{(1)} I_z^{(2)} + \dots\right)$$

$$+ \left(J^{(a,a)} + D^{(a,a)}\right)\left(I_z^{(4)} I_z^{(5)} + \dots\right) + \left(J^{(a,e)} + D^{(a,e)}\right)\left(I_z^{(4)} I_z^{(1)} + \dots\right) \tag{E-4}$$

In this case the requirement $\mathbf{X}^2 = \mathbf{I}$ is indeed satisfied. In the transformation of $\boldsymbol{\rho}$ into $\mathbf{X}\boldsymbol{\rho}\mathbf{X}$, the change in the indices can be schematically indicated as

$$\rho(1,2,3,4,5,6) \rightarrow \rho(4,5,6,1,2,3) = X\rho(1,2,3,4,5,6)X \tag{E-5}$$

****

Example III.2: Bond shift in cyclooctatetraene

In cyclooctatetraene (Fig. III.3) there are eight protons (hydrogen nuclei), and the backbone carbon atoms are connected by an alternating series of single and double bonds. The exchange process interchanges the single and double bonds, and at the same time "inverts" the molecule (see Figure). All protons are "magnetically equivalent" ,i.e., they have identical chemical shifts (or: all have zero chemical shift relative to their common resonance). Their interactions, however, depend on their relative positions, and these are not identical in the two configurations. As in s-trioxane, in isotropic solvents the interactions are weak, but in liquid crystalline solvents there are strong dipolar interactions. In this case, the transformation is formally described either as a shift to the right:

$$\boldsymbol{\rho}(1,2,3,4,5,6,7,8) \rightarrow \boldsymbol{\rho}(8,1,2,3,4,5,6,7) = \boldsymbol{X}\boldsymbol{\rho}(1,2,3,4,5,6,7,8)\boldsymbol{X}^{-1} \tag{E-1}$$

or as a shift to the left:

$$\boldsymbol{\rho}(1,2,3,4,5,6,7,8) \rightarrow \boldsymbol{\rho}(2,3,4,5,6,7,8,1) = \boldsymbol{X}^{-1}\boldsymbol{\rho}(1,2,3,4,5,6,7,8)\boldsymbol{X} \tag{E-2}$$

In this case, operating with the exchange operator twice does *not* bring the density matrix back to its original form. This is true both for the shift to the right:

$$\boldsymbol{X}^{2}\boldsymbol{\rho}(1,2,3,4,5,6,7,8)\boldsymbol{X}^{-2} = \boldsymbol{\rho}(7,8,1,2,3,4,5,6) \tag{E-3}$$

and for the shift to the left:

$$\boldsymbol{X}^{-2}\boldsymbol{\rho}(1,2,3,4,5,6,7,8)\boldsymbol{X}^{2} = \boldsymbol{\rho}(3,4,5,6,7,8,1,2) \tag{E-4}$$

In this case the condition $\mathbf{X}^2=\mathbf{I}$ is not satisfied, in spite of the fact that physically there are only two distinct configurations. This apparently counter-intuitive situation will be analysed in Example III.5 below.

****

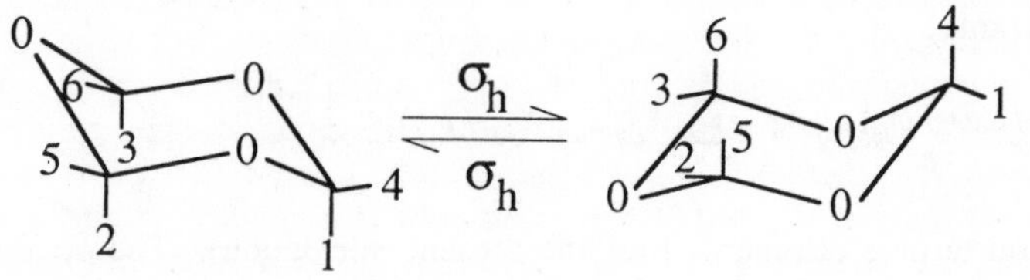

Fig. III.2. *The ring inversion process in s-trioxane.*

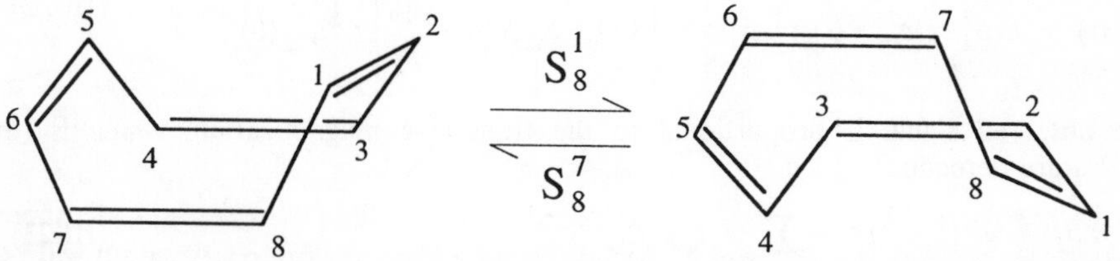

Fig. III.3. *The bond shift process in cyclooctatetraene.*

(b) Approximations for slow and fast exchange

When the exchange process is slow, τ^{-1} is small relative to typical frequencies in the Hamiltonian. Its effect in Eq. (III-18) may be considered as a small perturbative correction to the ordinary Liouville-von Neumann calculation. It is therefore natural to solve the equation in the basis in which **H** is diagonal. In principle such a solution involves finding all elements of the time dependent density matrix. However, as long as one is not concerned with multiple pulse experiments, there is no need to apply Eq. (III-18) to all matrix elements of $\boldsymbol{\rho}$. In a CW experiment or in a simple FID experiment (following a single pulse) one can only observe off-diagonal "single-quantum" elements of $\boldsymbol{\rho}$, i.e., those $\boldsymbol{\rho}_{ij}$ for which $|m_i - m_j| = 1$. Also, the assumption $\mathbf{X} = \mathbf{X}^{-1}$ will be made in the following calculations.

In the basis which diagonalizes **H**, the equation for off-diagonal elements is (in the rotating frame)

$$\frac{d}{dt}\rho_{i,j} = -i\left(\Delta\omega + \omega_{i,j}\right)\rho_{i,j} + \frac{1}{\tau}\sum_{m,n}\left(X_{i,m}\,\rho_{m,n}\,X_{n,j} - \rho_{i,j}\right) - \frac{1}{T_2}\rho_{i,j} \tag{III-20}$$

(the irradiation term is disregarded for the present calculation). The operator **X** only changes the magnetic environment of a spin, but not the quantum state of a spin, so the $\mathbf{I}_z$ projection value of each spin is unchanged by the transformation. Therefore **X** can only have non-zero matrix elements between states with the same total $\mathbf{I}_z$ projection. Out of the sum on m,n we shall only consider the diagonal term, which is the main effect of the "perturbation". The result is (with $k = \tau^{-1}$):

$$\frac{d}{dt}\rho_{i,j} = -\left\{ i\left(\Delta\omega + \omega_{i,j}\right) + k\left(1 - X_{i,i}X_{j,j}\right) + \frac{1}{T_2} \right\}\rho_{i,j} \tag{III-21}$$

which is solved by

$$\rho_{i,j}(t) = \exp\left\{-i\left(\Delta\omega + \omega_{i,j}\right)t\right\}\exp\left\{ -\left[k\left(1 - X_{i,i}X_{j,j}\right) + \frac{1}{T_2}\right]t \right\}\rho_{i,j}(0) \tag{III-22}$$

The observed signal is proportional to the transverse magnetization, which is (in quadrature detection)

$$\langle I_+\rangle(t) = Tr\{I_+\rho(t)\} = \sum_{i,j}(I_+)_{i,j}\,\rho(t)_{j,i} \tag{III-23}$$

Consequently, the density matrix elements in Eq. (III-22) result in a decaying signal, which is equivalent to a Lorentzian line in the frequency domain. The main effect of the exchange is to change the line width as

$$\frac{1}{T_2} \rightarrow \frac{1}{T_2} + \frac{1}{\tau}\left(1 - X_{i,i}X_{j,j}\right) \tag{III-24}$$

The main part of this change is simply related to the exchange rate. The smaller correction, proportional to $\mathbf{X}_{i,i}\,\mathbf{X}_{j,j}$ results from the possibility of states i,j remaining unchanged by the transformation of the molecule.

For fast exchange it was found in the Bloch-McConnell calculation that the system is effectively in an average position, because the jump rate is so high that the experiment cannot "see" the jumps. This indicates that in a quantum mechanical description it is natural to work in the basis of the average Hamiltonian

$$S \equiv \frac{1}{2}(H + XHX) \tag{III-25}$$

which is obtained by summing the Hamiltonians of all possible configurations, and dividing by the number of configurations. This Hamiltonian is, by construction, symmetric with respect to exchange:

$$XSX = \frac{1}{2}(XHX + X^2HX^2) = \frac{1}{2}(XHX + H) = S \tag{III-26}$$

One may now decompose the original Hamiltonian into a sum of two terms - the average Hamiltonian and the difference

$$A \equiv H - S = \frac{1}{2}(H - XHX) \tag{III-27}$$

which is found to be anti-symmetric with respect to exchange:

$$XAX = \frac{1}{2}(XHX - X^2HX^2) = \frac{1}{2}(XHX - H) = -A \tag{III-28}$$

Since the average Hamiltonian does not distinguish between the two states of the system, it obviously corresponds to the limit of infinitely fast exchange. Any observable effects of exchange at a finite rate are due to the deviation from average, i.e., due to **A**. Its magnitude is expected to be similar to that of S, but both have to be small relative to the exchange rate:

$$k \gg \frac{1}{\hbar}\|A\|,\ \frac{1}{\hbar}\|S\| \tag{III-29}$$

in the fast exchange regime.

The density matrix is decomposed in a similar manner:

$$\rho = \sigma + \alpha \tag{III-30}$$

with the definitions

$$\sigma \equiv \frac{1}{2}(\rho + X\rho X) \qquad \alpha \equiv \frac{1}{2}(\rho - X\rho X) \tag{III-31}$$

These parts are symmetric and anti-symmetric, respectively, under the exchange operation:

$$X\sigma X = \sigma \qquad X\alpha X = -\alpha \tag{III-32}$$

Since one observes all molecules simultaneously, the result of an observation is a sum over the magnetization of molecules in the two different states. This sum is determined by the average density matrix $\boldsymbol{\sigma}$. It is therefore desirable to isolate an equation for this part of the density matrix, in the basis of the average Hamiltonian. Other relaxation mechanisms will be neglected here, so the equation of motion for the density matrix (III-18) can be rewritten as

$$\frac{d}{dt}(\sigma+\alpha) = \frac{1}{i\hbar}\{[S,\sigma]+[A,\sigma]+[S,\alpha]+[A,\alpha]\} - \frac{2}{\tau}\alpha \tag{III-33}$$

Operating on this equation with the transformation **X...X**, the signs of the time derivative of $\boldsymbol{\sigma}$ and of the first and fourth commutators are unchanged. The signs of all other terms are inverted. From the sum and difference of Eq. (III-33) and its transformed version one obtains two coupled equations for the two parts of the density matrix:

$$\frac{d}{dt}\sigma = \frac{1}{i\hbar}\{[S,\sigma]+[A,\alpha]\} \tag{III-34a}$$

$$\frac{d}{dt}\alpha = \frac{1}{i\hbar}\{[A,\sigma]+[S,\alpha]\} - \frac{2}{\tau}\alpha \tag{III-34b}$$

An interaction picture can be defined through the transformation:

$$S^{(I)} \equiv e^{iSt} S e^{-iSt} = S \qquad A^{(I)} \equiv e^{iSt} A e^{-iSt} \tag{III-35a}$$

$$\sigma^{(I)} \equiv e^{iSt} \sigma e^{-iSt} \qquad \alpha^{(I)} \equiv e^{2t/\tau} e^{iSt} \alpha e^{-iSt} \tag{III-35b}$$

The factor of $e^{2t/\tau}$ in the definition of $\alpha^{(I)}$ is not part of the transformation, but is included for convenience in later calculations. It follows from the definitions that

$$\frac{d}{dt}\sigma^{(I)} = \frac{1}{i\hbar} e^{-2t/\tau} [A^{(I)}, \alpha^{(I)}] \tag{III-36a}$$

$$\frac{d}{dt}\alpha^{(I)} = \frac{1}{i\hbar} e^{2t/\tau} [A^{(I)}, \sigma^{(I)}] \tag{III-36b}$$

These equations are still coupled, but in a simpler way. Integrating over (III-36b):

$$\alpha^{(I)} = \frac{1}{i\hbar}\int_{-\infty}^{0} e^{2t'/\tau}\,[A^{(I)}(t'),\sigma^{(I)}(t')]\,dt' \qquad \text{(III-37)}$$

which can be substituted into (III-36a):

$$\frac{d}{dt}\sigma^{(I)} = \frac{1}{(i\hbar)^2}\,e^{-2t/\tau}\,[A^{(I)}(t),\int_{-\infty}^{0} e^{2t'/\tau}\,[A^{(I)}(t'),\sigma^{(I)}(t')]]\,dt' =$$

$$= -\frac{1}{\hbar^2}\int_{-\infty}^{0} e^{-2(t-t')/\tau}\,[A^{(I)}(t),[A^{(I)}(t'),\sigma^{(I)}(t')]]\,dt' \qquad \text{(III-38)}$$

This is a closed equation for the average density matrix, for which one only needs to use directly that part of the Hamiltonian which contains the information on the exchange effects. It is written in the same form as the Bloch-Redfield equation (II-90) , and the analogy extends also to most of its details . The average density matrix $\sigma^{(I)}$ here is the analog of the average density matrix $\langle\sigma\rangle$ there, and the anti-symmetric Hamiltonian **A** here is analogous to the stochastic Hamiltonian $\mathbf{H_1}$ there. The time dependent exponent, however, does not have its analog there. Writing (III-38) for specific matrix elements of $\sigma^{(I)}$, and doing integration by parts, one may obtain simple expressions for the elements of the Redfield operator, provided

$$\left|\frac{A_{i,j}A_{j,n}}{\hbar^2\left(i\omega_{i,j}+2/\tau\right)}\right| \ll |\omega_{i,n}| \qquad \text{(III-39)}$$

for general values of i,j,n. The result is:

$$\frac{d}{dt}\sigma^{(I)}(t)_{i,j} = -\sum_{m,n} R^{(exch.)}_{ij,mn}\,\sigma^{(I)}(t)_{m,n} \qquad \text{(III-40)}$$

with the following definition for the Redfield superoperator describing fast exchange:

$$R^{(exch.)}_{ij,mn} \equiv \frac{A_{i,m}A_{n,j}}{\hbar^2\left(i\omega_{i,m}+2/\tau\right)} + \frac{A_{i,m}A_{n,j}}{\hbar^2\left(-i\omega_{i,m}+2/\tau\right)}$$

$$+\ \delta_{i,m}\sum_{p}{}' \frac{A_{n,p}A_{p,j}}{\hbar^2\left(i\omega_{n,p}+2/\tau\right)} + \delta_{j,n}\sum_{p}{}' \frac{A_{i,p}A_{p,m}}{\hbar^2\left(-i\omega_{p,m}+2/\tau\right)} \qquad \text{(III-41)}$$

The primes on the summations indicate the restriction that $E_n = E_j$ for the first sum and $E_i = E_m$ for the second sum. Eq. (III-40) is exactly in the form of Eq. (II-95). If there are no degenerate levels, the restrictions on the sums can be replaced by multiplying each sum by $\delta_{im}\delta_{nj}$. Returning to the laboratory frame:

$$\frac{d}{dt}\sigma(t)_{i,j} = -i\omega_{i,j}\sigma(t)_{i,j} - \sum_{m,n} R^{(exch.)}{}_{ij,mn}\,\sigma(t)_{m,n} \qquad \text{(III-42)}$$

which is analogous to (II-98).

III.3 Invariance Properties of Exchanging Systems

In the Bloch-Redfield formalism the main complication is due to external effects, i.e., due to the stochastic influence of the environment. Thus having a larger spin system is not necessarily related to a more complicated solution of Redfield's equation. In chemical exchange, however, the complication is in a process taking place within the system. A larger size of the system will immediately manifest itself in a larger dimension of the relevant equation for the density matrix. For a single spin I, the total number of linearly independent wave functions is (2I+1). If the relevant system of spins has spin quantum numbers I_1, I_2,... then the total number of spin product wave functions is the product of the corresponding numbers of wave functions:

$$d_H = \Pi_k(2I_k + 1) \qquad \text{(III-43)}$$

where k goes over all spins in the molecule. Since d_H is the dimension of Hilbert space, the number of elements of $\boldsymbol{\rho}$ is d_H^2. The equation of motion of the density matrix is effectively in Liouville space, with the supervector $\boldsymbol{\rho}$ having dimension d_H^2 and the supermatrix having dimension $d_H^2 \times d_H^2$. The dimensions of the numerical problem therefore grow exponentially with the size of the system. For example, in the common case in which there is a system of protons (hydrogen nuclei) exchanging among them, $I_k = ½$ for each nucleus so for N spins: $d_H = 2^N$. It is thus very important to use all those properties of the system and of the exchange process which lead to a reduction in the dimensions of the calculations. Three types of such properties will be discussed in this section, each of them associated with some kind of invariance property of the chemically exchanging system. Two of them are useful for any rate of exchange, and the third is mainly useful for relatively fast exchange.

(a) *Invariance of the total spin I_z projection*

As mentioned above in connection with the approximation for slow exchange, the transformation operator **X** does not change the value of the total $\mathbf{I}_z$ projection. In other words, the eigenvalue of $\mathbf{I}_z$ which characterizes a wave function is invariant under the exchange process. This simple observation implies that the coupling of elements of the density matrix is restricted to "blocks" within the matrix, characterized by their I_z value, and the calculation can be broken up into small parts. This can be done both for CW and for FID magnetic resonance experiments.

Consider first a CW experiment. The Hamiltonian in the rotating frame can be written as

$$H = H_0 + \omega_1 I_x \tag{III-44}$$

where $\mathbf{H_0}$ contains the Zeeman term $\omega_0\mathbf{I_z}$ as well as intra-molecular interaction terms. Most magnetic resonance experiments are done in high field, where the Zeeman term is much larger than those interactions, so only their "secular" part, i.e., the part diagonal in $\mathbf{I_z}$ has to be included in (III-44). The irradiation term ω_1 $\mathbf{I_x}$ is not secular and is not necesarily large, but it is essential for observing non-vanishing transverse magnetization.

Neglecting the trivial T_1, T_2 terms in Eq. (III-18), the steady state equation (zero time derivative) is:

$$-\frac{1}{i\hbar}[\omega_1 I_x, \rho] = \frac{1}{i\hbar}[H_0, \rho] + \frac{1}{\tau}\left(X\rho X^{-1} - \rho\right) \tag{III-45}$$

In the basis which diagonalizes simultaneously $\mathbf{H_0}$ and the total $\mathbf{I_z}$, a matrix element of the equation is:

$$-\frac{\omega_1}{i\hbar}\sum_k \left((I_x)_{i,k}\rho_{k,j} - \rho_{i,k}(I_x)_{k,j}\right) = -i\,\omega_{i,j}\rho_{i,j} + \frac{1}{\tau}\sum_{k,l}\left(X_{i,k}\rho_{k,l}(X^{-1})_{l,j} - \rho_{i,j}\right) \tag{III-46}$$

The eigenvalues of $\hbar^{-1}\mathbf{I_z}$ are denoted as $m_i \equiv \langle i|\mathbf{I_z}|i\rangle$. Since $\mathbf{X_{ik}} \neq 0$ only if $m_i = m_k$, and $(\mathbf{X^{-1}})_{lj} \neq 0$ only if $m_l = m_j$, the only non-zero terms in the sum on the right-hand side of (III-46) must have $m_k - m_l = m_i - m_j \equiv \Delta m_{ij}$. Thus the exchange process couples elements of the density matrix only if they are of the same quantum multiplicity, since the value of Δm can be regarded as a number of elementary photons required for a transition containing Δm steps. It is true that different quantum multiplicities are coupled together on the left hand side of the equation, but it will now be shown that this coupling is not important in practice. As noted previously, the CW experiment can only probe those elements of the density matrix with $|\Delta m_{ij}| = 1$, so it is sufficient to consider those elements. Furthermore, in CW the irradiation is assumed to be weak: $\omega_1/\omega_0 \ll 1$, so the elements of the density matrix remain throughout the experiment close to their equilibrium values (without irradiation). At equilibrium with only a static field, only the diagonal elements of $\boldsymbol{\rho}$ are non-zero. Therefore in Eq. (III-46) any diagonal elements of $\boldsymbol{\rho}$ are much larger than off-diagonal elements of $\boldsymbol{\rho}$ occurring there. Since $(\mathbf{I_x})_{kl} \neq 0$ only if $|\Delta m_{kl}| = 1$, the sum on the left hand side of the equation is approximately equal to

$$-\frac{\omega_1}{i\hbar}\sum_k (I_x)_{i,j}\left(\rho_{j,j} - \rho_{i,i}\right) \approx \frac{i\,\omega_1\,\omega_0}{k_B T}(I_x)_{i,j} \tag{III-47}$$

This is a constant for each pair of states (i,j), so there is no coupling between the elements of $\boldsymbol{\rho}$ other than that created by the exchange term.

Now consider an FID experiment. The irradiation during the initial pulse creates a special initial condition for subsequent evolution of the system, but in the differential equation for that evolution there is no irradiation term. Thus the equation of motion is

$$\frac{d}{dt}\rho = \frac{1}{i\hbar}[H_0, \rho] + \frac{1}{\tau}\left(X\rho X^{-1} - \rho\right) \tag{III-48}$$

Following the same arguments as above, the equation does not couple elements of ρ with a different quantum multiplicity. In fact, the same conclusion could also be reached in a different way. It is known in magnetic resonance that the FID time domain behaviour of the magnetization is the Fourier transform of the CW frequency domain spectrum, provided the CW irradiation is weak, so that linear response theory can be used. Thus, calculating one of the two types of spectra is equivalent to calculating the other.

From the general result obtained here it follows that the equations for the "relevant" elements of ρ, i.e. those elements with $|\Delta m_{ij}| = 1$, can be divided into several sets. If the total spin operator of the system is $\mathbf{I} = \sum_k \mathbf{I}_k$, then the possible eigenvalues of the total $\mathbf{I}_z$ are: m = $-I_{max}$, $-I_{max}+1$,..., $I_{max}-1$, I_{max} where $I_{max} = \sum_k I_k$. Then the time evolution of those density matrix elements connecting, for example, m = 0 states with m = 1 states is completely independent of the evolution of those elements which connect m = 1 states with m = 2 states, etc. . Since the dimension of the Liouville space matrix in Eq. (III-19) is the square of the number of relevant states, the reduction in computations is very significant.

The dimensions of the Liouville supermatrix for cases of practical importance in chemical exchange are very large, and it is thus not easy to follow for such problems all the details of the invariance rules discussed here and in the next subsection. It is therefore convenient to illustrate these invariance conditions in a system of the smallest possible dimensions, so that the basic concepts can be clarified. The structure of the Liouville space operators will be studied here for a two-spin system, taking into account the $\mathbf{I}_z$ invariance. Such a system is any molecule in which there are two identical spins, which are involved in a chemical exchange process. The molecule may include other atomic nuclei as well, but these are not of the same species and have zero magnetic moment, so they are irrelevant for NMR experiments.

For simulating the effects of chemical exchange in NMR the molecule can be regarded as having only two nuclei. Each of them is assumed to have I = ½ , so each spin has (2·½+1) = 2 possible wave functions: $|\alpha\rangle \equiv |m{=}½\rangle$ and $|\beta\rangle \equiv |m{=}{-}½\rangle$. The total number of spins is N = 2, so there are 2^N = 4 independent wave functions for the system as a whole. The wave functions diagonal in $\mathbf{I}_{z1}$ and in $\mathbf{I}_{z2}$ are

$$|\alpha\alpha\rangle = |\alpha\rangle_1|\alpha\rangle_2 \qquad |\alpha\beta\rangle = |\alpha\rangle_1|\beta\rangle_2$$

$$|\beta\alpha\rangle = |\beta\rangle_1|\alpha\rangle_2 \qquad |\beta\beta\rangle = |\beta\rangle_1|\beta\rangle_2 \tag{III-49}$$

Here one wave function - $|\alpha\alpha\rangle$ - has a total $\mathbf{I}_z$ eigenvalue of 1, one wave function - $|\beta\beta\rangle$ - has an eigenvalue of -1, and the other two wave functions have an eigenvalue of 0.

An exchange process will effectively exchange the indices of the two nuclei as mentioned in the previous Section, so it will interchange $|\alpha\beta\rangle$ with $|\beta\alpha\rangle$ and will have no effect on $|\alpha\alpha\rangle$ and on $|\beta\beta\rangle$. A formal expression for the exchange operator **X** can be constructed in the following way. A general wave function is

$$|\psi\rangle = c_1|\beta\beta\rangle + c_2|\beta\alpha\rangle + c_3|\alpha\beta\rangle + c_4|\alpha\alpha\rangle \quad \text{(III-50)}$$

which can be written in vector form as:

$$|\psi\rangle = \begin{pmatrix} c_1 \\ c_2 \\ c_3 \\ c_4 \end{pmatrix} \quad \text{(III-51)}$$

where the indices 1, 2, 3, 4 refer to the functions $|\beta\beta\rangle$, $|\beta\alpha\rangle$, $|\alpha\beta\rangle$ and $|\alpha\alpha\rangle$ respectively. The exchange operator is represented in the current basis by the following matrix:

$$X = \begin{pmatrix} 1 & 0 & 0 & 0 \\ 0 & 0 & 1 & 0 \\ 0 & 1 & 0 & 0 \\ 0 & 0 & 0 & 1 \end{pmatrix} \quad \text{(III-52)}$$

so that it operates on a general wave function as

$$X\psi = \begin{pmatrix} 1 & 0 & 0 & 0 \\ 0 & 0 & 1 & 0 \\ 0 & 1 & 0 & 0 \\ 0 & 0 & 0 & 1 \end{pmatrix} \begin{pmatrix} c_1 \\ c_2 \\ c_3 \\ c_4 \end{pmatrix} = \begin{pmatrix} c_1 \\ c_3 \\ c_2 \\ c_4 \end{pmatrix} \quad \text{(III-53)}$$

A general Hilbert space operator **A** on this system is represented by a 4 × 4 matrix, with the elements $\mathbf{A}_{i,j} = \langle i|\mathbf{A}|j\rangle$. The matrix is transformed by the exchange operator through the relation:

$$A' = XAX^{-1} = \begin{bmatrix} 1 & 0 & 0 & 0 \\ 0 & 0 & 1 & 0 \\ 0 & 1 & 0 & 0 \\ 0 & 0 & 0 & 1 \end{bmatrix} \begin{bmatrix} A_{1,1} & A_{1,2} & A_{1,3} & A_{1,4} \\ A_{2,1} & A_{2,2} & A_{2,3} & A_{2,4} \\ A_{3,1} & A_{3,2} & A_{3,3} & A_{3,4} \\ A_{4,1} & A_{4,2} & A_{4,3} & A_{4,4} \end{bmatrix} \begin{bmatrix} 1 & 0 & 0 & 0 \\ 0 & 0 & 1 & 0 \\ 0 & 1 & 0 & 0 \\ 0 & 0 & 0 & 1 \end{bmatrix} =$$

$$= \begin{bmatrix} 1 & 0 & 0 & 0 \\ 0 & 0 & 1 & 0 \\ 0 & 1 & 0 & 0 \\ 0 & 0 & 0 & 1 \end{bmatrix} \begin{bmatrix} A_{1,1} & A_{1,3} & A_{1,2} & A_{1,4} \\ A_{2,1} & A_{2,3} & A_{2,2} & A_{2,4} \\ A_{3,1} & A_{3,3} & A_{3,2} & A_{3,4} \\ A_{4,1} & A_{4,3} & A_{4,2} & A_{4,4} \end{bmatrix} = \begin{bmatrix} A_{1,1} & A_{1,3} & A_{1,2} & A_{1,4} \\ A_{3,1} & A_{3,3} & A_{3,2} & A_{3,4} \\ A_{2,1} & A_{2,3} & A_{2,2} & A_{2,4} \\ A_{4,1} & A_{4,3} & A_{4,2} & A_{4,4} \end{bmatrix} \qquad \text{(III-54)}$$

This expression for **A'** could be obtained from the matrix of **A** simply by interchanging the indices 2,3 everywhere. This transformation formula applies, for example, to the density matrix.

The number of elements in the density matrix is $4^2 = 16$. Thus the basic equation (III-18) involves in principle the simultaneous treatment of 16 variables, and the Liouville supermatrix has $16^2 = 256$ elements. The first consideration which reduces the dimensions of the problem is that in the present context, the only relevant elements $\rho_{i,j}$ are those with $|\Delta m_{ij}| = 1$. Due to the hermiticity of $\boldsymbol{\rho}$, it is sufficient to calculate elements with $\Delta m_{ij} = -1$ (or $\Delta m_{ij} = 1$), because those with $\Delta m_{ij} = 1$ (or $\Delta m_{ij} = -1$) will then be found by complex conjugation. There are 2 such elements connecting the m = -1 manifold with the m = 0 manifold: $\langle\beta\beta|\boldsymbol{\rho}|\alpha\beta\rangle$ and $\langle\beta\beta|\boldsymbol{\rho}|\alpha\beta\rangle$, and 2 elements connecting the m = 0 manifold with the m = 1 manifold: $\langle\alpha\beta|\boldsymbol{\rho}|\alpha\alpha\rangle$ and $\langle\beta\alpha|\boldsymbol{\rho}|\alpha\alpha\rangle$. This amounts to a total of 4 relevant elements of $\boldsymbol{\rho}$, implying a relevant Liouville submatrix of dimensions 4×4.

At this point the separation into blocks can be applied. Since density matrix elements connecting the m = -1 and m = 0 manifolds are not coupled to those elements connecting the m = 0 and m = 1 manifolds, each of these sets of elements can be treated separately. Thus the 4×4 equation is broken up into two separate blocks along the diagonal, each of them of dimensions 2×2. The numerical problem is thus greatly reduced from the original calculation of 256 elements of **L**.

The implications of these simple considerations for the Liouville supermatrix can be illustrated with the diagram in Fig. III.4. This diagram describes the matrix structure of the complete Liouvillian operator $\mathbf{L}_{(c)}$, which is defined as the sum of **L** of Eq. (III-19) and of the exchange superoperator appearing in that equation. Fig. III.4 presents the general structure of the full Liouvillian, in terms of the basis described above, and taking into account the rules explained so far. The indices 1, 2, 3, 4 are used in the diagram for the functions $|\beta\beta\rangle$, $|\beta\alpha\rangle$, $|\alpha\beta\rangle$ and $|\alpha\alpha\rangle$ respectively. An element $\mathbf{L}_{(c)ij,mn}$ connects ρ_{ij} with ρ_{mn}, so the non-zero elements of $\mathbf{L}_{(c)}$ can be found by using Eq. (III-19) in the basis which diagonalizes **H**. Comparing **A** and **A'** in Eq. (III-54) it is clear that the exchange process connects only the following pairs of elements of the density matrix: $\rho_{1,2}$ with $\rho_{1,3}$; $\rho_{2,1}$ with $\rho_{3,1}$; $\rho_{2,4}$ with $\rho_{3,4}$; $\rho_{4,2}$ with $\rho_{4,3}$; $\rho_{2,2}$ with $\rho_{3,3}$; and $\rho_{2,3}$ with $\rho_{3,2}$. In other words, each element of $\boldsymbol{\rho}$ is connected by the exchange process only to one element of ρ - either to itself (in the case of $\rho_{1,1}$, $\rho_{1,4}$, $\rho_{4,1}$ and $\rho_{4,4}$) or to a different element. This connection scheme is consistent with the rule that only elements of ρ with the same Δm value can be coupled by the exchange process.

The complete Liouvillian $\mathbf{L}_{(c)}$, including both **L** of Eq. (III-19) and the exchange superoperator, thus couples each element of ρ with itself (at least due to **L**) and possibly with one other element of ρ. Fig. III.4 shows the structure of this 16 × 16 Liouvillian matrix. The dark squares are the only locations in the matrix where non-zero elements can be found. The small dark squares represent single elements of $\mathbf{L}_{(c)}$ and the bigger squares represent 2 × 2 blocks, each one containing 4 elements of $\mathbf{L}_{(c)}$. The dashed lines, based on the value of the first row index and the first column index (see Figure) conveniently divide the matrix into three regions along the diagonal.

The top left region includes three dark squares. The first from the top is the position of the element $\mathbf{L}_{(c)11,11}$, connecting $\rho_{1,1}$ with itself (this element is actually equal to $-i\omega_{1,1} = 0$). The second square in that region includes the positions of $\mathbf{L}_{(c)12,12}$ (coupling $\rho_{1,2}$ with itself, with $-i\omega_{1,2}-1/\tau$) and of $\mathbf{L}_{(c)13,13}$ (coupling $\rho_{1,3}$ with itself, with $-i\omega_{1,3}-1/\tau$), as well as $\mathbf{L}_{(c)12,13}$ and $\mathbf{L}_{(c)13,12}$ (each of them being equal to $1/\tau$, and both of them coupling $\rho_{1,2}$ with $\rho_{1,3}$) . These four elements form a 2 × 2 submatrix of $\mathbf{L}_{(c)}$, operating on the subvector of ρ containing only the elements $\rho_{1,2}$ and $\rho_{1,3}$. This is one of the two submatrices mentioned above, connecting elements with $\Delta m_{i,j} = -1$. The third square in that region is the position of $\mathbf{L}_{(c)14,14}$, connecting $\rho_{1,4}$ with itself (this element is equal to $-i\omega_{1,4}$).

The third region (in the bottom right corner of the full matrix) has a similar structure. The second region has a more complicated structure, but with the same basic pattern. The solid lines connect the positions of four elements of $\mathbf{L}_{(c)}$: $\mathbf{L}_{(c)24,24}$, $\mathbf{L}_{(c)24,34}$, $\mathbf{L}_{(c)34,24}$ and $\mathbf{L}_{(c)34,34}$. These four positions are separated by the positions of other matrix elements, creating the incorrect impression that they are unrelated. However, in practice they form a 2 × 2 submatrix of $\mathbf{L}_{(c)}$, which operates on the subvector of ρ containing only $\rho_{2,4}$ and $\rho_{3,4}$. In fact, this is the second of the two 2 × 2 submatrices of $\mathbf{L}_{(c)}$ connecting elements with $\Delta m_{i,j} = -1$.

Example III.3: *Implications of I_z invariance for s-trioxane*

An example of practical relevance is the case of s-trioxane mentioned above. This molecule has six relevant nuclei, all of which possess spin I = ½. The maximal total spin is thus I = 3, with m = -3, -2, ..., 2, 3 . In a spin product function each of the six factors has two possible values (m_k = ½ or m_k = -½), giving a total of 2^6 = 64 different wave functions. A simple calculation of permutations shows that there is 1 state with m = -3 or m = 3, there are 6 states with m = -2 or m = 2, 15 states with m = -1 or 1, and 20 states with m = 0. The number of elements in the density matrix is 64^2 = 4096. Thus Eq. (III-18) involves in principle the simultaneous treatment of 4096 variables. The relevant ones are those with $|\Delta m_{ij}| = 1$. Due to hermiticity, it is sufficient to calculate elements with $\Delta m_{ij} = -1$ (or $\Delta m_{ij} = 1$). There are 6 such elements connecting the m = -3 manifold with the m = -2 manifold or the m = 2 manifold with the m = 3 manifold; 90 elements connecting the m = -2 manifold with the m = -1 manifold or the m = 1 manifold with the m = 2 manifold; and 300 elements connecting the m = -1 manifold with the m = 0 manifold or the m = 0 manifold with the m = 1 manifold. This amounts to a total of 792 relevant elements of ρ, implying a relevant Liouville submatrix of dimensions 792 × 792.

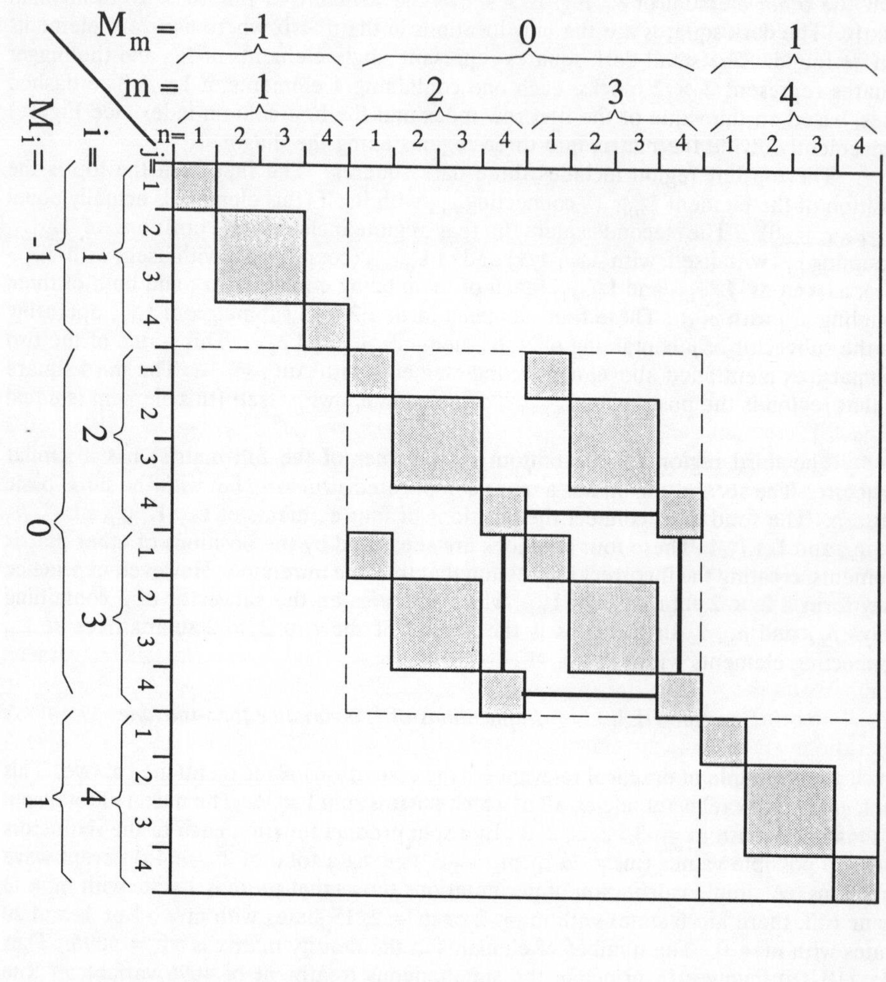

Fig. III.4. *The overall structure of the Liouville matrix for a two-spin system undergoing chemical exchange. Each small square represents a single matrix element, and each bigger square represents four elements. The second square from the top is a 2 × 2 matrix operating on* $\rho_{1,2}$ *and* $\rho_{1,3}$. *The four elements close to the center of the diagram, connected by heavy black lines, form a matrix coupling* $\rho_{2,4}$ *and* $\rho_{3,4}$. *For additional details see text.*

Now the separation into blocks can be carried out. Since density matrix elements connecting the m = -3 and m = -2 manifolds, for example, are not coupled to those elements connecting the m = -2 and m = -1 manifolds, each of these sets of elements can be treated separately. Consequently the 792 × 792 equation is broken up into six separate blocks along the diagonal, with dimensions 6, 90, 300, 300, 90 and 6 respectively. Now the largest matrix is only 300 × 300, rather than 4096 × 4096. The dimensions of the problem are thus much smaller than would seem from the simple calculation of d_H^2 = 4096 elements of $\boldsymbol{\rho}$.

****

(b) *Symmetries of the exchanging molecule*

Even when the invariance of $\mathbf{I}_z$ under chemical exchange is taken into account, the dimensions of the numerical problem may be quite large. More importantly, the question of selection rules for transitions is always significant, regardless of computational considerations. Group theory has been employed in many areas of spectroscopy in order to take advantage of molecular symmetry, and derive selection rules for quantum transitions. The basis of the method is the invariance of a Hamiltonian (or some other important function characterizing the system) under symmetry operations applied to the molecule. The same method can also be applied here, both for deriving selection rules and for reducing the amount of computations. The main problem in the present context is that the symmetry of a molecule in its static configuration (at the limit of infinitely slow exchange) differs from the symmetry of the average configuration, corresponding to the limit of infinitely fast exchange. A method to overcome this difficulty will now be presented. Some of the mathematical background is given in Appendix E. Only an outline of the method is sketched here, with an emphasis on the physical meaning of each stage in the calculation.

In order to make use of molecular symmetry, one has to find first those symmetry operations which leave the magnetic resonance Hamiltonian of the molecule unchanged. Each such operation can be a rotation by a certain angle about a particular molecular axis, or reflection in some plane, or inversion. The symmetry operations are found separately for the Hamiltonian of the static molecule and for the Hamiltonian of the average configuration. Each set of symmetry operations forms a group (see Appendix E). This means that doing two symmetry operations one after the other is equivalent to some other valid symmetry operation, which belongs to the same set. It also means that for every allowed symmetry operation, also the inverse operation belongs to the same set. Having found the group G of allowed symmetry operations $g_1, g_2, g_3, \ldots$, each element of the group (i.e., each symmetry operation) can be represented by a matrix $R(g_k) = R_k$, so that the matrices themselves form a group (isomorphic to the symmetry group). Such matrices operate directly on the wave functions, for example, so they express directly the geometrical content of these operations. Their dimension is the dimension of Hilbert space in the problem.

In addition to these operators, which have some simple "physical" significance, there are other matrices which form groups that are isomorphic to G ,i.e., they have exactly the same mathematical structure as G. In general there is more than one set of matrices representing the group in this manner. These matrices "know" only about the symmetries, and not about the physical problem. Each set of such matrices is called "a representation of the group". The important representations are the irreducible representations (see Appendix E), and it is only to these that we shall refer here. The term "irreducible representation" will be abbreviated as IRREP. The dimension of the IRREP, and even the trace (the sum of diagonal elements) of each matrix in the IRREP, depend on the specific group and the specific element involved. If the IRREP is denoted by Γ then the matrix representing the operation g_k in Γ can be denoted as $M^{\Gamma}(R_k)$. Its trace (also called its "character") is denoted as $\chi^{\Gamma}(R_k) \equiv Tr\{M^{\Gamma}(R_k)\}$. Using this "representation" of the group it is possible to construct projection operators, which operate on a given wave function of the system to project out that component of the function which is invariant under a particular symmetry operation. The formula for such projection operators is:

$$P^{\Gamma} = \frac{d^{\Gamma}}{N} \sum_{R} \chi^{\Gamma}(R) R \tag{III-55}$$

where N is the order of the group G (i.e., the number of elements in the group), d^{Γ} is the dimension of the IRREP Γ, and the sum extends over all operations R belonging to the group (the Hilbert space operators R_k are taken to be equivalent to the abstract operations g_k which they represent). Operating with such projection operators on the basis of Hilbert space consisting of spin product functions, one obtains in general linear combinations of the original basis functions. These linear combinations define another basis for Hilbert space. What is special about the new basis is that each function in it has a simple behaviour under symmetry operations. It is an eigenfunction (with eigenvalue 1) of the projection operator corresponding to one particular IRREP, whereas projection operators of other IRREPs will give zero (an eigenvalue of 0) when operating on such a function. The function is then said to belong to that IRREP, for which it has an eigenvalue of 1. Moreover, functions beloging to different IRREPs are orthogonal.

These new functions are known as "symmetry adapted wave functions". The Hamiltonian can have non-zero matrix elements between functions belonging to the same IRREP ("symmetry"), but not between functions belonging to different IRREPs ("symmetries"). This can be seen as follows. The Hamiltonian is invariant under any of its own symmetries, and therefore commutes with their projection operators:

$$P^{\Gamma} H (P^{\Gamma})^{-1} = H \quad \Rightarrow \quad [P^{\Gamma}, H] = 0 \tag{III-56}$$

Now suppose $\psi_i^{\Gamma a}$ is a function generated by the projection operator for IRREP Γ_a. Then one may construct many different functions $\phi_{i,\alpha}$ (including $\psi_i^{\Gamma a}$ itself) from which the symmetry adapted $\psi_i^{\Gamma a}$ can be obtained by operating with the projection operator $\mathbf{P}^{\Gamma a}$. As a result, one may write matrix elements between functions belonging to two (possibly equal) IRREPs as:

$$\langle \psi_i^{\Gamma a} | H | \psi_j^{\Gamma b} \rangle = \langle \phi_{i,\alpha} | P^{\Gamma a} H P^{\Gamma b} | \phi_{j,\beta} \rangle = \langle \phi_{i,\alpha} | P^{\Gamma a} P^{\Gamma b} H | \phi_{j,\beta} \rangle \ \alpha \ \delta_{a,b} \tag{III-57}$$

The proportionality relation on the extreme right results from the orthogonality of functions belonging to different representations. The commutation of the projectors $\mathbf{P}^{\Gamma a}$ and $\mathbf{P}^{\Gamma b}$ with **H** thus leads to the result that **H** does not mix different symmetries.

Transitions between states having different symmetries are therefore impossible, and they are known as forbidden transitions.

In the previous Section the Liouville matrix of an exchanging two-spin system was discussed in detail. Such a physical system is too simple to have both non-trivial symmetry and an exchange process as described above. Note that the exchange process cannot be a symmetry operation, otherwise the exchange will not affect the Hamiltonian. In order to demonstrate the use of symmetry considerations on the simplest possible case, a *non-exchanging* two-spin system will be analysed here. This system is assumed to have a symmetry group G = C_2 , containing only two symmetry elements (geometrical symmetry operations): the identity operation E, and the operation of rotation by 180° , commonly denoted by C_2 . Thus G = {E, C_2} in the notation of the present Section. The order of G is N = 2 . This group has two IRREPs, commonly denoted by A, B respectively. Each of these IRREPs consists of two 1 × 1 matrices, representing E and C_2 . The trace of a 1 × 1 matrix is just the single element of that matrix. Therefore the characters of the two IRREPs are just the elements of the corresponding matrices. From the character table of this group:

$$\chi^A(E) = 1 \qquad \chi^A(C_2) = 1 \tag{III-58a}$$

$$\chi^B(E) = 1 \qquad \chi^B(C_2) = -1 \tag{III-58b}$$

The projection operators for the two IRREPs are, from Eq. (III-55):

$$\mathbf{P}^A = \frac{1}{2}\left\{\chi^A(E)\,E + \chi^A(C_2)\,C_2\right\} = \frac{1}{2}\{E + C_2\} \tag{III-59a}$$

$$\mathbf{P}^B = \frac{1}{2}\left\{\chi^B(E)\,E + \chi^B(C_2)\,C_2\right\} = \frac{1}{2}\{E - C_2\} \tag{III-59b}$$

Assuming C_2 interchanges the two spins (as a symmetry operation, having no effect on the Hamiltonian) one would get

$$\mathbf{P}^A\,|m_1 m_2\rangle = \frac{1}{2}\{E + C_2\}\,|m_1 m_2\rangle = \frac{1}{2}\{|m_1 m_2\rangle + |m_2 m_1\rangle\} \tag{III-60a}$$

$$\mathbf{P}^A\,|m_1 m_2\rangle = \frac{1}{2}\{E - C_2\}\,|m_1 m_2\rangle = \frac{1}{2}\{|m_1 m_2\rangle - |m_2 m_1\rangle\} \tag{III-60b}$$

Now assume for simplicity that the spins have I = ½ , and use the indices 1, 2, 3, 4 for the wave functions as in the case considered in the previous Section. Operating with $\mathbf{P}^A$, $\mathbf{P}^B$ on each of the four product functions, the following results would be obtained:

$$P^A\,|\beta\beta\rangle = |\beta\beta\rangle \qquad P^A\,|\alpha\alpha\rangle = |\alpha\alpha\rangle \tag{III-61a}$$

$$P^A\,|\beta\alpha\rangle = P^A|\alpha\beta\rangle = \frac{1}{2}\{|\alpha\beta\rangle + |\beta\alpha\rangle\} \equiv |2'\rangle \tag{III-61b}$$

$$P^B\,|\beta\beta\rangle = P^B\,|\alpha\alpha\rangle = 0 \tag{III-62a}$$

$$P^B\,|\beta\alpha\rangle = -P^B|\alpha\beta\rangle = \frac{1}{2}\{|\beta\alpha\rangle - |\alpha\beta\rangle\} \equiv |3'\rangle \tag{III-62b}$$

The projection operators generate in this case a transformation from the basis functions 1, 2, 3, 4 to the symmetry basis functions 1, 2', 3', 4. Three of the symmetry adapted functions (1, 2' and 4) belong to A, the symmetric representation, and only one belongs to B, the anti-symmetric representation. The Hamiltonian can only have non-zero elements between the functions belonging to A. The allowed transitions are: 1 → 2' and 2' → 4 (single quantum) and 1 → 4 (double quantum), but 1 → 3' and 3' → 4 are forbidden.

It should be emphasized again that the analysis here applied to a *static* two-spin system, unlike the exchanging system analysed in the previous Section. A more complicated system is needed for demonstrating the combination of symmetry with dynamic effects. Such a system will be treated in Example III.4 below.

Continuing with the general case, also the transverse magnetization operator $\mathbf{I}_x \equiv \sum_k (\mathbf{I}_x)_k$ (sum over all spins in the molecule) will only have non-zero elements between functions belonging to the same IRREP. This is because it is a completely symmetric combination of the spin operators of different spins in the molecule, so it is not affected by any exchange, which simply amounts to a permutation of the terms in this sum. Being completely symmetric under all symmetry operations for the Hamiltonian, $\mathbf{I}_x$ thus commutes with the symmetry projection operators, in the same manner as the Hamiltonian itself.

Therefore different IRREPs can be coupled in Eq. (III-45) or (III-48) only by the exchange operator **X**. If this operator mixes all IRREPs, no simplification of the problem is possible. Usually, however, this is not the case, and then the problem can be reduced in size in the following way. First one defines the average Hamiltonian as a sum over the Hamiltonians of all mutually exchanging configurations, divided by the number of these configurations:

$$H_{ave} \equiv \frac{1}{N}\sum_{k=0}^{N-1} X^k H X^{-k} \tag{III-63}$$

If $\mathbf{X}^{-1} = \mathbf{X}$, so that $\mathbf{X}^2 = \mathbf{I}$, the only terms in the sum are those with k = 0 and k = 1, exactly as in Eq. (III-25).

Now construct the group G_{ave} of symmetry operations of the average Hamiltonian. Then G must be a subgroup of G_{ave}, and the operator **X** belongs to G_{ave} but not to G. The central question to be addressed is how the IRREPs of G are related to those of

G_{ave}. There is a systematic way to generate IRREPs of a group from those of its subgroup, and vice versa. The first process is called induction of representations, and the second is called subduction of representations. Studying these mathematical techniques one can reach conclusions which have significant implications for the physical problem discussed here. If an IRREP Γ^a of G induces one or more IRREPs of G_{ave}, but none of them is induced also by another IRREP of G, then Γ^a is invariant with respect to the exchange process. Functions belonging to Γ^a will not be mixed by **X** with functions belonging to different IRREPs of G. On the other hand, if two IRREPs of G induce the same IRREP of G_{ave}, then the dynamic process will mix functions belonging to these IRREPs. Correlation tables contain the information on the relation between the IRREPs of groups and their subgroups, so using a correlation table one can determine what kind of symmetry mixing occurs in a particular problem.

Even with such mixing, elements of $\boldsymbol{\rho}$ corresponding to allowed transitions (such as $\rho_{1,2'}$ and $\rho_{2',4}$ in the two-spin case discussed above) will not be mixed with elements corresponding to symmetry-forbidden transitions (such as $\rho_{1,3'}$ in the case mentioned above). This is intuitively obvious, because the exchange process only permutes the spins in a valid configuration of the molecule, and does not change the symmetry of the molecule. The formal mathematical proof of this rule in the general case is not simple, and will not be given here (see Ref. 11, Ch. 14 and Ref. 13).

Example III.4: *Symmetry factorization for s-trioxane*

In s-trioxane the symmetry group of the Hamiltonian is G = C_{3v} and that of the average Hamiltonian is G_{ave} = D_{3h}. The correlation table of these two groups, shown in Fig. III.5, happens to be a very simple one:

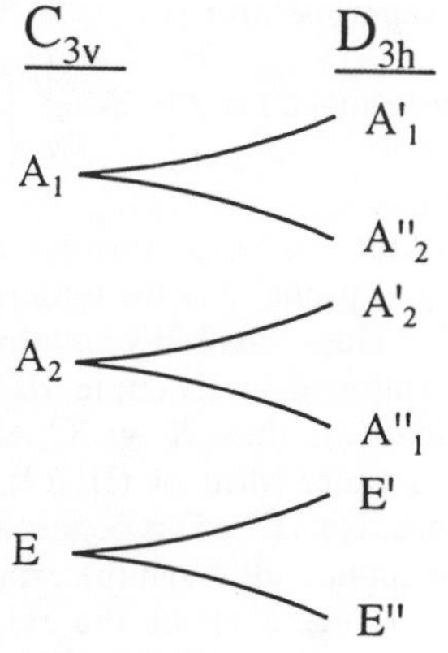

Fig. III.5. *The correlation diagram of C_{3v} and D_{3h}*

Every IRREP of G_{ave} is correlated with only one IRREP of G, and therefore the dynamic process does not cause any "mixing" of IRREPs, and the symmetry factorization is maximal. Applying the projection operators for IRREPs A_1, A_2 and E one obtains: in the m = 3 or m = -3 manifolds, a single symmetry adapted function beloging to A_1; in the m = 2 or m = -2 manifolds, 2 symmetry adapted functions belonging to A_1 and two pairs of functions belonging to E (which is two dimensional, so its functions always come in pairs); in the m = 1 or m = -1 manifolds, 4 functions in A_1, 1 function in A_2 and 5 pairs of functions in E; and in the m = 0 manifold, 6 functions in A_1, 2 in A_2 and 6 pairs in E. Allowed transitions are those which satisfy two selection rules: $|\Delta m| = 1$ and unchanged symmetry (IRREP) in the transition. Thus the equations are factored significantly, the largest block involving 30 transitions (30 elements of $\boldsymbol{\rho}$) belonging to E between m = 0 and m = 1 (or m = -1) manifolds. The largest matrices in Liouville space are thus 30 × 30, rather than 300 × 300 as found from $\mathbf{I}_z$ invariance.

****

(c) *Symmetries of the exchange process*

It was seen above that for fast exchange there is an approximate method for calculating the density matrix, provided that $\mathbf{X} = \mathbf{X}^{-1}$. Here a method will be described for making such an approximation under more general circumstances. The idea is to use symmetries in Liouville space in order to examine the problem in the more general case. When working in Liouville space it is convenient to use the supervector and superoperator notation, so that Eq. (III-18) is replaced by a slightly modified version of (III-19):

$$\frac{d}{dt}\boldsymbol{\rho} = \frac{1}{i\hbar}\boldsymbol{L}\boldsymbol{\rho} + \boldsymbol{C}^{(exch.)}\boldsymbol{\rho} + \boldsymbol{R}^{(phen.)}\boldsymbol{\rho} \qquad \text{(III-64)}$$

where the chemical exchange superoperator is

$$\boldsymbol{C}^{(exch.)} = \frac{1}{\tau}\left[\frac{1}{2}\left(\boldsymbol{X}\otimes\boldsymbol{X}^{-1} + \boldsymbol{X}^{-1}\otimes\boldsymbol{X}\right) - \boldsymbol{I}\otimes\boldsymbol{I}\right] \equiv \frac{1}{\tau}\left[\frac{1}{2}\left(\boldsymbol{Y} + \boldsymbol{Y}^{-1}\right) - \boldsymbol{J}\right] \qquad \text{(III-65)}$$

The modification here is that the exchange process is expressed as a sum of two contributions, reflecting the equal probability for undergoing the same type of process in two opposite directions. This possibility materializes, for example, in the cyclooctatetraene molecule mentioned in Example III.2 above. If there is only one possible direction (as in s-trioxane), then $\mathbf{X} = \mathbf{X}^{-1}$ and $\mathbf{Y} = \mathbf{Y}^{-1}$, so the exchange superoperator reduces to the simpler form of (III-19). If there are more than two possible "directions" for exchange, Eq. (III-65) is generalized in an obvious manner, and the method to be described still applies, although the expressions become more involved. In Eq. (III-64) no assumption is made about the relative magnitude of τ^{-1}, so the equation is just the Liouville space version of Eq. (III-18) with a general $\mathbf{X}$.

The exchange process is formally equivalent to a permutation of the spins. Since the number of spins is finite, performing the same permutation repeatedly one must return to the original configuration after a finite number of steps. Operating on a wave function in Hilbert space:

$$X^N \psi(1,2,3,...,n) = \psi(1,2,3,...,n) \quad \Rightarrow \quad X^N = I \tag{III-66}$$

for some number N. Consequently, operating on an arbitrary Liouville space supervector, or Hilbert space operator **A** results in:

$$Y^N A(1,2,3,...,n) = A(1,2,3,...,n) \quad \Rightarrow \quad Y^N = J \tag{III-67}$$

In general the finite number of spins implies only: N ≤ n! (because n! is the number of possible permutations of the sequence (1,2,...,n)), but in practical cases N is much smaller. Then T ≡ {$\mathbf{J} = \mathbf{Y}^0$, $\mathbf{Y}^1$, $\mathbf{Y}^2$,..., $\mathbf{Y}^{N-1}$} is a cyclic group of order N, consisting of superoperators which operate in Liouville space. Their operation is defined through

$$Y^k(A) = X^k A X^{-k} \tag{III-68}$$

where **A** is any Hilbert space operator. As a cyclic group it is abelian, i.e., all its elements commute with one another. As a result of the cyclic property: $\mathbf{Y}^{k+N} = \mathbf{Y}^k$. This is described by saying that the powers of **Y** are only defined **modulo N.** The chemical exchange superoperator is expressed in terms of these powers as:

$$C^{(exch.)} \equiv \frac{1}{\tau}\left[\frac{1}{2}\left(Y + Y^{-1}\right) - J\right] = \frac{1}{\tau}\left[\frac{1}{2}\left(Y + Y^{N-1}\right) - Y^0\right] \tag{III-69}$$

Therefore all the elements of T commute with the exchange superoperator:

$$Y^k\left\{C^{(exch.)}(A)\right\} \equiv \frac{1}{\tau}\left\{\frac{1}{2}\left(Y^{k+1} + Y^{k-1}\right) - Y^k\right\} A = \frac{1}{\tau}\left[\frac{1}{2}\left(Y + Y^{-1}\right) - Y^0\right]\left\{Y^k(A)\right\}$$

$$\Rightarrow \quad [Y^k, C^{(exch.)}] = 0 \tag{III-70}$$

Physically the last equation means that doing the chemical exchange operation, and then doing k steps of pure permutation (without the full structure of the chemical exchange process) is the same as doing the k permutation steps first, and then the actual chemical exchange. Since each permutation step defines a new (permuted) basis for Hilbert space, one may regard the N steps as defining a sequence of N Hilbert spaces, differing only in their labeling of spins in the wave functions (or in operators).

Mathematically these commutation relations imply that T is a symmetry group for the form of the chemical exchange superoperator $\mathbf{C}^{(exch.)}$. This is analogous to the situation with molecular symmetry, where the Hamiltonian of a molecule commutes with the symmetry operators under which the molecular configuration is physically unchanged. The labeling of atoms does change, but only in such a way that when the labeling of identical atoms is removed the effect of the operation cannot be seen. The wave function of such a molecule changes (its labels are permuted) under the symmetry operation, but because the change is not physically meaningful, the new Hamiltonian is physically unchanged. This means it has the same eigenfunctions and energy levels, and thus also the same transition frequencies and transition intensities. In the present case the elements of T change the labels in the supervectors on which they operate, but this does not change the form of the chemical exchange supeoperator. Just as in the molecular symmetry case the Hamiltonian has the same form for each (permuted) configuration, in the present case the exchange superoperator has the same form for each (permuted) molecular configuration. Molecular symmetry is geometrical, relating to operations in real three dimensional space. The symmetry of the exchange superoperator is algebraic, relating to formal operations in an abstract space (Hilbert space). Nevertheless, the group theoretical results applicable to geometrical symmetries are also relevant to abstract symmetries.

If the IRREPs of T are denoted by λ, then the projection operator for the general IRREP is:

$$P^{(\lambda)} = \frac{1}{N}\sum_{k=0}^{N-1} \exp(2\pi ik\lambda/N) \cdot Y^k \tag{III-71}$$

where the number λ determines the representation λ (see Appendix E). The allowed values for λ are the integers λ = 0,1,2,...,N-1 . For any of these values one may add an integer multiple of N, but this will not change the representation or the projection operator. Thus the values of λ are only defined **modulo N** . Therefore the Hamiltonian and the density matrix can be decomposed as

$$H = \sum_{\lambda} P^{(\lambda)} H \equiv \sum_{\lambda} H^{(\lambda)} \qquad \rho = \sum_{\lambda} P^{(\lambda)} \rho \equiv \sum_{\lambda} \rho^{(\lambda)} \tag{III-72}$$

Each of the components $\mathbf{H}^{(\lambda)}$ or $\boldsymbol{\rho}^{(\lambda)}$ belongs to a particular symmetry λ, being the projection on IRREP λ of the original operator, **H** or $\boldsymbol{\rho}$. From Eq. (III-71) for the projection operator one may verify for any Hilbert space operator **A** that the λ-components are indeed eigenfunctions of **Y**:

$$YA^{(\lambda)} = Y\{P^{(\lambda)}A\} = \frac{1}{N}\exp(-2\pi i\lambda/N)\sum_{k=0}^{N-1} \exp(2\pi i(k+1)\lambda/N) \cdot Y^{k+1}A =$$

$$= \exp(-2\pi i\lambda/N)\, A^{(\lambda)} \tag{III-73}$$

It follows that any given symmetry component of the arbitrary operator **A** is an eigenfunction (an "eigen-supervector") of the chemical exchange superoperator:

$$C^{(exch.)}A^{(\lambda)} = \frac{1}{\tau}\left[\frac{1}{2}\left(e^{-2\pi i\lambda/N} + e^{2\pi i\lambda/N}\right) - 1\right]A^{(\lambda)} = \frac{1}{\tau}\left(\cos(2\pi\lambda/N) - 1\right)A^{(\lambda)} \tag{III-74}$$

The eigenvalue is:

$$-\gamma_\lambda = -\frac{1}{\tau}\left(1 - \cos(2\pi\lambda/N)\right) \tag{III-75}$$

In particular, $\gamma_0 = 0$. All other eigenvalues are non-zero. The definition of γ is chosen so that γ_λ is always positive or zero, as is usually chosen for decay coefficients.

On the basis of the expressions derived above one may extract equations of motion for each (algebraic) symmetry component of ρ. We assume for simplicity that the longitudinal and transverse relaxation times due to other mechanisms are equal: $T_1 = T_2$. This assumption is valid at relatively high temperatures, where also the exchange is usually relatively fast, which is just the domian for which the current method can be applied. In any case, even if $T_1 \neq T_2$ the equations to be presented will only become slightly more complicated, but their essential structure will not change.

The first stage is writing **H** and ρ in terms of their symmetry components, and using the eigenvalue equation (III-74) for the chemical exchange superoperator. Then the equation of motion (III-64) becomes:

$$\sum_{\lambda=0}^{N-1} \frac{d}{dt}\rho^{(\lambda)} = \frac{1}{i\hbar}\sum_{\mu,\nu=0}^{N-1} [H^{(\mu)},\rho^{(\nu)}] - \sum_{\lambda=0}^{N-1} \gamma_\lambda \rho^{(\lambda)} - \frac{1}{T_2}\sum_{\lambda=0}^{N-1} \rho^{(\lambda)} \qquad \text{(III-76)}$$

In this form of the equation every component of the density matrix, and every component of the Hamiltonian, is an eigenfunction of the chemical exchange superoperator. Operating on both sides of the equation with $\mathbf{P}^{(\lambda)}$ one obtains an equation of motion for such components of ρ:

$$\frac{d}{dt}\rho^{(\lambda)} = \frac{1}{i\hbar} P^{(\lambda)} \sum_{\mu,\nu=0}^{N-1} [H^{(\mu)},\rho^{(\nu)}] - \left(\gamma_\lambda + \frac{1}{T_2}\right)\rho^{(\lambda)} \qquad \text{(III-77)}$$

To calculate the projection of the commutator, note that for any two operators **A, B** the product $\mathbf{A}^{(\mu)}\,\mathbf{B}^{(\nu)}$ belongs to an IRREP which is the direct product of the two IRREPs μ, ν ; this product is equal to the IRREP $\mu+\nu$ (see Appendix E). Operating with $\mathbf{P}^{(\lambda)}$ the projected part of the product belongs to $\lambda = \mu+\nu$, so that $\nu = \lambda-\mu$. Therefore Eq. (III-77) can be rewritten as

$$\frac{d}{dt}\rho^{(\lambda)} = \frac{1}{i\hbar}\sum_{\mu=0}^{N-1} [H^{(\mu)},\rho^{(\lambda-\mu)}] - \left(\gamma_\lambda + \frac{1}{T_2}\right)\rho^{(\lambda)} \qquad \text{(III-78)}$$

in terms of symmetry components with respect to the group T (of *algebraic* symmetries). The original equation, Eq. (III-64) is completely equivalent to the set of equations (III-78) for all λ. This set of equations is a generalization of Eq. (III-34), applying to any rate of exchange. From these equations one may derive a relaxation equation for fast chemical exchange, similar to Eqs. (III-38)-(III-40) above. Before indicating how this is done, it is important to know how Eq. (III-78) can be simplified by taking into account also the geometrical symmetries of the molecule.

If a wave function $\psi(1,2,...,n)$ is represented by the ordered sequence of labels (1,2,...,n) of individual spins, then a symmetry operation on the molecule simply permutes the labels. The Hamiltonian is invariant under such an operation. The point group of geometrical symmetry operations is isomorphic to a group of permutations, which is in general a subgroup (usually a very small one !) of the group of all possible permutations of n objects. This restricted group of permutations will be denoted by $G = \{g_1, g_2, ...g_d\}$, and its dimension by d. It is convenient to assume that $g_1 = e$ where e is the identity permutation (corresponding to the unit operator **I**). These permutations represent the allowed set of configurations of a single chemical isomer. This set of configurations will be called a "permutational isomer".

Now the exchange operator **X** also permutes labels in wave functions, but it obviously does not belong to the symmetry group of the molecule. The corresponding permutation will be denoted by x; operating with it on all elements of G, one obtains a **left coset** of G: $xG \equiv \{xg_1, xg_2,..., xg_d\}$. This coset represents the possible configurations of a second chemical isomer (or "permutational isomer"), which can be reached from the original one only through the chemical exchange operation. Operating again with x one obtains another coset: x^2G, representing yet another chemical isomer, and so on. From Eq. (III-66) it is clear that for some number N: $x^N = e$ (e is the identity permutation). Thus $x^NG = \{x^Ng_1, x^Ng_2,... x^Ng_d\} = G$. In other words, x^N carries G into itself, by transforming each of its elements into itself. Thus N is the maximal number of different isomers which can be reached by performing the chemical exchange operation. This is consistent with the relation $\mathbf{X}^N\,\mathbf{H}\,\mathbf{X}^{-N} = \mathbf{H}$ which follows directly from Eq. (III-67).

However, the definition of a group (or a coset) does not refer to the order of elements in it. It may occur that a lower power of x will carry G into G, except for a rearrangement of its elements: $x^MG = G$, where $M < N$. If this is the case, then $x^M = x^Me$ is an element of G, so the coset $x^{M+r}G$ is equal to x^rG, for any r. In particular: $x^NG = x^MG$, so that N must be divisible by M. Thus in the sequence of cosets $xG, x^2G,...$ each coset is repeated exactly $L \equiv N/M$ times. This means that the exchange operation is related in a special way to the symmetry operations of the system, so that out of the N "isomers" only M are really different isomers, all the others being repetitions of the same isomers but with a physically irrelevant rearrangement of the configurations belonging to each isomer. In such a case the Hamiltonian should remain unchanged. A different way to look at this is to observe that x^M simply generates a configuration that is symmetry related to the original one. By definition, such a configuration must have the same Hamiltonian as the original configuration.

It follows that if $x^M \in G$, then $\mathbf{Y}^M\,\mathbf{H} = \mathbf{X}^M\,\mathbf{H}\,\mathbf{X}^{-M} = \mathbf{H}$, and therefore in the decomposition of **H** into symmetry components with respect to T there are only M different terms. Applying the projection operator $\mathbf{P}^{(\lambda)}$ and using the notation $q = \exp(2\pi i\lambda/N)$ one finds:

$$P^{(\lambda)}H = \frac{1}{N}\sum_{k=1}^{N} q^k Y^k H = \frac{1}{N}\sum_{m=1}^{M}\left[\sum_{l=1}^{L} q^{(l-1)M}\right] q^m Y^m H \tag{III-79}$$

Summing over the geometrical series enclosed by parentheses, one obtains:

$$P^{(\lambda)}H = \frac{1}{M}\sum_{m=1}^{M} q^m Y^m H \qquad (if\ q^M = 1) \qquad or \tag{III-80a}$$

$$P^{(\lambda)}H = 0 \qquad (if\ q^M \neq 1) \tag{III-80b}$$

In other words, the Hamiltonian has only M non-zero components, with $\lambda = 0, L, 2L, ..., (M\text{-}1)L$, and all other symmetry components of **H** are zero. This means that in Eq. (III-78) one does not need to sum over all values of μ from 0 to N-1 (or 1 to N), but only over the values $\mu = mL$, with $m = 0,1,2,..., M\text{-}1$.

It is now possible to derive the relaxation equation for fast exchange. The derivation will not be carried out here (see Ref. 14 for the calculation), but the simplified form of Eq. (III-78) will be written in a way which emphasizes the analogy to Eq. (III-34), which served as the basis for such a derivation in a special case. The line shape is the trace of the density matrix with some operator ($\mathbf{I}_x$, $\mathbf{I}_y$, or some combination of them) of the total spin. These operators are symmetric under spin exchanges, and thus belong to the totally symmetric IRREP, $\lambda = 0$. Therefore only $\rho^{(0)}$ will give a non-zero contribution to the line shape, so one only needs to calculate the evolution of this component of the density matrix. Eq. (III-78) with the restriction on values of μ is:

$$\frac{d}{dt}\rho^{(0)} = \frac{1}{i\hbar}[H^{(0)},\rho^{(0)}] + \frac{1}{i\hbar}\sum_{m=1}^{M-1}[H^{(mL)},\rho^{(N-mL)}] - \left(\frac{1}{T_2}\right)\rho^{(0)} \qquad \text{(III-81a)}$$

$$\frac{d}{dt}\rho^{(mL)} = \frac{1}{i\hbar}\{[H^{(0)},\rho^{(mL)}] + [H^{(mL)},\rho^{(0)}]\} + \frac{1}{i\hbar}\sum_{k=1}^{M-1}[H^{(kL)},\rho^{((m-k)L)}]$$

$$- \left(\gamma_\lambda + \frac{1}{T_2}\right)\rho^{(mL)} \qquad (m = 1, 2, ..., M-1) \qquad \text{(III-81b)}$$

The summation in (III-81b) is only over $k \neq m$. The greatest simplification occurs if $M = 2$, because then there are only two equations:

$$\frac{d}{dt}\rho^{(0)} = \frac{1}{i\hbar}\{[H^{(0)},\rho^{(0)}] + [H^{(L)},\rho^{(L)}]\} - \left(\frac{1}{T_2}\right)\rho^{(0)} \qquad \text{(III-82a)}$$

$$\frac{d}{dt}\rho^{(L)} = \frac{1}{i\hbar}\{[H^{(0)},\rho^{(L)}] + [H^{(L)},\rho^{(0)}]\} - \left(\gamma_L + \frac{1}{T_2}\right)\rho^{(L)} \qquad \text{(III-82b)}$$

If $1/T_2$ is ignored, Eqs. (III-82) are just Eqs. (III-34) in a more general notation. Thus the fast exchange approximation which was discussed above (in Eq. (III-34), for example) for the special case $\mathbf{X}^2 = \mathbf{I}$, is also valid for more general cases, as long as $\mathbf{X}^2\,\mathbf{H}\,\mathbf{X}^{-2} = \mathbf{H}$. In terms of the parameters defined here, one does not need $N = 2$ for this set of two equations; $M = 2$ is the only requirement. Moreover, a similar approximation scheme can be constructed also for larger values of M, by applying to Eqs. (III-81) the same considerations for the fast exchange. The resulting equations are in the form of Eq. (III-40), and the exchange superoperator is a generalization of the superoperator of Eq. (III-41). The end result is that the symmetry of the exchange process combines with the symmetry of the molecule to simplify the problem and reduce its dimensions considerably in the regime of fast exchange.

Example III.5: Symmetry and exchange operations on configurations of cyclooctatetraene

In Example III.2 above it was found for cyclooctatetraene that there seem to be only two physically distinct isomers, and yet the chemical exchange process is not equal to its inverse. This can be understood using the concepts introduced in the present subsection. If the molecule on the left hand side of Fig. III.3 is taken as a reference configuration, it will be represented as: [1 2 3 4 5 6 7 8] . Each of the numbers 1-8 represents a specific hydrogen atom, and each of the eight positions in the sequence represents a specific physical site in the molecular "skeleton". Applying the symmetry operations of D_{2d} , which is the symmetry group of the Hamiltonian, the following eight configurations are obtained:

By E (identity):

c_1 = [1 2 3 4 5 6 7 8]

By C_2 (rotation by $2\pi/2$ about vertical axis):

c_2 = [5 6 7 8 1 2 3 4]

By S_4 (rotation by $2\pi/4$ and then reflection in a perpendicular plane):

c_3 = [3 4 5 6 7 8 1 2] and c_4 = [7 8 1 2 3 4 5 6]

By C_2 (rotation by $2\pi/2$ about horizontal axes):

c_5 = [4 3 2 1 8 7 6 5] and c_6 = [8 7 6 5 4 3 2 1]

By σ_d (reflection through vertical planes)

c_7 = [2 1 8 7 6 5 4 3] and c_8 = [6 5 4 3 2 1 8 7]

The symmetry group of this molecule is equivalent to the group of permutations G = {g_1, g_2, ...,g_8} , in which each g_i is defined as transforming the sequence (1 2 3 4 5 6 7 8) to the sequence c_i . For example,

$$g_2 = \begin{pmatrix} 1 & 2 & 3 & 4 & 5 & 6 & 7 & 8 \\ 5 & 6 & 7 & 8 & 1 & 2 & 3 & 4 \end{pmatrix} \qquad g_4 = \begin{pmatrix} 1 & 2 & 3 & 4 & 5 & 6 & 7 & 8 \\ 7 & 8 & 1 & 2 & 3 & 4 & 5 & 6 \end{pmatrix} \qquad etc.$$

In particular, the identity operation E is represented by the identity permutation:

$$e = \begin{pmatrix} 1 & 2 & 3 & 4 & 5 & 6 & 7 & 8 \\ 1 & 2 & 3 & 4 & 5 & 6 & 7 & 8 \end{pmatrix}$$

Therefore, instead of working directly with the sets of numbers representing geometrical configurations of the molecule, one may work with the permutation group G to obtain the result of any operation on the molecule. The ("right") exchange permutation for this molecule is

$$x = \begin{bmatrix} 1 & 2 & 3 & 4 & 5 & 6 & 7 & 8 \\ 8 & 1 & 2 & 3 & 4 & 5 & 6 & 7 \end{bmatrix}$$

Then:

$$x^2 = \begin{bmatrix} 1 & 2 & 3 & 4 & 5 & 6 & 7 & 8 \\ 7 & 8 & 1 & 2 & 3 & 4 & 5 & 6 \end{bmatrix} \qquad x^3 = \begin{bmatrix} 1 & 2 & 3 & 4 & 5 & 6 & 7 & 8 \\ 6 & 7 & 8 & 1 & 2 & 3 & 4 & 5 \end{bmatrix} \qquad etc.$$

so that $x^8 = e$. Thus $N = 8$ is the lowest power such that $\mathbf{X}^N = \mathbf{I}$. This is certainly much lower than the bound of $8! = 40{,}320$. On the other hand, $x^2 = g_4$, so x^2 is one of the elements in the group G ! This is the cause for the result to be obtained below, that there are only two distinct cosets formed by powers of x acting on G.

Operating on the group G with the ("right") exchange permutation, one obtains the following coset of permutations xG, equivalent to the configurations:

$xc_1 = [8\ 1\ 2\ 3\ 4\ 5\ 6\ 7]$ $\qquad xc_2 = [4\ 5\ 6\ 7\ 8\ 1\ 2\ 3]$

$xc_3 = [2\ 3\ 4\ 5\ 6\ 7\ 8\ 1]$ $\qquad xc_4 = [6\ 7\ 8\ 1\ 2\ 3\ 4\ 5]$

$xc_5 = [5\ 4\ 3\ 2\ 1\ 8\ 7\ 6]$ $\qquad xc_6 = [1\ 8\ 7\ 6\ 5\ 4\ 3\ 2]$

$xc_7 = [3\ 2\ 1\ 8\ 7\ 6\ 5\ 4]$ $\qquad xc_8 = [7\ 6\ 5\ 4\ 3\ 2\ 1\ 8]$

Operating again with x, the following configurations are obtained:

$x^2c_1 = [7\ 8\ 1\ 2\ 3\ 4\ 5\ 6] = c_4$ $\qquad x^2c_2 = [3\ 4\ 5\ 6\ 7\ 8\ 1\ 2] = c_3$

$x^2c_3 = [1\ 2\ 3\ 4\ 5\ 6\ 7\ 8] = c_1$ $\qquad x^2c_4 = [5\ 6\ 7\ 8\ 1\ 2\ 3\ 4] = c_2$

$x^2c_5 = [6\ 5\ 4\ 3\ 2\ 1\ 8\ 7] = c_8$ $\qquad x^2c_6 = [2\ 1\ 8\ 7\ 6\ 5\ 4\ 3] = c_7$

$x^2c_7 = [4\ 3\ 2\ 1\ 8\ 7\ 6\ 5] = c_5$ $\qquad x^2c_8 = [8\ 7\ 6\ 5\ 4\ 3\ 2\ 1] = c_6$

The resulting configurations are identical to the original configurations, except for their order. For example, $x^2c_1 = c_4$, $x^2c_3 = c_1$ etc. Therefore, if we continue applying powers of x to G, until we reach x^8G, each of the two cosets G and xG will simply be repeated four times. This means that in this particular system $M = 2$ is the number of physically distinct isomers reachable by the exchange operation, although the exchange itself has to be repeated $N = 8$ times in order to generate the identity transformation. As a result, for this molecule $\mathbf{X}^2\,\mathbf{H}\,\mathbf{X}^{-2} = \mathbf{H}$, which was found above to be a sufficient condition for using Eq. (III-34). Using the fast exchange approximation, the Liouville matrix is broken into smaller blocks than before, making the calculations much easier (see Ref.

14).

****

Suggested References

* On the Bloch-McConnell equations:

1. H.S. Gutowsky, D.W. McCall and C.P. Slichter, *J. Chem. Phys.* **21**, 279 (1953).
2. H.M. McConnell, *J. Chem. Phys.* **28**, 430 (1958).
3. A. Carrington and A.D. McLachlan, *"Introduction to Magnetic Resonance"* (Harper and Row, 1967), Ch. 12.

* On the Kaplan-Alexander density matrix formalism:

4. J. Kaplan, *J. Chem. Phys.* **28**, 278; **29**, 462 (1958).
5. S. Alexander, *J. Chem. Phys.* **37**, 967; **37**, 974 (1962).
6. G. Binsch in *"Dynamic Nuclear Magnetic Resonance Spectroscopy"*, edited by L.M. Jackman and F.A. Cotton (Academic Press, New York, 1975), Ch. 3.
7. J.I. Kaplan and G. Fraenkel, *"NMR of Chemically Exchanging Systems"* (Academic Press, New York, 1980).

* On the general theory of groups and their applications:

8. M. Hamermesh, *"Group Theory and Its Application to Physical Problems"* (Addison Wesley, Reading, MA , 1962).
9. F.A. Cotton , *"Chemical Applications of Group Theory"*, third edition (Wiley, New York, 1990).

* On aspects of group theory relevant to the discussion of symmetries in this chapter:

10. L. Jansen and M. Boon, *"Theory of Finite Groups. Applications in Physics"* (North Holland, Amsterdam, 1967).
11. S.L. Altmann, *"Induced Representations in Crystals and Molecules"* (Academic Press, London, 1977).

* On the use of symmetry in the density matrix formalism:

12. Z. Luz and R. Naor, *Mol. Phys.* **46**, 891 (1982).
13. S. Szymanski, *Mol. Phys.* **55**, 763 (1985).
14. D. Gamliel, Z. Luz and S. Vega, *J. Chem. Phys.* **85**, 2516 (1986).

CHAPTER IV

STOCHASTIC RELAXATION THEORY

IV.1 Relaxation Due to Random Modulations of the Motion

(a) Random modulation of a classical oscillator

The harmonic oscillator model is very useful for analogies with many physical systems, including magnetic resonance systems. It is therefore instructive to study the effect of stochastic processes on a simple oscillator, and then generalize the results to the relevant cases. For an ordinary one-dimensional classical oscillator the position at time t is given by the real part of x(t), where $x(t) = A\exp\{i(\omega t+\phi)\}$ so that

$$\frac{d}{dt}x(t) = i\omega\, x(t) \tag{IV-1}$$

In this equation it is understood that the real part of each side has to be taken for describing the actual physical variables. Here ω is the constant frequency of the oscillations and ϕ is their initial phase at time t=0. Now suppose there are some disturbances which make the frequency time dependent in a a random manner. Thus the frequency ω becomes a stochastic function of time, and the equation is changed to

$$\frac{d}{dt}x(t) = i\omega(t)x(t) \tag{IV-2}$$

The stochastic modulation is assumed to be stationary, so that the probability $p(\omega,t)$ for having at time t the value ω for the frequency is independent of the time t (but not of the frequency ω !). The frequency is assumed to have some arbitrary average value

$$\omega_0 \equiv \frac{\int_0^T \omega(t)\,p(\omega)\,dt}{\int_0^T p(\omega)\,dt} = \frac{1}{T}\int_0^T \omega(t)\,dt \tag{IV-3}$$

where T is a sufficiently long time. The averaging is done here without the probability function $p(\omega)$, because $p(\omega)$ constant in time and therefore irrelevant for this calculation. Then $\omega(t) = \omega_0 + \omega_1(t)$, where the time average of $\omega_1(t)$ is zero. The solution of Eq. (IV-2) is:

$$x(t) = x(0)\exp\left[i\int_0^t \omega(t')\,dt'\right] = x(0)\exp\left[i\omega_0 t + i\int_0^t \omega_1(t')\,dt'\right] \tag{IV-4}$$

If a measurement is performed, one can only expect to observe some average value of x(t). For example, if there is an ensemble of oscillators, one has to average over all oscillators in order to predict the observed result. This obviously corresponds to the situation in magnetic resonance, where one observes simultaneously a whole ensemble of spins. The expectation value is:

$$\langle x(t)\rangle = \int x(t)p(\omega)\,d\omega = x(0)\exp(i\omega_0 t)\int d\omega\, p(\omega)\exp\left[i\int_0^t \omega_1(t')\,dt'\right] =$$

$$= x(0)\exp(i\omega_0 t)\langle\exp(i\int_0^t \omega_1(t')\,dt')\rangle \tag{IV-5}$$

Then the auto-correlation function between the values of x at time 0 and time t is equal to

$$\langle x(t)x^*(0)\rangle = |x(0)|^2\exp(i\omega_0 t)\langle\exp(i\int_0^t \omega_1(t')\,dt')\rangle \tag{IV-6}$$

The non-trivial part of this expression is the average

$$\Phi(t) = \langle\exp(i\int_0^t \omega_1(t')\,dt')\rangle = \int d\omega_1\, p(\omega_1)\exp(i\int_0^t \omega_1(t')\,dt') \tag{IV-7}$$

which is called the **relaxation function** of the oscillator, because it governs the non-trivial part in the time evolution of the oscillator; this part of the evolution will be shown below to describe a relaxation behaviour. The relaxation function is closely related to whatis defined in statistics as the **characteristic function** of the random variable $\omega(t)$, which is

$$\phi(t) = \langle\exp(i\omega t)\rangle = \int_{-\infty}^{\infty}\exp(i\omega t)p(\omega)\,d\omega \tag{IV-8}$$

A **spectral density** matrix may be defined by the Fourier transform of the time average of $x^*(t)x(t+\tau)$:

$$S(\omega) \equiv \int_{-\infty}^{\infty} d\tau\, e^{-i\omega\tau} \left(\lim_{T\to\infty} \frac{1}{T} \int_{-T}^{T} dt\, x^*(t)\, x(t+\tau) \right) \qquad \text{(IV-9)}$$

For an ergodic system the time average can be replaced by an average over the probability distribution of x, and this average is independent of t due to the assumption of stationarity. Thus the spectral density is the Fourier transform of the correlation function:

$$S(\omega) \equiv \int_{-\infty}^{\infty} d\tau\, e^{-i\omega\tau} \langle x(t)\, x^*(0) \rangle \qquad \text{(IV-10)}$$

This relation is known as the **Wiener-Khinchin theorem.** It determines the connection between fluctuations in the value of x(t) and the frequency distribution of the time variations of x(t). The spectral density is closely related to the linear response of the system, a quantity which can be measured experimentally. For example, in magnetic resonance the magnetic susceptibility χ represents the response of the system to an external magnetic field:

$$\boldsymbol{M} = \chi \boldsymbol{B} \quad \Rightarrow \quad M_i = \sum_{i,j} \chi_{i,j} B_j \qquad \text{(IV-11)}$$

where **M** is the magnetization vector induced by the external magnetic field **B**, and χ is a tensor (a matrix) in the general case. It can be shown that χ', the real part of χ, describes scattering or refraction phenomena, whereas the imaginary part χ'' describes absorption. In magnetic resonance the measured quantity is absorption, so χ'' is the interesting variable. This variable is a function of the frequency ω.

A general result in statistical mechanics, the **fluctuation-dissipation theorem,** states that the linear response of a system to a weak perturbing force is proportional to the Fourier transform of the correlation function of the system; under theabove-mentioned conditions, this is equivalent to the spectral density introduced above. More specifically, using $\chi(\omega)$ for the response function of the oscillator, it is proven that

$$\chi(\omega) = \frac{\omega}{k_B T} \int_{-\infty}^{\infty} dt \cos(\omega t)\, \langle x(t)\, x^*(0) \rangle \qquad \text{(IV-12)}$$

where T is the temperature and k_B is Boltzmann's constant. The energy absorbed by the system is proportional to the imaginary part of the response function, and the factor of ω is to be expected there, since it is the energy (in units of $\hbar$) of one quantum of radiation. The non-trivial part of the expression is the integral, which expresses the "number of photons". In practical cases in magnetic resonance this integral has a sharp resonance in one position or one region, and is negligibly small outside this frequency region. Ignoring proportionality constants, one can define the line shape of the absorption intensity as a normalized spectral density function:

$$I(\omega-\omega_0) = \frac{\int_{-\infty}^{\infty} d\tau\, e^{-i\omega\tau}\langle x(t)\, x^*(0)\rangle}{2\pi\cdot\langle x(0)\, x^*(0)\rangle} = \frac{1}{2\pi}\int_{-\infty}^{\infty} dt \exp(-i(\omega-\omega_0)t)\,\Phi(t) \qquad \text{(IV-13)}$$

It is convenient to define it as a function of the difference ω-ω_0 (rather than of ω), since the fluctuations center around ω_0 .

The stochastic process may be characterized by two parameters, one measuring the amplitude of the modulations and one measuring the rate of decay of the autocorrelation function. The **modulation amplitude** Δ is defined as:

$$\Delta = \left\{\int d\omega_1\, p(\omega_1)\,(\omega_1)^2\right\}^{1/2} = \langle(\omega_1)^2\rangle^{1/2} \qquad \text{(IV-14)}$$

Since the process is stationary, the calculation of Δ does not involve the time t. The **correlation time** τ_C is defined as:

$$\tau_C = \frac{\int_0^{\infty} d\tau\,\langle\omega_1(t)\,\omega_1(t+\tau)\rangle}{\langle\omega_1(t)\,\omega_1(t)\rangle} = \frac{1}{\Delta^2}\int_0^{\infty} d\tau\,\langle\omega_1(t)\,\omega_1(t+\tau)\rangle \qquad \text{(IV-15)}$$

The integrand in this formula is the correlation function of the modulation. In this expression the time does appear formally as an argument of the correlation function, but the value of the correlation function depends only on the time difference τ. It is also possible to show that the value of this correlation function is always less than or equal to 1. Using these two parameters it is possible to define the ranges of fast modulation and slow modulation by the inequalities

$$\Delta\tau_C \gg 1 \qquad (\textit{slow modulation}) \qquad \text{(IV-16)}$$

when the correlation time is relatively long, so that dephasing is slow, and

$$\Delta\tau_C \ll 1 \qquad (\textit{fast modulation}) \qquad \text{(IV-17)}$$

when the correlation time is relatively short, so that dephasing is fast.

(b) Solution for the case of Markovian modulation

In order to solve Eq. (IV-2) a model is needed for the stochastic process. Finding an appropriate model for the process may be difficult, so in many cases one only assumes a model for the correlation function of Eq. (IV-15). The solution here will be based on the assumption that ω_1 undergoes Markovian modulation, which is relatively easy to treat mathematically and is of great practical importance. It is assumed for simplicity that the oscillator can be in one of a discrete set of states E_j , and $p_j(t) \equiv p(\omega_1=\omega_j)$ is the probability that at time t the oscillator is in state E_j , so that the random frequency ω_1

is equal at that moment to ω_j . The stochastic process consists of random "jumps" of the oscillator between these states. For a Markovian process, the evolution of the system is described by

$$P(E_k,t\,|\,E_j,0) = \sum_m P(E_k,t\,|\,E_m,t-\tau)\,P(E_m,t-\tau\,|\,E_j,0) \tag{IV-18}$$

Here $P(\beta,t_2\,|\,\alpha,t_1)$ is the *conditional probability* that the system is in state β at time t_2 when it is known that it was in state α at time t_1. If the process is assumed to be stationary, the conditional probability depends only on the time difference t_2-t_1 and not on the value of t_1 (see Chapter II). One can therefore define $P_{\alpha\beta}(s)$ as the *conditional probability* $P(\beta,s+t_0\,|\,\alpha,t_0)$, and then the above equation becomes

$$P_{jk}(t) = \sum_m P_{jm}(t-\tau)\,P_{mk}(\tau) \tag{IV-19}$$

We assume that for short time intervals this probability depends linearly on time:

$$P_{jj}(t) = 1 - c_j t \tag{IV-20a}$$

$$P_{jk}(t) = c_j W_{jk} t \qquad (for\ \ k \neq j) \tag{IV-20b}$$

where $c_j = \tau_j^{-1}$ is the inverse of the average lifetime of state j, and W_{jk} is the transition probability from state E_j to state E_k. These transition probabilities are normalized so that

$$\sum_k W_{jk} = 1 \qquad (k \neq j) \tag{IV-21}$$

In other words, the factor of c_j determines the total probability per unit time for depleting state E_j . Given that the system is initially in state E_j , and that it makes a transition (with probability per unit time c_j), there is a total probability of 1 for having a transition to some other state. Then the equation of motion can be rewritten as

$$P_{jk}(t) = P_{jk}(t-\tau)\,P_{kk}(\tau) + \sum_{m\neq k} P_{jm}(t-\tau)\,P_{mk}(\tau) =$$

$$= P_{jk}(t-\tau)\,(1 - c_k\tau) + \sum_{m\neq k} P_{jm}(t-\tau)\,c_m W_{mk}\,\tau \tag{IV-22}$$

Now subtract $P_{jk}(t-\tau)$ from both sides of the equation, divide by τ and take the limit $\tau \rightarrow 0$, so that the intermediate time t-τ tends to t. The result is simply the time derivative at time t:

$$\frac{d}{dt} P_{jk}(t) = -c_k P_{jk}(t) + \sum_{m \neq k} P_{jm}(t)\, c_m W_{mk} \tag{IV-23}$$

For a fixed j value this is a set of differential equations coupling the conditional probabilities P_{jk} (for all k), which may be written as a vector equation:

$$\frac{d}{dt} (\Pi_j(t))_k = - \sum_m (\Pi_j(t))_m D_{m,k} \tag{IV-24}$$

Here Π_j is the row vector having $P_{jk}(t)$ as the element in column k, and D is a matrix defined by:

$$D_{k,k} = c_k \qquad and \qquad D_{m,k} = -c_m W_{mk} \qquad (m \neq k) \tag{IV-25}$$

Now a matrix P may be constructed by arranging the row vectors Π_j as rows of a matrix. Thus Eq. (IV-24), which now ranges over all values of both j and k, becomes a matrix equation:

$$\frac{d}{dt} P = -P D \tag{IV-26}$$

In a steady state the time derivative vanishes, and then Eq. (IV-24) implies that: $\Pi_j = \xi_0 \equiv N^{-1} (P_1^{(0)}, P_2^{(0)},...,P_N^{(0)})$ for all j, where N is the number of states and $P_i^{(0)}$ is the steady state population of the state i. This corresponds to the equilibrium situation. On the other hand, the equation: $D\,\eta_0 = 0$ is satisfied by the column vector η_0 in which all elements are equal to N^{-1}. Thus ξ_0 is the left eigenvector of D, and η_0 is the right eigenvector of D, both with the eigenvalue 0.

With this model for the stochastic process it is possible to calculate $\Phi(t)$ and the line shape. The definition of $\Phi(t)$ implies that the initial and final conditions are arbitrary. It is thus possible to calculate the relaxation function as a sum over its values in all possible cases, with the appropriate weight factor for each initial state:

$$\Phi(t) = \sum_{j,k} (\xi_0)_j\, Q_{jk} \tag{IV-27}$$

where $Q_{jk}(t) \equiv \{\Phi(t)\}_{jk}$ is the relaxation function for the case in which the system is initially in state E_j and finally in state E_k. One may now assume a coupled evolution of the $Q_{jk}(t)$ with the same behaviour as the time evolution of the $P_{jk}(t)$, in order to derive an equation of motion analogous to Eq. (IV-26). However, exactly the same result can be obtained by the simplifying assumption that $Q_{jk}(t)$ is proportional to the probability $P_{jk}(t)$:

$$Q_{jk}(t) = P_{jk}(t)\, \Phi(t) \quad \Rightarrow \quad \frac{d}{dt} Q_{jk}(t) = i\,(\omega_1)_k t\, Q_{jk}(t) + \left[\frac{d}{dt} P_{jk}(t)\right] \Phi(t) \tag{IV-28}$$

so the following equation couples the different $Q_{jk}(t)$:

$$\frac{d}{dt} Q = i Q \Omega - Q D \tag{IV-29}$$

where Ω is a diagonal matrix, with $\Omega_{j,j} = (\omega_1)_j$. Given the initial condition Q(0) = I (the unit matrix), the solution is

$$Q(t) = Q(0) \exp\left\{ (i\Omega - D) t \right\} \tag{IV-30}$$

which leads to

$$\Phi(t) = \xi_0 \cdot \exp\left\{ (i\Omega - D) t \right\} \cdot \eta_0 \equiv u(t) \cdot \eta_0 \tag{IV-31}$$

In this equation the scalar function $\Phi(t)$ is obtained by calculating the exponential of a matrix, multiplying the column vector η_0 with it, and then multiplying the resulting column vector with the row vector ξ_0. Here u(t) is a row vector, defined by the product of the row vector ξ_0 with the exponential of the matrix $(i\Omega\text{-}D)t$. It satisfies the equation

$$\frac{d}{dt} u(t) = u(t) \cdot \{ i\Omega - D \} \quad ; \quad u(0) = \xi_0 \tag{IV-32}$$

This equation is mathematically the same as the equations of motion for a set of coupled damped harmonic oscillators. The oscillation frequencies are contained in Ω, and the damping results from the matrix D. The Fourier transform of this vector is

$$v(\omega) = \int_0^\infty dt \exp(-i\omega t) \, u(t) \tag{IV-33}$$

so from Eq. (IV-13) the line shape is

$$I(\omega - \omega_0) = \frac{1}{2\pi} \int_{-\infty}^{\infty} dt \exp\left(-i(\omega - \omega_0)t\right) \Phi(t) =$$

$$= \frac{1}{\pi} Re \left\{ \int_0^\infty dt \exp\left(-i(\omega - \omega_0)t\right) u(t) \cdot \eta_0 \right\} = \frac{1}{\pi} Re \left\{ v(\omega - \omega_0) \cdot \eta_0 \right\} \tag{IV-34}$$

"Re" stands here for the real part of the complex number. From the above equations it follows that

$$\nu(\omega) = \int_0^\infty dt\, \xi_0 \cdot \exp\{(i\Omega - D - i\omega I)t\} \;\Rightarrow\; \nu(\omega) = \xi_0 \cdot \{i\Omega - D - i\omega I\}^{-1} \quad \text{(IV-35)}$$

The equation on the right is the result of carrying out formally the integration in the equation on the left. In this equation, the expression $\{i\Omega\text{-}D\text{-}i\omega I\}^{-1}$ is the inverse of the matrix $\{i\Omega\text{-}D\text{-}i\omega I\}$. Therefore the line shape can be written as

$$I(\omega - \omega_0) = \frac{1}{\pi} Re\left\{ \xi_0 \cdot \{i\Omega - D - i(\omega - \omega_0) I\}^{-1} \cdot \eta_0 \right\} \quad \text{(IV-36)}$$

The inverse of the matrix which appears in this formula can be calculated by first diagonalizing it as

$$T^{-1}\{i\Omega - D\} T = \Lambda \quad \text{(IV-37)}$$

where Λ is a diagonal matrix, because then also

$$T^{-1}\left\{i\Omega - D - i(\omega - \omega_0) I\right\} T = \Lambda - i(\omega - \omega_0) I \equiv \Lambda'$$

$$\Rightarrow \quad \{i\Omega - D - i\omega I\} = T\Lambda' T^{-1} \quad \text{(IV-38)}$$

Therefore the inverse is given by

$$\{i\Omega - D - i\omega I\}^{-1} = T\,(\Lambda')^{-1}\, T^{-1} \quad \text{(IV-39)}$$

and the inverse of the diagonal matrix Λ' is also a diagonal matrix, in which each element is the inverse of the corresponding element in Λ':

$$\left\{(\Lambda')^{-1}\right\}_{i,j} = \delta_{i,j} \frac{1}{\Lambda_{i,i} - i(\omega - \omega_0)} \quad \text{(IV-40)}$$

Thus the line shape is a sum of Lorentzian lines:

$$I(\omega - \omega_0) = \frac{1}{\pi} Re\left\{ \sum_{i,j,k} (\xi_0)_i \, T_{i,j} \frac{1}{\Lambda_{j,j} - i(\omega - \omega_0)} (T^{-1})_{j,k} (\eta_0)_k \right\} =$$

$$= \frac{1}{\pi} Re\left\{ \sum_j A_j \frac{1}{\Lambda_{j,j} - i(\omega - \omega_0)} \right\} \quad \text{(IV-41)}$$

in which the j'th line has the intensity coefficient

$$A_j \equiv \sum_{i,k} (\xi_0)_i T_{i,j} (T^{-1})_{j,k} (\eta_0)_k \tag{IV-42}$$

As noted in Chapter II, the Fourier transform of such a line is a function of the form $\exp\{-(ia+b)t\}$, corresponding to damped oscillations. Since the line shape was defined as the Fourier transform of the relaxation function, the relaxation function behaves as a sum of decaying terms, each of them oscillating at its characteristic frequency.

Example IV.1: *A two state resonance spectrum*

Suppose one measures the nuclear magnetic resonance spectrum of atom A, which is part of a system in which the following chemical reaction is taking place:

$$AB + C \rightleftarrows AC + B \tag{E-1}$$

The magnetic environment of A in the molecule AB is assumed to differ from its environment in molecule AC, so A will have different resonance frequencies in the two molecules. This is actually one of the types of a "chemical exchange" process, studied in the previous chapter. It is formally similar to the oscillator problem, because for a single spin the Bloch equations lead to: $M_x + iM_y \;\alpha\; \exp(i(\omega-\omega_0)t)$, so the observed magnetization behaves in the same manner as the position of an oscillator. The two possible frequencies correspond to two possible values of ω_1. The damping matrix is

$$D = \begin{pmatrix} c_B & -c_B \\ -c_C & c_C \end{pmatrix} \tag{E-2}$$

where $\tau_B = c_B^{-1}$ and $\tau_C = c_C^{-1}$ are the lifetimes of the molecules AB and AC, respectively. The total probability for being in one of the states is $c_T = c_B + c_C$, and the relative probabilities for being in the two states are: $x_1 \equiv (\xi_0)_B = c_C/c_T$ and $x_2 \equiv (\xi_0)_C = c_B/c_T$. These occupation probabilities are directly proportional to the lifetimes, and inversely proportional to the depletion coefficients. Applying the line shape theory developed above, the magnetic resonance line shape can be calculated (from Eq. (IV-41)) if the matrix $i\Omega$-D is inverted. Here

$$i\Omega - D = \begin{pmatrix} i\omega_B - c_B & c_B \\ c_C & i\omega_C - c_C \end{pmatrix} \tag{E-3}$$

and instead of diagonalizing it, one may invert it directly using the rule: $A^{-1} = (\det A)^{-1}\,\mathrm{adj}(A)$ where det A is the determinant of A and adj(A) is the adjoint matrix of A. Thus the inverse of $i\Omega - i\Delta\omega - D$ (where $\Delta\omega = \omega-\omega_0$) is

$$\frac{1}{-\mu\nu - i(\gamma\mu + \beta\nu)} \begin{bmatrix} i\nu - \gamma & -\beta \\ -\gamma & i\mu - \beta \end{bmatrix} \tag{E-4}$$

using the shorter notation: $\beta \equiv c_B$, $\gamma \equiv c_C$, $\mu \equiv \omega_B - \Delta\omega$ and $\nu \equiv \omega_C - \Delta\omega$. The line shape is equal to the real part of

$$\frac{1}{\pi(\beta+\gamma)} \frac{1}{-\mu\nu - i(\gamma\mu + \beta\nu)} (\gamma \;\; \beta) \begin{bmatrix} i\nu - \gamma & -\beta \\ -\gamma & i\mu - \beta \end{bmatrix} \begin{bmatrix} 1 \\ 1 \end{bmatrix} =$$

$$= \frac{1}{\pi(\beta+\gamma)} \frac{\gamma(i\nu - \gamma - \beta) + \beta(-\gamma + i\mu - \beta)}{-\mu\nu - i(\gamma\mu + \beta\nu)} =$$

$$= \frac{-1}{\pi(\beta+\gamma)} \frac{\mu\nu - i(\gamma\mu + \beta\nu)}{(\mu\nu)^2 + (\gamma\mu + \beta\nu)^2} \times \left(i(\beta\mu + \gamma\nu) - (\beta+\gamma)^2\right) \tag{E-5}$$

Therefore

$$I(\Delta\omega) = \frac{\mu\nu(\beta+\gamma)^2 - (\beta\mu + \gamma\nu)(\gamma\mu + \beta\nu)}{\pi(\beta+\gamma)\left\{(\mu\nu)^2 + (\gamma\mu + \beta\nu)^2\right\}} =$$

$$= \frac{\mu\nu(2\beta\gamma) - \beta\gamma(\mu^2 + \nu^2)}{\pi(\beta+\gamma)\left\{(\mu\nu)^2 + (\gamma\mu + \beta\nu)^2\right\}} =$$

$$= \frac{\beta\gamma(\mu - \nu)^2}{\pi(\beta+\gamma)\left\{(\mu\nu)^2 + (\gamma\mu + \beta\nu)^2\right\}} = \frac{c_T x_1 x_2 (\omega_B - \omega_C)^2}{\pi\left\{(\mu\nu)^2 + c_T^{\,2}\,\omega_{ave}^{\,2}\right\}} \tag{E-6}$$

where $\omega_{ave} \equiv x_1\,\mu + x_2\,\nu$ is the average frequency. This result is the same as the standard solution of the classical equations for chemical exchange. The formalism developed so far is thus applicable to chemical exchange, at least in its classical formulation. It is also applicable, as will be seen later on, to problems of rotational diffusion, where the jump variable is continuous rather than discrete.

****

(c) Cumulant expansion for the relaxation function

The relaxation function, defined in Eq. (IV-7), was calculated in the previous subsection for an oscillator, in the case of Markovian modulation of the oscillation frequency. For more general types of modulation it is important to have some systematic way to treat the basic equation of the oscillator, even when it is impossible to derive an exact solution of the problem. Since the experimentally observed quantity is the line shape, which is a Fourier transform of the relaxation function, one needs a method for calculating the relaxation function in the general case. Here we shall work with the frequency ω rather than with its non-trivial part ω_1. The calculation is done using the **cumulant function** K(t) and the **cumulants** or **semi-invariants** κ_n which are defined for the characteristic function (Eq. (IV-8)) through the following relations:

$$\phi(t) \equiv \exp\{K(t)\} \quad ; \quad K(t) \equiv \sum_{n=1}^{\infty} \frac{(i\,t)^n}{n!}\,\kappa_n \tag{IV-43}$$

This means the cumulant function is defined as the natural logarithm of the characteristic function (assuming the logarithm is well defined) and one assumes it can be expanded as shown here, where the cumulants are the coefficients for the expansion as a function of t. The cumulants are closely related to the **moments** μ_n of the distribution of the random variable ω, defined as:

$$\mu_n \equiv \langle \omega^n \rangle \equiv \int_{-\infty}^{\infty} \omega^n\, p(\omega)\, d\omega \tag{IV-44}$$

Writing the exponential in the characteristic function as a series (which is the definition of the exponential), ϕ(t) is seen to be equal to

$$\phi(t) = \sum_{n=0}^{\infty} \frac{(i\,t)^n}{n!}\,\mu_n \tag{IV-45}$$

Comparing terms with equal powers of t in the two expansions of the relaxation function, one finds

$$\mu_0 = I \tag{IV-46a}$$

$$\mu_1 = \kappa_1 \tag{IV-46b}$$

$$\mu_2 = \kappa_2 + (\kappa_1)^2 \tag{IV-46c}$$

$$\mu_3 = \kappa_3 + 3\,\kappa_1\,\kappa_2 + (\kappa_1)^3 \tag{IV-46d}$$

etc. These relations can be inverted to give the cumulants in terms of the moments:

$$\kappa_1 = \mu_1 \tag{IV-47a}$$

$$\kappa_2 = \mu_2 - (\mu_1)^2 \tag{IV-47b}$$

$$\kappa_3 = \mu_3 - 3\,\mu_1\,\mu_2 + 2(\mu_1)^3 \tag{IV-47c}$$

and so on. In general, the n'th moment is expressed in terms of cumulants of order not higher than n, and the n'th cumulant is expressed in terms of the moments of order not higher than n. In particular, the first cumulant is equal to the average or expectation value, and the second cumulant is equal to the variance of the distribution.

From the equations presented above the characteristic function may be calculated once a model for the stochastic process, with a distribution function $p(\omega)$, has been determined. Moreover, if the variable ω has a distribution of values but does not change with time (e.g., frozen molecular orientations in a glass) the relaxation function is equal to the characteristic function. Thus the above equations specify the method for calculating the relaxation function of this type of systems. However, in general the relaxation function of Eq. (IV-7) involves the calculation of a non-trivial integral in the exponent, and the definitions given above are not sufficient. In addition, as mentioned in connection with the harmonic oscillator above, it is not always possible to specify the process by giving explicitly its distribution function. It may be that only correlation functions can be found for the stochastic variable. In order to deal also with such cases, a more general method for calculating the cumulant function is needed. Such a method can be derived by generalizing the definition of the moments and cumulants to multi-variable distributions. Instead of the single variable ω there is a set of n variables $\omega_1(s_1)$, $\omega_2(s_2)$,..., $\omega_n(s_n)$ with a distribution function $p(\omega_1,...,\omega_n)$. The arguments s_k are the analogs of: $i\cdot t$ where t is the time parameter. The generalized relaxation function is equal, on the one hand, to:

$$\Phi(t) \equiv \exp\{K(t)\} \quad ; \quad K(t) \equiv \sum_{n=1}^{\infty} \frac{(s_1)^{\nu_1}(s_2)^{\nu_2}\ldots(s_n)^{\nu_n}}{(\nu_1!)(\nu_2!)\ldots(\nu_n!)}\,\kappa(\nu_1,\nu_2,\ldots,\nu_n) \tag{IV-48}$$

which serves as a definition of the generalized cumulants $\kappa(\nu_1, \nu_2, ..., \nu_n)$. On the other hand, the relaxation function is equal to:

$$\Phi(t) \equiv \langle \exp\left\{\sum_{j=1}^{n} s_j\,\omega_j\right\}\rangle = \sum_{n=0}^{\infty} \frac{(s_1)^{\nu_1}(s_2)^{\nu_2}\ldots(s_n)^{\nu_n}}{(\nu_1!)(\nu_2!)\ldots(\nu_n!)}\,\mu(\nu_1,\nu_2,\ldots,\nu_n) \tag{IV-49}$$

where

$$\mu(\nu_1,\nu_2,\ldots,\nu_n) \equiv \langle (\omega_1)^{\nu_1}(\omega_2)^{\nu_2}\ldots(\omega_n)^{\nu_n}\rangle \tag{IV-50}$$

Thus, for example, $\mu(1,1,...,1) = \langle\omega_1\ \omega_2\ ...\ \omega_n\rangle$ and $\kappa(\nu_k\!=\!n)$ is the coefficient κ_n for the single variable ω_k. Again, comparing the two expansions, one obtains general relations between the two types of expansion coefficients. As found above, the cumulants are formed from combinations of the moments, which are averaged expressions. Because of this relationship the cumulants are often denoted as:

$$\langle (\omega_1)^{\nu_1} (\omega_2)^{\nu_2} \ldots (\omega_n)^{\nu_n} \rangle_c \equiv \kappa(\nu_1, \nu_2, \ldots, \nu_n) \tag{IV-51}$$

where the "c" subscript indicates **cumulant averages** or **connected averages.** Using this notation, the results of comparing the two expansions are, for example:

$$\langle \omega_1 \rangle_c = \langle \omega_1 \rangle \tag{IV-52a}$$

$$\langle \omega_1 \omega_2 \rangle_c = \langle \omega_1 \omega_2 \rangle - \langle \omega_1 \rangle \langle \omega_2 \rangle \tag{IV-52b}$$

$$\langle \omega_1 \omega_2 \omega_3 \rangle_c = \langle \omega_1 \omega_2 \omega_3 \rangle - \langle \omega_1 \rangle \langle \omega_2 \omega_3 \rangle - \langle \omega_2 \rangle \langle \omega_3 \omega_1 \rangle - \langle \omega_3 \rangle \langle \omega_1 \omega_2 \rangle$$

$$+ 2 \langle \omega_1 \rangle \langle \omega_2 \rangle \langle \omega_3 \rangle \tag{IV-52c}$$

It is convenient to use simple diagrams to visualize the relationship between the different average quantities involved in these expressions. Fig. IV.1 is such a diagram, in which ordinary averages, or moments, are represented by sets of unlinked points, whereas cumulant averages are represented by sets of linked points. The "linked diagrams" are expressed in terms of the "unlinked diagrams".

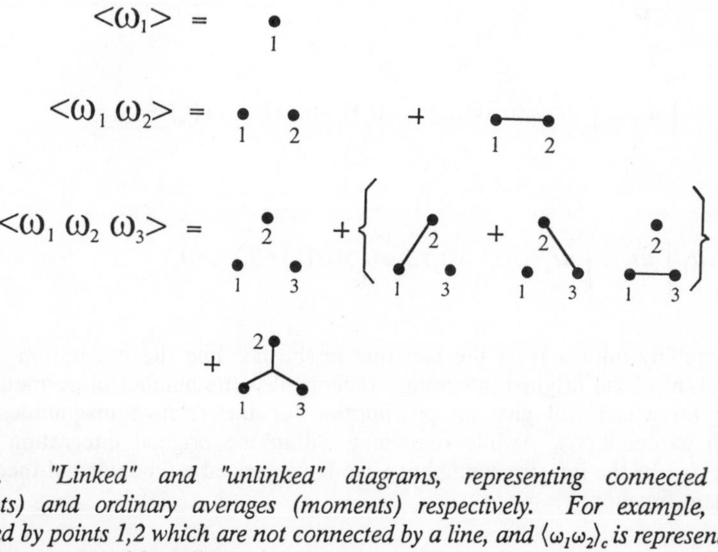

Fig. IV.1. "Linked" and "unlinked" diagrams, representing connected averages (cumulants) and ordinary averages (moments) respectively. For example, $\langle \omega_1 \omega_2 \rangle$ is represented by points 1,2 which are not connected by a line, and $\langle \omega_1 \omega_2 \rangle_c$ is represented by the same points connected by a line.

The cumulant average has the property of being equal to zero if one of the variables in it is uncorrelated with the others. This can be seen by using the definition of the generalized relaxation function, and writing it as a product of two averages, one average of the exponent involving the uncorrelated variable and the other average of the exponent involving all other variables. The result is that no cumulants appear in which the uncorrelated variable appears together with the other variables.

Finally, extending the above derivations to a continuous set of variables, the sum over $s_j\omega_j$ is transformed to an integral:

$$\sum_{j=1}^{n} s_j\,\omega_j \quad \rightarrow \quad \int_0^t dt'\,s(t')\,\omega(t') \qquad \text{(IV-53)}$$

When the sum is exponentiated, mixed products appear - products of different variables, each of them to some power. When the limiting expression leading to the integral is involved, mixed products involve mixed products of differentials, like $dt'_i dt'_j$ (and higher order products). Such products are smaller than first order, and therefore negligible in the integration if i = j . Thus the only products which remain are those in which each power ν_i is either 1 or 0. This leads to the result

$$\left\langle \exp\left\{ \int_0^t \omega(t')\,s(t')\,dt' \right\} \right\rangle = \exp\{\Sigma_c\} \qquad \text{(IV-54)}$$

where Σ_c , the sum over integrals with cumulants, is defined as:

$$\Sigma_c \equiv \int_0^t \langle\omega(t_1)\rangle_c\,s(t_1)\,dt_1 + \frac{1}{2!}\int_0^t dt_1\int_0^t dt_2\,\langle\omega(t_1)\,\omega(t_2)\rangle_c\,s(t_1)\,s(t_2) + \ldots$$

$$+ \frac{1}{n!}\int_0^t dt_1\int_0^t dt_2\ldots\int_0^t dt_n\,\langle\omega(t_1)\,\omega(t_2)\ldots\omega(t_n)\rangle_c\,s(t_1)\,s(t_2)\ldots s(t_n) + \ldots$$

$$= \sum_{n=1}^{\infty}\int_0^t dt_1\int_0^{t_1} dt_2\ldots\int_0^{t_{n-1}} dt_n\,\langle\omega(t_1)\,\omega(t_2)\ldots\omega(t_n)\rangle_c\,s(t_1)\,s(t_2)\ldots s(t_n) \qquad \text{(IV-55)}$$

The last equality follows from the fact that in the last line the integration region is limited to 1/n! of the original integration region (n! is the number of permutations of the integrand which will give all possibilities for the relative magnitudes of the integration parameters t_k , while remaining within the original integration region). Replacing $s(t_k)$ by the imaginary number i, we have derived a fundamental theorem for the cumulant function:

$$\left\langle \exp\left\{ i \int_0^t \omega(t')\,dt' \right\}\right\rangle = \exp\left\{ i\langle\omega(t)\rangle t - \int_0^t dt_1 \int_0^{t_1} dt_2 \langle\omega(t_1)\,\omega(t_2)\rangle_c + \ldots \right.$$

$$\left. + (i)^n \int_0^t dt_1 \int_0^{t_1} dt_2 \ldots \int_0^{t_{n-1}} dt_n \langle\omega(t_1)\,\omega(t_2)\ldots\omega(t_n)\rangle_c + \ldots \right\} \qquad \text{(IV-56)}$$

This formula makes it possible to calculate systematically the relaxation function for the most general case. In particular, one may approximate the relaxation function by calculating the low order cumulants and then substituting them in a truncated version of Eq. (IV-56).

(d) Relaxation behaviour as a function of the modulation rate

In Eq. (IV-13) the line shape for a randomly modulated oscillator was expressed in terms of the correponding relaxation function. This was solved in subsection I.(b) above for the special case of Markovian modulation. For more general types of modulation one can invoke the cumulant theorem derived above, obtaining:

$$\left\langle \exp\left\{ i \int_0^t \omega(t')\,dt' \right\}\right\rangle = \exp\{i\,\omega_0 t\} \exp\left\{ -\frac{1}{2!}\int_0^t dt_1 \int_0^t dt_2 \langle\omega(t_1)\,\omega(t_2)\rangle_c + \ldots \right\} =$$

$$\equiv \exp\{i\,\omega_0 t\} \exp\{K(t)\} \qquad \text{(IV-57)}$$

The cumulant function K(t) is re-defined here as being related to the non-trivial ω_1 rather than to the full frequency value $\omega = \omega_0 + \omega_1$. The average of ω_1 is zero, so the two-term cumulant in the exponential of Eq. (IV-57) is equal to the correponding moment (see Eq. (IV-52b)). Since it has been assumed that ω_1 is a stationary random variable, this moment (which is just its auto-correlation function) depends only on the time difference $\tau \equiv t_2 - t_1$. Thus $\langle\omega(t_1)\omega(t_2)\rangle$ in the integrand can be replaced by $\langle\omega(0)\omega(\tau)\rangle \equiv G(\tau)$, and the integration over t_2 is replaced by integration over τ in the limits τ: $-t_1 \rightarrow t-t_1$. Here τ is expressed as a function of t_1, which is inconvenient since one wants to integrate over t_1 first. If one examines the integration region which is shown in Fig. IV.2, one finds that the integral can be written as a sum over two integrals, in the following way:

$$-\frac{1}{2}\int_0^t dt_1 \int_{-t_1}^{t-t_1} d\tau \,\langle\omega_1(0)\,\omega_1(\tau)\rangle = -\frac{1}{2}\left(\int_{-t}^{0} d\tau \int_{-\tau}^{t} dt_1\, G(\tau) + \int_0^t d\tau \int_0^{t-\tau} dt_1\, G(\tau)\right) =$$

$$= -\int_0^t d\tau\,(t-\tau)\,G(\tau) \approx -t\int_0^\infty G(\tau)\,d\tau = -t\left(\Delta^2\,\tau_c\right) \qquad \text{(IV-58)}$$

The next to last step in the equation is based on the assumption that the calculation is done for times t which are much longer than the correlation time τ_C, which measures the time during which the auto-correlation $G(\tau)$ has an appreciable value. In a similar manner one finds that the higher order cumulants are also proportional to t. The result is that for times t which satisfy the condition $t \gg \tau_C$:

$$K(t) \approx -(a+ib)\,t \quad \Rightarrow \quad \Phi(t) \approx \exp\{-(a+ib)\,t\} \qquad \text{(IV-59)}$$

where

$$a = \left(\Delta^2\,\tau_C\right) + \ldots \quad ; \quad b = -\omega_0 + \ldots \qquad \text{(IV-60)}$$

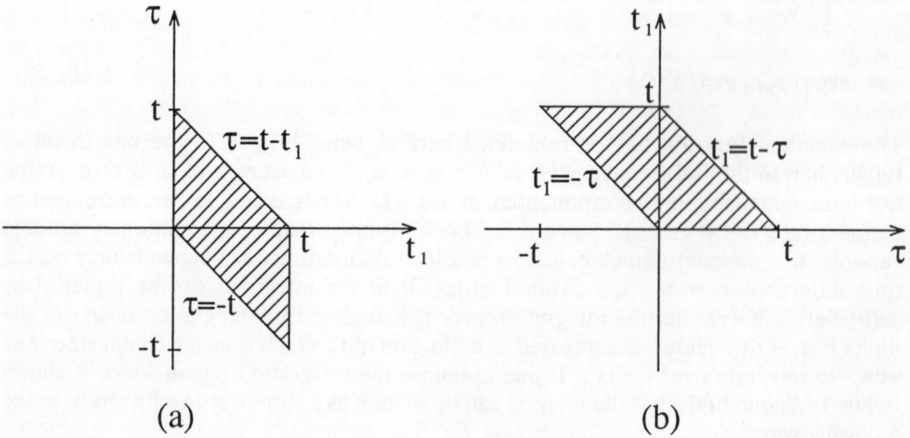

Fig. IV.2. *(a) The integration region for equation (IV-58), as viewed in the first part, first line of the equation. (b) The same integration region, as viewed in the second part, first line of the equation.*

It is interesting to study the line shape for the two extreme cases, in which the stochastic process is either fast or slow, as defined in subsection IV.1.(a). This may be done by looking at the behaviour of the relaxation function for very short times and for very long times. For very short times, i.e., for t $\ll \tau_C$:

$$\Phi(t) \approx \int d\omega_1 \, p(\omega_1) \exp\{i\,\omega_1(0)\,t\} \equiv \Phi_0(t) \tag{IV-61}$$

In this case the exponent is almost constant and equal to 1, so the line shape, which is the Fourier transform of the relaxation function, is essentially equal to the distribution of $(\omega\text{-}\omega_0)$. This is the situation as long as: $\tau_C\,\Delta \gg 1$, because the times t relevant to the calculation are of the order of $1/\Delta$.

For very long times, i.e., for t $\gg \tau_C$ Eq. (IV-59) applies, and the function may be approximated by a single decaying oscillator:

$$\Phi(t) \approx \exp\left\{-(a+ib)t\right\} \equiv \Phi_\infty(t) \tag{IV-62}$$

This is Fourier transformed to give a Lorentzian line shape in the frequency domain. The width of this line can be shown to be small relative to Δ, so this dynamic regime is called the motional narrowing regime.

IV.2 The Stochastic Liouville Equation - a Relaxation Function Approach

(a) The stochastic Liouville equation in classical mechanics

The method of using a relaxation function to solve an equation of motion can be extended to more general equations of motion than that discussed above. This will be demonstrated here both for classical mechanics and for quantum mechanics. In classical mechanics one may follow the dynamics of a physical system by following the time evolution of the density function ρ in phase space. This is done using Liouville's equation for the density function:

$$\frac{\partial \rho}{\partial t} = \sum_j \left(\frac{\partial H}{\partial q_j}\frac{\partial}{\partial p_j} - \frac{\partial H}{\partial p_j}\frac{\partial}{\partial q_j} \right) \rho = [H,\rho] \equiv \boldsymbol{L}\,\rho \tag{IV-63}$$

The square brackets denote here Poisson's brackets, q_j and p_j represent coordinates and momenta, respectively, and the sum is over all relevant degrees of freedom. The operator **L** is Liouville's operator for this classical equation. In cases which involve some random modulations of the system, the Hamiltonian can be divided into two parts:

$$H = H_0 + V(t) \tag{IV-64}$$

where H_0 includes the constant interactions in the system, and V(t) the time dependent perturbations which change with time in a random manner. Thus in Liouville's equation above one may separate the effects of the two parts of the Hamiltonian:

$$\frac{\partial \rho}{\partial t} = [H_0 + V(t), \rho] \equiv (L_0 + L_1(t))\,\rho \qquad \text{(IV-65)}$$

This is Liouville's classical equation for the case in which stochastic processes appear, so it may be called a stochastic Liouville equation. It is similar to Eq. (IV-2) for the one-dimensional oscillator in its general form, suggesting that a similar method of soilution may be feasible. Moreover, it is also similar in the separation into a large constant term , analogous to the ω_0 term in the case of the oscillator, and a time-dependent term, analogous to the $\omega_1(t)$ term in the case of the oscillator. However, here one deals with operators and not with simple scalar functions, and therefore the analogous operations in the present case are not so simple. In order to focus on the effects of the random processes, it is useful to transform to an "interaction representation", in which the effects of the large and constant Hamiltonian do not mask the smaller time-dependent effects. The transformation is defined by:

$$\sigma(t) = \exp\{-L_0 t\}\,\rho(t) \quad ; \quad \sigma(0) = \rho(0) \qquad \text{(IV-66)}$$

Then the transformed density function in phase space has the following equation of motion:

$$\frac{\partial \sigma}{\partial t} = -L_0 \exp\{-L_0 t\}\rho(t) + \exp\{-L_0 t\}(L_0 + L_1(t))\rho(t) = \exp\{-L_0 t\}L_1(t)\rho(t) =$$

$$= \exp\{-L_0 t\}\, L_1(t) \exp\{L_0 t\}\, \sigma(t) \equiv \Omega(t)\,\sigma(t) \qquad \text{(IV-67)}$$

This is a generalization of Eq. (IV-2), and the method of solution used for the oscillator can thus be extended to the present problem, provided one takes into account the possible non-commutativity of Ω(t) and Ω(t') for t' $\neq$ t. The formal solution of the equation involves a time-ordered exponential:

$$\sigma(t) = T\left\{ \exp\left[i\int_0^t \Omega(t')\,dt' \right] \right\} \sigma(0) = \left\{ I + \int_0^t dt_1\,\Omega(t_1) + \int_0^t dt_1 \int_0^{t_1} dt_2\,\Omega(t_1)\,\Omega(t_2) + \dots \right.$$

$$\left. + \int_0^t dt_1 \int_0^{t_1} dt_2 \dots \int_0^{t_{n-1}} dt_n\,\Omega(t_1)\,\Omega(t_2)\dots\Omega(t_n) + \dots \right\} \sigma(0) \qquad \text{(IV-68)}$$

(**T** is the time ordering operator). Since V(t) has a random time dependence, it is necessary to average these expressions over the random variables in order to obtain a result that can be measured in practice. The result of the averaging is written formally as:

$$\langle \sigma(t) \rangle = \langle T \left\{ \exp \left[i \int_0^t \Omega(t') dt' \right] \right\} \rangle \sigma(0) \equiv \Phi(t) \sigma(0) \tag{IV-69}$$

or, transforming back to the original laboratory frame of reference:

$$\langle \rho(t) \rangle = \exp\{L_0 t\} \, \Phi(t) \, \rho(0) \tag{IV-70}$$

The operator Φ(t) is called the **relaxation operator** for the process Ω(t), and is clearly analogous to the relaxation function defined previously for the problem of a one-dimensional oscillator. If the initial state is known exactly then the initial distribution in phase space is a delta function:

$$\rho(0) = \Pi_j \, \delta(q_j - q_{j0}) \, \delta(p_j - p_{j0}) \tag{IV-71}$$

where q_{j0} and p_{j0} are the initial values of the coordinates and momenta, and the product is over all degrees of freedom. The solution obtained for $\langle\rho(t)\rangle$ in such a case represents the transition probability to go from the initial configuration to the final configuration in the time interval t.

As for the relaxation function in the case of the oscillator, it is desirable to have a general method for calculating the relaxation operator here. This can be done by employing a cumulant expansion as in Eqs. (IV-48) and (IV-55), where the scalar functions $\omega(t_i)$ are replaced by the generally non-commuting operators $\Omega(t_i)$, so the exponential is time ordered as in Eq. (IV-68). It is possible to generalize to this case all the concepts discussed above, and obtain generally applicable formal solutions for the problem of calculating the relaxation operator. However, the expressions are complicated, and will not be given here. As will be shown in the next chapter, there is an alternative method of solving the stochastic Liouville equation, which is less general but easier to apply in practice.

Example IV.2: *A randomly modulated magnetization vector*

Suppose a spin system is placed in a constant magnetic field of magnitude B_0, the direction of which is defined as the **z** direction. Suppose further that there are local perturbing electromagnetic fields, changing randomly with time, which also act on the spins. The classical equations for the magnetization vector of the system are

$$\frac{d}{dt} M_x = -\omega M_y + F_x \tag{E-1a}$$

$$\frac{d}{dt} M_y = \omega M_x + F_y \tag{E-1b}$$

and the longitudinal component of the vector is unimportant for the present discussion. Here $\omega = \omega_0 + \omega' = \gamma(B_0 + B^p{}_z)$, where ω' represents the contribution of the **z**-

component of the local perturbing field B^p. The "driving force" $F_{x,y}$ terms result from the action of the transverse components of the local fields, similar to the $\omega_1 M_z$ term in Eq. (II-5b). They are defined as

$$F_x = \gamma M_z B^p{}_y \qquad F_y = -\gamma M_z B^p{}_x \tag{E-2}$$

Equations (E-1) closely resemble the canonical equations of Hamilton for the coordinate q and the conjugate momentum p:

$$\frac{\partial}{\partial t} q = \frac{\partial H}{\partial p} \qquad \frac{\partial}{\partial t} p = -\frac{\partial H}{\partial q} \tag{E-3}$$

in which H is the Hamiltonian function. The equations for the magnetization can be written in the form (E-3) if one defines: $q \equiv M_y$, $p \equiv M_x$ and

$$H \equiv \frac{1}{2}\omega\left(p^2 + q^2\right) + pF_y - qF_x = \frac{1}{2}\omega\left(M_x^2 + M_y^2\right) - \gamma M_z\left(M_x B^p{}_x + M_y B^p{}_y\right) \tag{E-4}$$

The main part of this Hamiltonian is formally similar to that of a one-dimensional harmonic oscillator, and is proportional to the classical expression for the energy contained in a magnetic field. The additional part is assumed to describe the perturbation. It is true that both magnetization components are physically the same type of variables, but it is mathematically convenient here to regard one of them as a coordinate and the other as its conjugate momentum. With this identification, one may calculate Poisson's brackets as in Eq. (IV-63), with

$$\frac{\partial H}{\partial p} = \omega p + F_y = \omega M_x + F_y \qquad \frac{\partial H}{\partial q} = \omega q - F_x = \omega M_y - F_x \tag{E-5}$$

Liouville's equation for the density function is therefore

$$\frac{\partial \rho}{\partial t} = \left(\omega\left(M_y \frac{\partial}{\partial M_x} - M_x \frac{\partial}{\partial M_y}\right) - F_x \frac{\partial}{\partial M_x} - F_y \frac{\partial}{\partial M_y}\right)\rho = \boldsymbol{L}\rho \tag{E-6}$$

This is a stochastic Liouville equation, since ω, F_x and F_y are assumed to be stochastic variables. Now define the operator

$$\boldsymbol{J}_z \equiv M_y \frac{\partial}{\partial M_x} - M_x \frac{\partial}{\partial M_y} \tag{E-7}$$

This operator is formally like the **z**- component of an angular momentum operator, if q and p defined above are regarded as a coordinate and its momentum. It is now useful to transform away the dependence of the magnetization on the constant frequency ω_0, by the definition

$$\sigma(t) = \exp\{-\omega_0 \boldsymbol{J}_z t\}\,\rho(t) \tag{E-8}$$

as in Eq. (IV-66). Differentiating this "interaction representation" density function, the

Liouville equation becomes:

$$\frac{\partial \sigma}{\partial t} = \left(-\omega_0 J_z + \omega J_z - e^{-\omega_0 J_z t} \left(F_x \frac{\partial}{\partial M_x} + F_y \frac{\partial}{\partial M_y} \right) e^{\omega_0 J_z t} \right) \sigma =$$

$$= \left(\omega'(t) J_z - (F_x(t)\cos(\omega_0 t) + F_y(t)\sin(\omega_0 t)) \frac{\partial}{\partial M_x} - (-F_x(t)\sin(\omega_0 t) + F_y(t)\cos(\omega_0 t)) \frac{\partial}{\partial M_y} \right) \sigma$$

$$\equiv i\,\Omega'(t)\,\sigma \tag{E-9}$$

Specific assumptions about the stochastic processes involved make it possible to solve this stochastic Liouville equation, obtaining an explicit description of the relaxation behaviour of the system.

****

(b) *The stochastic Liouville equation in quantum mechanics*

In the quantum mechanical treatment of a random modulation problem one may also construct an equation which is similar to the classical stochastic Liouville equation. One starts by writing down the Liouville - von Neumann equation of motion for the density matrix:

$$\frac{\partial \rho}{\partial t} = \frac{1}{i\hbar} [\mathbf{H}_0 + \mathbf{V}(t), \rho] \equiv (\mathbf{L}_0 + \mathbf{L}_1(t))\,\rho \tag{IV-72}$$

This equation looks similar to Eq. (IV-65), but here the square brackets denote the quantum commutator and not Poisson's brackets. The Hamiltonian is the quantum mechanical Hamiltonian, ρ(t) is the quantum mechanical density operator, and **V**(t) is the stochastic perturbation term. Notice that the factor of $1/i\hbar$ has been included here in the definition of the operators $\mathbf{L}_0$ and $\mathbf{L}_1$, in order to emphasize the analogy to the classical case. As in the classical case, one may define an interaction representation:

$$\sigma(t) = \exp\{-L_0 t\}\,\rho(t) = \exp\{i H_0 t/\hbar\}\,\rho(t)\,\exp\{-i H_0 t/\hbar\} \tag{IV-73}$$

with the initial condition: $\sigma(0) = \rho(0)$, and the right hand side of the equation is obtained using the definition of the Liouville super-operator (compare Eqs. (I-61),(I-69)). The equation of motion in the interaction representation is:

$$\frac{\partial \sigma}{\partial t} = \exp\{-L_0 t\}\, L_1(t)\, \rho(t) = \frac{1}{i\hbar} \exp\{i H_0 t/\hbar\} \left(V(t)\rho(t) - \rho(t) V(t)\right) \times$$

$$\times \exp\{-i H_0 t/\hbar\} = \frac{1}{i\hbar}[\, V^{(I)}(t), \sigma\,] \equiv \Omega(t)\, \sigma(t) \qquad \text{(IV-74)}$$

where $V^{(I)} \equiv \exp\{iH_0t/\hbar\}\ V(t)\ \exp\{-iH_0t/\hbar\}$ is the perturbation in the interaction representation. In analogy to the classical case, this equation is also called a stochastic Liouville equation. It may be formally integrated as done in Eq. (IV-68) above for the classical case. Following the time ordered integration and exponentiation, one needs to average over the stochasic process, as in Eq. (IV-69) above. The result is:

$$\langle \sigma(t) \rangle = \langle T \left\{ \exp\left[\int_0^t \Omega(t')\, dt' \right] \right\} \rangle\, \sigma(0) \equiv \Phi(t)\, \sigma(0) \qquad \text{(IV-75)}$$

which is the formal solution for the quantum mechanical stochastic Liouville equation. It may be practically developed using cumulants, as done for the classical case. This and other methods of solution will be discussed in the next chapter.

IV.3 The Stochastic Liouville Equation - a Distribution Function Approach

(a) Derivation of the equation in classical mechanics

The stochastic Liouville equation (SLE) has been derived above, both in its classical and in its quantum versions, with the relaxation function or operator as the main tool for dealing with the equation. The same physical problem may also be treated in a different approach, in which the central concept is that of the distribution function of the stochastic process. The main idea in this approach is to construct an equation describing the time dependence of the distribution function, and use this equation in order to obtain an equation of motion for the relevant physical variables. In fact, this is more intuitively appealing, and also leads to a form of the SLE which can be solved for practical cases more easily than by the cumulants method, which is the natural method of solution in the relaxation function approach.

In the previous derivation of the stochastic Liouville equation it was clear that two types of variables appear in the equation. One type includes the "interesting" physical variable, which one actually measures, like the oscillator cooordinate **x**, or the magnetization vector, etc. The other type is that of the stochastic variable, like the frequency ω, which influences through its modulation the interesting variable. The relevant physical variable will be denoted here by: **u**, and the stochastic variable by: **λ**. The time behaviour of the stochastic variable is obviously governed only by some probability distribution. However, **u** may also depend in some way on some other random variables, not known to us. For example, in a macroscopic ensemble there is also a probability distribution of possible values for the relevant physical variable, due to standard considerations of statistical mechanics. In order to take into account both causes of randomness, one may define the joint probability distribution W(**u**,λ,t) for finding these two variables with the values u, λ at time t. If the time dependence of this function is known, it can be used to construct an equation of motion for the observable physical variable, which is an average of **u**:

$$\langle u(t) \rangle_\lambda = \int W(u,\lambda,t)\, u\, du \tag{IV-76}$$

The variable **u** may be multi-dimensional (e.g., a three dimensional vector), and in that case the average is calculated by a multi-dimensional integration. This average is calculated for a particular value of the stochastic variable λ, but for the result of a measurement one obviously needs to average also over that variable.

The rate of change of W is, to first order in the changes in **u** and in λ, a sum of the changes caused by them separately:

$$\frac{\partial W}{\partial t} = \left(\frac{\partial W}{\partial t}\right)_\lambda + \left(\frac{\partial W}{\partial t}\right)_u \tag{IV-77}$$

where a subscript $)_\alpha$ indicates that the variable α is held constant. Thus the first term on the right hand side results from the change in **u** , and the second term results from the change in λ. Now the joint distribution W can be expressed in terms of the probability distribution for one variable and the conditional probability for the other variable, provided that the first variable has a known value:

$$W(u,\lambda,t) = p_\lambda(\lambda,t) \cdot P_{u|\lambda}(u,t\,|\,\lambda,t) \tag{IV-78a}$$

$$W(u,\lambda,t) = p_u(u,t) \cdot P_{\lambda|u}(\lambda,t\,|\,u,t) \tag{IV-78b}$$

Here p_λ, p_u are the single variable probability distributions at time t for λ, **u** respectively, and $P_{u|\lambda}$, $P_{\lambda|u}$ are the conditional probabilities for **u** (or λ) having a certain value at time t when λ (or **u**) has a given value at that instant. It will now be assumed that

$$\frac{\partial}{\partial t} p_\lambda(\lambda,t) = \Gamma_\lambda\, p_\lambda(\lambda,t) \tag{IV-79}$$

where Γ_λ is, in general, dependent on λ. This assumption is fairly general, and may be regarded as a generalization of the master equation which applies to Markovian processes (see Chapter II). Therefore

$$\left(\frac{\partial W}{\partial t}\right)_u = \left(\frac{\partial p_\lambda(\lambda,t)}{\partial t}\right)_u P_{u|\lambda}(u,t\,|\,\lambda,t) = \Gamma_\lambda\, p_\lambda(\lambda,t)\, P_{u|\lambda}(u,t\,|\,\lambda,t) =$$

$$= \Gamma_\lambda\, W(u,\lambda,t) \tag{IV-80}$$

As for the other contribution to Eq. (IV-77), one may write an equation for the rate of change in $p_u(\mathbf{u},t)$ on the basis of the assumption of conservation of probability. The probability function $p_u(u,t)$ may be regarded as the density function in a phase space, in which **u** is the coordinate. In general **u** is not one-dimensional, so it will be regarded as a vector. Suppose first that is is a three dimensional vector (e.g., a magnetization vector). Then the mathematical divergence theorem can be used to express the fact that the probability to find specific values for the components of the vector can only "flow" in phase space without the overall probability (for having any values) becoming smaller or larger. Thus the rate at which probability leaves a volume element in phase space is equal, by the divergence theorem, to

$$\frac{\partial}{\partial t} p_u(u,t) = -\nabla\cdot\big(v\, p_u(u,t)\big) = -\nabla\cdot\left(\frac{du}{dt} p_u(u,t)\right) \tag{IV-81}$$

where **v** is the "velocity" of motion in phase space, or: **v** = d**u**/dt . This is the same as the continuity equation in hydrodynamics, the only difference being that there p_u is the mass density and **v** is the velocity of the fluid motion. In the present context, the equation can be intuitively justified in the following manner. The change in p_u at the "point" (in phase space) u during a short time interval Δt is $\Delta p_u/\Delta t$. On the other hand, the total amount of change in the density p_u in the close environment of the point **u** is equal to the change in the "current density" around that point: $\Delta(p_u \cdot (d\mathbf{u}/dt))$. Thus the rate of change of p_u per unit length of **u** is: $\Delta(p_u \cdot (d\mathbf{u}/dt))/\Delta\mathbf{u}$ which is equal, in the limit of infinitely small changes, to the divergence:

$$\nabla_u\cdot\left(\frac{du}{dt}p_u\right) = \frac{\partial}{\partial u_x}\left(\frac{du_x}{dt}p_u\right) + \frac{\partial}{\partial u_y}\left(\frac{du_y}{dt}p_u\right) + \frac{\partial}{\partial u_z}\left(\frac{du_z}{dt}p_u\right) \tag{IV-82}$$

It thus follows that

$$\left(\frac{\partial W}{\partial t}\right)_\lambda = \left(\frac{\partial p_u(u,t)}{\partial t}\right)_\lambda P_{\lambda|u}(\lambda,t\,|\,u,t) = -\nabla\cdot\left(\frac{du}{dt}p_u(u,t)\right)P_{\lambda|u}(\lambda,t\,|\,u,t) =$$

$$= -\nabla\cdot\left(\frac{du}{dt}W(u,\lambda,t)\right) \tag{IV-83}$$

Combining the two results, the rate of change of the joint distribution is equal to:

$$\frac{\partial W}{\partial t} = -\nabla_u\cdot\left(\frac{du}{dt}W(u,\lambda,t)\right) + \Gamma_\lambda W(u,\lambda,t) \tag{IV-84}$$

If the time rate of change of **u**, namely $\mathbf{F} \equiv d\mathbf{u}/dt$ is known then it is possible to obtain an equation of motion for the observed variable $\langle \mathbf{u}\rangle$, which is an average over all possible values of that variable. The equation is derived as follows:

$$\frac{\partial}{\partial t}\langle u\rangle = \int du \frac{\partial W}{\partial t} u = \int du\Big(-u\nabla_u\cdot(F(u,t)\,W(u,\lambda,t)) + u\Gamma_\lambda W(u,\lambda,t)\Big) =$$

$$= -\Big(u\cdot(F\,W)\big| - \int du\,(\nabla_u\cdot u)(F\,W)\Big) + \Gamma_\lambda \int du\,u\,W(u,\lambda,t) = \langle F\rangle + \Gamma_\lambda\langle u\rangle \quad \text{(IV-85)}$$

In the second line integration by parts was used, and the last step is based on the assumption that at the limits of integration the product: $\mathbf{u}\cdot(\mathbf{FW})$ tends to zero. Thus the desired equation of motion is:

$$\frac{\partial}{\partial t}\langle u\rangle_\lambda = \langle\frac{du}{dt}\rangle + \Gamma_\lambda\langle u\rangle \quad \text{(IV-86)}$$

The first term on the right hand side is analogous to the $\mathbf{L_0}$ term with Poisson's brackets in Eqs. (IV-63),(IV-65). The second term replaces the $\mathbf{L_1}$ term in those equations, and it is the most prominent feature of the new equation. The operator Γ_λ represents the influence of the stochastic modulations on the behaviour of the interesting physical variable. In order to solve the equation one needs a model for the time dependence of the related probability distribution (Eq. (IV-79)), and then the equation can be handled either with a cumulants expansion or otherwise. As will be seen later on, a very useful approach is to expand every part of the equation in eigenfunctions of the stochastic operator Γ_λ , if such eigenfunctions can be found.

Example IV.3: A spin in a randomly changing magnetic field

Suppose a magnetic moment (a "spin") is under the influence of two kinds of magnetic fields. On the one hand, there is a constant external magnetic field of magnitude B_0, and its direction is choen as the **z**-axis of the laboratory frame of reference. On the other hand, there is a local field $B^p(t)$ due to other magnetic moments in the neighborhood of the relevant spin, and this local field changes with time in a random manner due to molecular motions. One may define the following Larmor frequencies:

$$\Omega_x = \gamma B^p{}_x \qquad \Omega_y = \gamma B^p{}_y \qquad \Omega_z = \gamma B^p{}_z + \gamma B_0 = \gamma B^p{}_z + \omega_0 \quad \text{(E-1)}$$

From the standard equations for a magnetic moment precessing in a magnetic field, which are the same as the Bloch equations without the relaxation terms, one obtains for the magnetization vector **m**:

$$\frac{d}{dt} m_x = \Omega_y m_z - \Omega_z m_y \tag{E-2a}$$

$$\frac{d}{dt} m_y = \Omega_z m_x - \Omega_x m_z \tag{E-2b}$$

$$\frac{d}{dt} m_z = \Omega_x m_y - \Omega_y m_x \tag{E-2c}$$

These equations can be summarized by the vector equation:

$$\frac{d}{dt} \boldsymbol{m} = \left(\boldsymbol{\Omega}_0 + \Omega^p(t)\right) \times \boldsymbol{m} \tag{E-3}$$

where $\boldsymbol{\Omega}_0 \equiv (0, 0, \omega_0)$ and $\boldsymbol{\Omega}^p(t) = (\Omega_x, \Omega_y, \Omega_z - \omega_0)$. In order to apply Eq. (IV-84), one can employ the general vector identity

$$-\boldsymbol{A} \cdot (\phi \boldsymbol{\nabla} \times \boldsymbol{r} - \boldsymbol{r} \times \boldsymbol{\nabla} \phi) = \boldsymbol{A} \cdot (\boldsymbol{r} \times \boldsymbol{\nabla} \phi) \tag{E-4}$$

for a scalar function ϕ and vectors **A**, **r**, where the differentiation is according to **r**, and A is independent of **r**. The stochastic equation for the probability distribution now becomes

$$\frac{\partial}{\partial t} W(\boldsymbol{m},\lambda,t) = -\left(\boldsymbol{\Omega}_0 + \Omega^p(t)\right) \times \boldsymbol{m} \frac{\partial}{\partial \boldsymbol{m}} W(\boldsymbol{m},\lambda,t) + \boldsymbol{\Gamma} W(\boldsymbol{m},\lambda,t) \tag{E-5}$$

In practice one can only measure the average over this magnetization vector:

$$\boldsymbol{M} \equiv \langle \boldsymbol{m} \rangle = \int W(\boldsymbol{m},\lambda,t)\, \boldsymbol{m}\, d\boldsymbol{m} \tag{E-6}$$

This average magnetization evolves in time according to

$$\frac{d}{dt} \boldsymbol{M} = \left(\boldsymbol{\Omega}_0 + \Omega^p(t)\right) \times \boldsymbol{M} + \boldsymbol{\Gamma}_\lambda \boldsymbol{M} \tag{E-7}$$

Thus the averaging process modifies the equation of motion by adding a relaxation term, which is effectively a line width term.

****

(b). Derivation of the equation in quantum mechanics

In the relaxation function approach the SLE was derived quantum mechanically in a manner completely analogous to that of the classical derivation, the main difference being that the classical Poisson's brackets were replaced by the quantum commutator brackets. In the distribution function approach one can also repeat the classical derivation *mutatis mutandis* for the quantum mechanical case, but there is an additional difficulty in the quantum treatment. In the classical treatment it was assumed that the relevant physical variable **u** has a value, or (for a vector) a finite set of values, at time t. One could thus differentiate according to **u** or, in the multi-dimensional case, calculate the gradient of a function according to **u**. In quantum mechanics one deals with a wave function or a density matrix, which have an infinite number of values in the general case. The concept of a derivative then needs to be generalized, to the concept of a **functional derivative** (or **variational derivative**). This concept will be defined here, and then it will be used in the derivation of the SLE in the quantum case.

In many physical contexts one encounters quantities which depend not on a single value of a variable, but on an infinite set of values - on a whole range of values of a function. A very simple example is the integral in a particular domain of some three dimensional function:

$$I = \int_a^b dx \int_c^d dy \int_e^f dz \, \phi(x,y,z) \tag{IV-87}$$

The value of the integral obviously depends on the value of the function $\phi(x,y,z)$ at all points (x,y,z) within the specified region of integration. A more interesting example is the expression

$$\xi(x,y,z;m) = H\,\psi(x,y,z;m) \tag{IV-88}$$

where **H** is a quantum mechanical Hamiltonian operator, and $\psi(x,y,z;m)$ is a wave function which depends on three space coordinates and on one spin "coordinate". Since the equation is defined for all values of x,y,z and m, it defines the function ξ over the whole range of values of these variables. Changing the functional form of ψ will necessarily change ξ. In general, there are situations in which a quantity $L(\psi,\partial\psi/\partial x,...)$ is defined, which depends linearly on a function and possibly also on its derivatives. In such cases it may be interesting to find out the change in L when the function ψ is changed as a whole. This is the kind of change which is measured by a **variation**, or by a functional derivative. It generalizes the concept of an ordinary derivative, which measures the change in a certain quantity when its single argument changes. The analog of a differential of an ordinary function is the variation

$$\delta L \equiv L\left(\psi + \delta\psi, \frac{\partial\psi}{\partial x} + \delta\frac{\partial\psi}{\partial x}, ...\right) - L\left(\psi, \frac{\partial\psi}{\partial x}, ...\right) \tag{IV-89}$$

In analogy to the derivative, which measures the rate of change of the differential, the functional derivative measures the rate of change of the variation. The variation can be expressed in terms of such derivatives:

$$\delta L = \frac{\delta L}{\delta\psi}\delta\psi + \frac{\delta L}{\delta\left(\frac{\partial\psi}{\partial x}\right)}\delta\left(\frac{\partial\psi}{\partial x}\right) + \ldots \tag{IV-90}$$

where

$$\delta\left(\frac{\partial\psi}{\partial x}\right) = \frac{\partial}{\partial x}(\psi + \delta\psi) - \frac{\partial}{\partial x}(\psi) = \frac{\partial}{\partial x}(\delta\psi) \tag{IV-91}$$

For example, if $L = H\psi = -(\hbar^2/2m)\nabla^2\psi + V\psi$, then

$$\delta L = -\frac{\hbar^2}{2m}\delta\left(\nabla^2\psi\right) + V\delta\psi \tag{IV-92}$$

With these definitions it is possible to construct an SLE from Schrödinger's equation. Starting from the equation

$$\frac{\partial\psi}{\partial t} = \frac{1}{i\hbar}H\psi \tag{IV-93}$$

the time derivative of the average wave function can be calculated from

$$\frac{\partial}{\partial t}\langle\psi\rangle_\lambda = \int\frac{\partial W}{\partial t}\psi\,d\psi \tag{IV-94}$$

where the time derivative of the joint distribution function is given by

$$\frac{\partial}{\partial t}W(\psi,\lambda,t) = -\frac{\delta}{\delta\psi}\left(\frac{\partial\psi}{\partial t}W(\psi,\lambda,t)\right) + \mathbf{\Gamma}_\lambda W(\psi,\lambda,t) \tag{IV-95}$$

Thus the average wave function evolves according to the equation

$$\frac{\partial}{\partial t}\langle\psi\rangle_\lambda = \int d\psi\left(-\frac{1}{i\hbar}\psi\frac{\delta}{\delta\psi}(H\psi W) + \psi\mathbf{\Gamma}_\lambda W\right) =$$

$$= -\frac{1}{i\hbar}\left(\psi(H\psi W)\,| - \int d\psi\frac{\delta\psi}{\delta\psi}(H\psi W)\right) + \mathbf{\Gamma}_\lambda\int d\psi\,\psi W =$$

$$= \frac{1}{i\hbar}\int d\psi(H\psi W) + \mathbf{\Gamma}_\lambda\langle\psi\rangle = \frac{1}{i\hbar}H\langle\psi\rangle + \mathbf{\Gamma}\langle\psi\rangle \tag{IV-96}$$

This result for the average wave function is completely analogous to Eq. (IV-86) obtained above for a classical physical variable. In a similar manner one can start from the Liouville - von Neumann equation

$$\frac{d}{dt}\rho = \frac{1}{i\hbar}[H,\rho] \tag{IV-97}$$

and use the appropriate functional derivative $\delta/\delta\rho$, which is a sum over the functional derivatives related to the individual matrix elements. The result is Kubo's SLE for the quantum mechanical density matrix

$$\frac{\partial}{\partial t}\langle\boldsymbol{\rho}\rangle = \frac{1}{i\hbar}[\boldsymbol{H},\langle\boldsymbol{\rho}\rangle] + \boldsymbol{\Gamma}\langle\boldsymbol{\rho}\rangle \tag{IV-98}$$

This is the version of the stochastic Liouville equation with which we shall be concerned in the rest of this book. It is similar in form to the Bloch-Redfield perturbative relaxation equation (Eq. (II-99)), although the averages in the two equations are not calculated over the same variables. The stochastic operator $\boldsymbol{\Gamma}$ appearing in the SLE is directly related to the stochastic variable λ, unlike the Bloch-Redfield operator which results from a sequence of operations related to the stochastic process. Moreover, the equations differ not just in the computational steps needed to apply them, but in their domain of applicability. The SLE applies equally well to fast and slow random processes, whereas the perturbative equation applies only for random motion which is fast relative to the inverse line width of the spectrum. As for the averaging appearing in this equation, the average sign will be omitted from here on, with the understanding that the equation refers only to the suitably averaged density matrix.

Example IV.4: *Modulation of one spin by another spin*

This example is in fact a special case of the problem considered in Example IV.3. However, here it is treated quantum mechanically, so frequency terms are imaginary (because of the $1/i\hbar$ factor in the Liouville - von Neumann equation), and the relaxation term remains real.

Suppose a magnetic resonance experiment is carried out on a system with two spin species, for which the spin operators are denoted by **I**, **S** respectively. The spins are assumed to interact with a scalar interaction: $J\mathbf{I}\cdot\mathbf{S}$, and spin S is assumed to have a fast relaxation mechanism, not shared by spin I. For example, two nuclear spin I and S can be coupled in this manner, and spin S may have fast relaxation due to a quadrupole interaction on that nucleus, whereas spin I does not have such an interaction. If the magnetic field is scanned in a range in which the irradiation frequency ω is close to the resonance (Larmor) frequency of spin I and far from that of spin S, the Hamiltonian in the rotating frame is

$$H = (\omega_I - \omega)I_z + (\omega_S - \omega)S_z + J\boldsymbol{I}\cdot\boldsymbol{S} \approx (\omega_I - \omega)I_z + (\omega_S - \omega)S_z + JI_zS_z \tag{E-1}$$

The approximate equality on the right results from first order perturbation theory. The term far from resonance has a negligible effect, so the experiment is effectively described by

$$H \approx \left(\omega_I - \omega + J S_z\right) I_z \tag{E-2}$$

If spin S is in a certain quantum state $|S,m_S\rangle$ then the effective resonance frequency of spin I is

$$\omega_{eff} = \omega_I + J m_S \tag{E-3}$$

The additional interaction causes spin S to jump "randomly" (from the viewpoint of spin I) between (2S + 1) states having different values of m_S, so the effective resonance frequency of spin I jumps randomly between the different values allowed to it by Eq. (E-3). As usual, one measures the transverse magnetization $M_+ = M_x + iM_y = \langle I_x + iI_y \rangle = \mathrm{Tr}\{(I_x + iI_y)\rho\}$, but this has to be found for each of the (2S + 1) states. The values of M_+ for these states are arranged in a vector, denoted by **x**. For each state the equation of motion for the density matrix is very simple in the spin basis, because **H** is diagonal in that basis. If no time dependent irradiation is applied, these magnetizations change in time according to the Bloch equations, with the addition of a stochastic term. The magnetization vector satisfies the equation

$$\frac{d}{dt}\boldsymbol{x} = i\boldsymbol{\Omega}\boldsymbol{x} + \boldsymbol{\Gamma}\boldsymbol{x} \tag{E-4}$$

where $\boldsymbol{\Omega}$ and $\boldsymbol{\Gamma}$ are matrices. In the simplest case there are only two different states. This may occur if S = ½ , but then spin S cannot have a quadrupole moment so it must have some other relaxation mechanism. In such a situation the equation becomes:

$$\frac{d}{dt}\begin{pmatrix} x_1 \\ x_2 \end{pmatrix} = i\begin{pmatrix} \omega_1 & 0 \\ 0 & \omega_2 \end{pmatrix}\begin{pmatrix} x_1 \\ x_2 \end{pmatrix} + \frac{\gamma}{2}\begin{pmatrix} -1 & 1 \\ 1 & -1 \end{pmatrix}\begin{pmatrix} x_1 \\ x_2 \end{pmatrix} \tag{E-5}$$

It is easiest to understand the equation for the two extreme cases, in which only one of the two terms on the right-hand side appears. If there are no "jumps", $\gamma = 0$ and then one has two separate equations:

$$\frac{d}{dt}x_j = i\omega_j x_j \qquad (j = 1,2) \tag{E-6}$$

These equations, formally identical to the equations of two uncoupled simple harmonic oscillators, describe two free spins precessing in the external magnetic field with (possibly) different Larmor frequencies. In the opposite case, one assumes $\gamma \neq 0$ and $\omega_1 = \omega_2 = 0$. The essential point in the second assumption is the equality of the two frequencies; equating them to zero is done in practice by transforming to an interaction representation (rotating frame). Then one has simple rate equations, with a source term and a sink term in each equation, and with a jump rate of $k = \gamma/2$:

$$\frac{d}{dt}x_1 = kx_2 - kx_1 \qquad \frac{d}{dt}x_2 = kx_1 - kx_2 \tag{E-7}$$

These equations describe a process of exchange between two magnetic sites. In fact, when both parts of the equation (the precession and the exchange) are present, Eq. (E-5) is just a different form of the Bloch-McConnell equations for chemical exchange.

For the general case the equation will be solved in terms of the eigenvectors of the matrix $\mathbf{\Gamma}$. Diagonalizing the matrix one finds that its eigenvalues are

$$\kappa_1 = 0 \qquad \kappa_2 = -\gamma \tag{E-8}$$

with the corresponding eigenvectors

$$v_1 = \frac{1}{\sqrt{2}}\begin{pmatrix}1\\1\end{pmatrix} \qquad v_2 = \frac{1}{\sqrt{2}}\begin{pmatrix}1\\-1\end{pmatrix} \tag{E-9}$$

Then instead of using the standard basis for two dimensional vector space one could use the basis $\{\mathbf{v}_1, \mathbf{v}_2\}$. The eigenvector $\mathbf{v}_1$ corresoponds to equal magnetization in the two states, and therefore to a situation in which there is no net transfer of magnetization between the two states. This is the physical reason that the corresponding eigenvalue is $\kappa_1 = 0$, zero rate of magnetization transfer. Also in more general problems, the equilibrium state is represented by an eigenvector of $\mathbf{\Gamma}$ with the eigenvalue 0. The other eigenvector corresponds to a state of maximum difference between the magnetizations at the two states, and thus to a transfer of magnetization between them, represented by the non-zero rate $\kappa_2 = -\gamma$.

The present problem is simple enough that it is not necessary to transform explicitly to this basis. The solution of Eq. (E-4) is:

$$\boldsymbol{x}(t) = \exp\left\{\left(i\,\mathbf{\Omega} + \mathbf{\Gamma}\right)t\right\}\boldsymbol{x}(0) \tag{E-10}$$

which is calculated by diagonalizing the matrix in the exponent. Its eigenvalues are, assuming for convenience $\omega_2 = -\omega_1$:

$$\xi_{1,2} = -k \pm \left(k^2 - \omega_1^2\right)^{1/2} \equiv -k \pm r \qquad \left(k \equiv \frac{\gamma}{2}\right) \tag{E-11}$$

with corresponding eigenvectors

$$w_1 = \frac{1}{\sqrt{2k}}\begin{pmatrix}-k\\ i\omega_1 - r\end{pmatrix} \qquad w_2 = \frac{1}{\sqrt{2k}}\begin{pmatrix}-k\\ i\omega_1 + r\end{pmatrix} \tag{E-12}$$

The solution in Eq. (E-10) can now be written explicitly as done in Chapter II, subsection II.2.(b).

****

(c) *Modification of the equation for a correct equilibrium solution*

In the version of the SLE obtained above the emphasis was on the process of decay from some initial state, by the gradual loss of coherence. A complete description of the decay requires attention also to the final result of decay, which is normally assumed to be an equilibrium state. However, in the form of the equation derived above the decay is to an artificial equilibrium state, in which all possible states of the system are equally populated. In other words, all energy levels are equally populated in the final state. If it is a standard equilibrium state, in which the relative populations are given by the Boltzmann weights $\exp\{-E/k_BT\}$, this implies an infinite temperature. In fact, this is exactly the problem encountered in Chapter II in connection with the perturbative relaxation equation obtained by a semi-classical derivation. Using quantum mechanical arguments it is possible to show that the correct equilibrium state is:

$$\rho_0 = \frac{\exp\{-H(\Omega)/k_BT\}}{Tr\left(\exp\{-H(\Omega)/k_BT\}\right)} \approx p_0(\Omega)\left\{I - \frac{H(\Omega)}{k_BT} + \ldots\right\} \qquad \text{(IV-99)}$$

where Ω is the stochastic variable on which the Hamiltonian depends, and p_0 is the equilibrium distribution of this variable. It is further possible to show that the SLE obtained above should be modified as:

$$\frac{\partial}{\partial t}\boldsymbol{\rho}(\Omega,t) = \frac{1}{i\hbar}[H(\Omega,t),\boldsymbol{\rho}(\Omega,t)] + \boldsymbol{\Gamma}\left(\boldsymbol{\rho}(\Omega,t) - \boldsymbol{\rho}_0(\Omega,t)\right) \qquad \text{(IV-100)}$$

If the Hamiltonian is time dependent, also the equilibrium state is in principle time dependent.

Using the modified SLE one may re-derive the line shape expression, or the expression for the magnetic susceptibility, for the case of negligible saturation (weak irradiation). It is found to be the same as the standard line shape expression for this case, obtained without the modification term. However, the value of the density matrix which is used in that calculation, does depend on the new term through Eq. (IV-100).

An important question is the importance of the dependence of ρ_0 on the stochastic variable. In typical experiments in magnetic resonance, the magnetic field is high, i.e., the magnitude of its interaction with the spins is large relative to the magnitude of interactions within the system. Thus the terms in the Hamiltonian which depend on the stochastic variables are small relative to the Zeeman term, which is constant and independent of the stochastic variables. Under such circumstances, the effect of the stochastic variable on Eq. (IV-100) is quite small, and the ρ_0 term is negligible. However, if the stochastic terms are not very small relative to the Zeeman term, one has to use the equation as it is. The latter situation occurs, for example, in X-band EPR of triplet states, where the Zeeman term is typically larger by only one order of magnitude than the electron-electron dipolar interaction (ZFS, zero field splitting intercation). The ZFS term depends on the orientation of the molecule relative to the external field, which changes randomly with time in a non-crystalline environment. In that case one needs to use Eq. (IV-100) with the equilibrium term.

Finally it should be noted that using eigenfunctions of the relaxation operator $\mathbf{\Gamma}$ one may construct from the modified SLE an approximate theory for fast random motions. The approximation is based on the assumption of rapid stochastic motion, as in the derivation of the perturbative relaxation theory. The result is effectively the same as the Bloch-Redfield result for such a case, which is appropriate since for fast motions the perturbative theory applies.

Suggested References

* On the general treatment of random motions and their spectra:

1. L.E. Reichl, *"A Modern Course in Statistical Physics"* (University of Texas Press, Austin, TX 1980).
2. L. Landau and E.M. Lifshitz, *"Statistical Physics"* , 3rd edition, part 1, revised by E.M. Lifshitz and L.P. Pitaevski (Pergamon Press, Oxford, 1980).

* On Markov processes:

3. W. Feller, *"An Introduction to Probability Theory and Its Applications"*, third edition, revised printing (Wiley, New York,1970), Vol. I, Ch. XVII.
4. N.G. van Kampen, *"Stochastic Processes in Physics and Chemistry"* (North Holland, Amsterdam, 1981), Chs. IV, V and VIII.

* On random modulation of paramagnetic systems and the effect on the line shape:

5. P.W. Anderson, *J. Phys. Soc. Japan* **9**, 316 (1954).
6. R. Kubo, *J. Phys. Soc. Japan* **9**, 935 (1954).
7. R. Kubo in *"Fluctuation, Relaxation and Resonance in Magnetic Systems"*, edited by D. ter Haar (Oliver and Boyd, Edinburgh, 1962), p. 23.
8. R. Kubo, *J. Math. Phys.* **4**, 174 (1962).
9. R. Kubo in *"Stochastic Processes in Chemical Physics" - Adv. Chem. Phys.* vol. **XV**, edited by K.E. Shuler (1969), p. 101.
10. R. Kubo, *J. Phys. Soc. Japan*, vol. **26,** Supplement (1969), p. 1.
11. R. Kubo, *Nuovo Cimento* **6** , Supplement (1957), p. 1063; M. Suzuki and R. Kubo, *Mol. Phys.* **7**, 201 (1964).

* On functional (or variational) derivatives:

12. R. Courant and D. Hilbert, *"Methods of Mathematical Physics"* (Wiley-Interscience, New York, 1966), Ch. IV.
13. H. Goldstein, *Classical Mechanics* , second edition (Addison-Wesley, New York, 1980), Ch. 2.
14. J.D. Bjorken and S.D. Drell, *Relativistic Quantum Fields* (McGraw-Hill, New York, 1965), Ch. 11.

* On the modification of the SLE for a correct equilibrium:

18. A.J. Vega and D. Fiat, *J. Chem. Phys.* **60**, 579 (1974); A.J. Vega and D. Fiat, *J. Mag. Reson.* **13**, 260 (1974).
19. A.J. Vega and D. Fiat, *J. Mag. Reson.* **19**, 21 (1975).

CHAPTER V

METHODS FOR SOLVING THE STOCHASTIC LIOUVILLE EQUATION

V.1 The Method of Cumulants

(a) General form of the solution

In the previous chapter the Liouville - von Neumann equation for the density matrix was written as (Eq. (IV-72))

$$\frac{\partial \rho}{\partial t} = \frac{1}{i\hbar}[H_0 + V(t), \rho] \equiv (L_0 + L_1(t))\,\rho \tag{V-1}$$

where V(t) is the stochastic perturbation to the system. It was then shown that transforming to the interaction representation one obtains the stochastic Liouville equation (SLE) in the following form

$$\frac{\partial \sigma}{\partial t} = \frac{1}{i\hbar}[V^{(I)}(t), \sigma] \equiv \Omega(t)\,\sigma(t) \tag{V-2}$$

where $V^{(I)} \equiv \exp\{iH_0t/\hbar\}\, V(t)\, \exp\{-iH_0t/\hbar\}$ is the perturbation in the interaction representation. This was formally integrated as:

$$\langle \sigma(t) \rangle = \left\langle T\left\{ \exp\left[\int_0^t \Omega(t')\,dt' \right] \right\} \right\rangle \sigma(0) \equiv \Phi(t)\,\sigma(0) \tag{V-3}$$

On the other hand, it was also shown there that using the distribution function approach, one obtains for the suitably averaged density matrix the folowing equation of motion (eq. (IV-98)):

$$\frac{\partial}{\partial t}\rho = \frac{1}{i\hbar}[H, \rho] + \Gamma\rho \equiv (L_0 + L_1)\,\rho \tag{V-4}$$

Therefore also in this approach one can transform to the interaction representation to get equations (V-2), (V-3).

Our purpose here is to construct analytically a systematic solution for the SLE, using a cumulant expansion of the relaxation operator Φ(t). This is a very general method of solution, applicable for any model for the random motion. It also makes it possible to solve for the operator up to any given order, and thus the expansion can be used not only for fast random motions (where very few terms would be needed), but also for slow random motions, where a relatively large number of terms is necessary. The basic equation for the cumulant expansion is (see Eq. (IV-68)):

$$\Phi(t) = \left\langle T\left\{ \exp\left[i\int_0^t \Omega(t')\,dt' \right]\right\}\right\rangle = \left\langle \left\{ I + \int_0^t dt_1\,\Omega(t_1) + \int_0^t dt_1 \int_0^{t_1} dt_2\,\Omega(t_1)\,\Omega(t_2) \right.\right.$$

$$\left.\left. + \ldots + \int_0^t dt_1 \int_0^{t_1} dt_2 \ldots \int_0^{t_{n-1}} dt_n\,\Omega(t_1)\,\Omega(t_2)\ldots\Omega(t_n) + \ldots \right\}\right\rangle =$$

$$= \left\{ I + \int_0^t dt_1 \langle \Omega(t_1)\rangle_c + \int_0^t dt_1 \int_0^{t_1} dt_2 \langle \Omega(t_1)\,\Omega(t_2)\rangle_c + \ldots \right.$$

$$\left. + \int_0^t dt_1 \int_0^{t_1} dt_2 \ldots \int_0^{t_{n-1}} dt_n \langle \Omega(t_1)\,\Omega(t_2)\ldots\Omega(t_n)\rangle_c + \ldots \right\} \qquad \text{(V-5)}$$

The subscript "c" indicates cumulant averages (see Eq. (IV-52)), which occur here because the expansion here is a direct generalization of the expansion in Eq. (IV-56). The integration limits for each time variable are chosen so that in the n'th term the following inequalities are satisfied:

$$t_n \le t_{n-1} \le \ldots \le t_2 \le t_1 \qquad \text{(V-6)}$$

which ensures that the proper time ordering is kept in each term in this expansion. Due to this time ordering, in the average $\langle\Omega(t_1)\ \Omega(t_2)\ \ldots\ \Omega(t_n)\rangle_c$ the time variables must be related by this set of inequalities. The ordinary relation between cumulant averages on the one hand, and moments (ordinary averages) on the other hand, is the same as for c-number variables (i.e., variables which are not operators) except that here the time ordering is imposed on all such formulas. Thus, for example:

$$\langle \Omega(t_1)\rangle_c = \langle \Omega(t_1)\rangle \qquad \text{(V-7a)}$$

$$\langle \Omega(t_1)\,\Omega(t_2)\rangle_c = T\langle \Omega(t_1)\,\Omega(t_2)\rangle - \langle \Omega(t_1)\rangle\langle \Omega(t_2)\rangle \qquad \text{(V-7b)}$$

$$\langle \Omega(t_1)\Omega(t_2)\Omega(t_3)\rangle_c = T\langle \Omega(t_1)\Omega(t_2)\Omega(t_3)\rangle - \langle \Omega(t_1)\rangle T\langle \Omega(t_2)\Omega(t_3)\rangle - \langle \Omega(t_2)\rangle T\langle \Omega(t_3)\Omega(t_1)\rangle$$

$$- \langle \Omega(t_3)\rangle T\langle \Omega(t_1)\Omega(t_2)\rangle + 2\langle \Omega(t_1)\rangle\langle \Omega(t_2)\rangle\langle \Omega(t_3)\rangle \qquad \text{(V-7c)}$$

where the time ordering operator **T** ensures that the time variables are related by the same inequalities as written above. It should be born in mind that these equations assume the existence of the relevant cumulants and moments. It is possible, as will be seen in Example V.1 below, that the moments are not defined but the cumulants are well defined.

The formalism presented here is thus a particular form of time dependent perturbation theory in quantum mechanics. It is therefore an alternative to the perturbation expansion used in Chapter II, which led to the Bloch-Redfield equation. The two methods are similar in accounting for the effect of a random perturbation by systematic approximation schemes, based on series expansions. Both the iterative perturbation expansion of Chapter II and the cumulant expansion involve in the n'th order term products of n factors, each factor being of the form "$\mathbf{H}_1$"/"$\mathbf{H}_0$" where "$\mathbf{H}_1$" is a matrix element of the stochastic time dependent perturbation and "$\mathbf{H}_0$" is some energy difference in the main, constant Hamiltonian. In practice one obviously works only with the first few orders in each expansion. It is therefore clear that both approaches rely on the assumption that the stochastic effect is not too strong relative to the constant interactions. This is reasonable, since the stochastic interaction usually represents local intra-molecular or inter-molecule effects. These are expected to be smaller than the Zeeman interaction with the external magnetic field in ordinary magnetic resonance experiments.

On the other hand, there is a very important difference between the dependence on the time variable t in the two expansions. In the iterative expansion, an integral from the initial time to t appears in the differential equation for the density matrix. Since such an integro-differential equation is very difficult to solve, one makes the assumption that the relevant times t are long compared with the correlation time τ_C of the random motion, so the upper integration limit can be taken to infinity. This means that the random process must be fast relative to the interesting periods of time over which the system is examined. Moreover, even if one eliminates this restriction by working with the density matrix (Eq. (II-87)) rather than with its derivative, one still needs the assumption of a fast random process. This is because the magnitude of n'th order terms in the expansion can be shown to be of the order of $\{\langle \mathbf{H}_1^2\rangle \tau_C^2\}^{n/2}$, so high order terms can only be neglected if the changes with time of the random interaction are fast compared with the inverse magnitude of that interaction. The simplicity of the Bloch-Redfield equation resulted from several approximations which were made specifically for a second order perturbation result, and going beyond the second order would make the problem much more complicated. The most significant limitation in that approximation was the assumption of fast random motions, an assumption which is often not satisfied in the context of EPR.

In the cumulant expansion, the upper integration limit does not have to be "infinite", so the random process does not have to be fast relative to the relevant time periods in the experiment. The magnitude of the n'th order term in the general case is of the order of the n'th order moment of the relaxation operator $\Omega(t)$ times t^n , where t is a typical time value, which is of the order of τ_C. Thus on the one hand longer correlation times will require taking more terms in the expansion, just as for the perturbation expansion. On the other hand, if the moments μ_n of the distribution decrease with the order n sufficiently fast, the cumulants will decrease in a similar manner. In that case one may not need many of the cumulants even if the correlation time is long, i.e., the stochastic process is slow. Therefore, even working only to second order in the cumulant expansion does not necessarily imply a restriction to the fast motional regime. The example of Gaussian random processes, to be treated in the next subsection, demonstrates this point most clearly.

A rigorous treatment of the convergence of the cumulant expansion is quite complicated and will not be given here. The result of such a treatment is that the expansion is formally valid for any rate of the stochastic motion, and in this respect is preferable to the perturbative relaxation equation. Nevertheless, in practice one would need many terms in the expansion for slow stochastic motion. Thus one is not restricted to the very fast motional regime as in the Bloch-Redfield approach, but the regime of very slow motions is still beyond the domain of applicability of the technique. For some types of stochastic processes, however, the cumulant expansion is applicable to any rate of the motion. These are processes for which all high-order terms in the cumulant expansion must be equal to zero. This occurs in the cases of Gaussian modulation, which will be discussed in the next subsection.

Example V.1: Cumulant expansion for a Lorentzian distribution function

The purpose of this example is *not* to show a typical case of a cumulant expansion. Rather, it is to point out the good convergence properties of the cumulant expansion even in a problematic case. Suppose a stochastic variable ω has a Lorentzian distribution function (see Fig. (V.1)):

$$p(\omega) = \frac{1}{\pi} \frac{\gamma}{\gamma^2 + (\omega - \omega_0)^2} \qquad \text{(E-1)}$$

The function is normalized, i.e.,

$$\int_{-\infty}^{\infty} p(\omega)\, d\omega = 1 \qquad \text{(E-2)}$$

It is well known that the moments of this function are not defined, because the integrals $\mu_n \equiv {}_{-\infty}\int^{\infty} \omega^n\, p(\omega)\, d\omega$ (for $n \geq 1$) diverge. For this reason one does not usually choose such a distribution function to start with (the divergence can be eliminated by choosing some cutoff frequency ω_{cutoff} , and making $p(\omega)$ equal to zero for $|\omega - \omega_0| > \omega_{cutoff}$, but this will not be discussed here). Nevertheless, the cumulant expansion can be carried out for this case, and results in very simple expressions.

The basis for the cumulant expansion here is the characteristic function $\phi(t)$ of Eq. (IV-7), which is the Fourier transform of $p(\omega)$. This is equal to:

$$\phi(t) = \exp\left(i\,\omega_0 t - \gamma\,|t|\right) \tag{E-3}$$

This is a continuous function of t, which is differentiable almost everywhere, except for the point $t = 0$. From the definition of the cumulant expansion (Eq. (IV-43)) there is only one non-zero cumulant, κ_1, but its value depends on the sign of t. For $t > 0$:

$$\kappa_1 = \omega_0 + i\,\gamma \tag{E-4}$$

whereas for $t < 0$:

$$\kappa_1 = \omega_0 - i\,\gamma \tag{E-5}$$

For $t = 0$ both are formally possible, so the cumulant is not well defined there. Thus the cumulant expansion depends on the sign of t, and is discontinuous at $t = 0$, but if one works only with $t > 0$ (or only with $t < 0$) it can be used in the ordinary way.

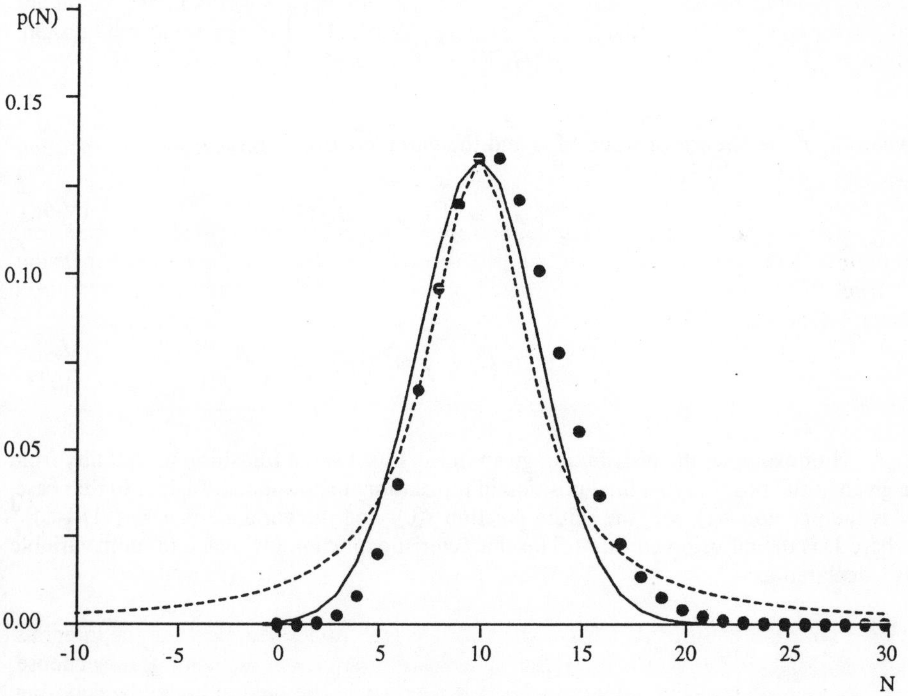

Fig. V.1. *A comparison between three types of distribution functions. The solid line represents a Gaussian distribution (see Eq. (V-8)) with $\mu = 10$ and $\sigma = 3$. The dashed line represents a Lorentzian distribution (Eq. (E-1)) with $\omega_0 = 10$ and $\gamma = 3$. The isolated points represent a (discrete) Poisson distribution (Eq. (V-35)), with $\langle N \rangle = 10$.*

The convergence of the cumulant expansion in a case in which moments diverge indicates that the cumulant expansion has better convergence properties. This can be seen directly when such an expansion is used in statistical mechanics or in many body perturbation theory, for which diagrammatic methods are used to sum over expansions of this kind. The cumulants are then related to "linked cluster" diagrams (rather than to arbitrary diagrams, which may be "linked" or "unlinked"), which have better convergence properties than ordinary diagrams.

****

(b) *Solving with the cumulant expansion for Gaussian modulation*

An important class of stochastic processes is that of Gaussian processes. In the simplest case, a single stochastic variable ω characterizes the behavior of the system. The variable can be, e.g., the oscillation frequency of a one dimensional oscillator, or the position of a particle undergoing one dimensional translational diffusion. A Gaussian (or: normal) distribution for that variable is defined by (see Fig. (V.1))

$$p(\omega) = \frac{1}{\sqrt{2\pi\sigma^2}} \exp\left(-\frac{(\omega-\mu)^2}{2\sigma^2} \right) \tag{V-8}$$

where μ, σ^2 are the mean value of ω and the variance, respectively:

$$\mu = \langle \omega \rangle = \int_{-\infty}^{\infty} \omega\, p(\omega)\, d\omega \tag{V-9a}$$

$$\sigma^2 = \langle \omega^2 \rangle - \langle \omega \rangle^2 = \int_{-\infty}^{\infty} \omega^2 p(\omega)\, d\omega - \mu^2 \tag{V-9b}$$

For example, the distribution given in Eq. (V-8) is the transition probability from a given initial position to a final position in translational Brownian diffusion. In that case, ω is the position x(t), μ is the initial position x(0), and the variance is $\sigma = \{2D(t-t_0)\}^{1/2}$ where D is the diffusion constant. The characteristic function for such a random variable is calculated as:

$$\phi(t) \equiv \int_{-\infty}^{\infty} \exp(i\omega t) p(\omega)\, d\omega = \frac{1}{\sqrt{2\pi\sigma^2}} \int_{-\infty}^{\infty} \exp\left(-\frac{(\omega-\mu)^2}{2\sigma^2} + i\omega t \right) d\omega =$$

$$= \frac{1}{\sqrt{2\pi\sigma^2}} \int_{-\infty}^{\infty} \exp\left(-\frac{(\omega-\mu-i\sigma^2 t)^2}{2\sigma^2} - \frac{\sigma^2 t^2}{2} + i\mu t \right) d\omega \qquad \text{(V-10)}$$

The terms independent of ω can be taken out of the integral, and the integration variable can be changed to $y=\omega/\sigma$. Then, defining $a = \mu/\sigma + i\sigma t$:

$$\phi(t) = \frac{\sigma}{\sqrt{2\pi\sigma^2}} \exp\left(i\mu t - \frac{\sigma^2 t^2}{2} \right) \int_{-\infty}^{\infty} \exp\left(-\frac{(y-a)^2}{2} \right) dy \qquad \text{(V-11)}$$

The remaining integral can be calculated by standard integration in the complex plane, and is equal to $(2\pi)^{1/2}$. Therefore the characteristic function for the Gausssian distribution is:

$$\phi(t) = \exp\left(i\mu t - \frac{\sigma^2 t^2}{2} \right) \qquad \text{(V-12)}$$

Comparing this result with the cumulant expansion of Eq. (IV-43), it is clear that only the first two cumulants are non-zero:

$$\langle \omega \rangle_c = \mu \quad ; \quad \langle \omega^2 \rangle_c = \sigma^2 \quad ; \quad \langle \omega^n \rangle_c = 0 \quad (n \geq 2) \qquad \text{(V-13)}$$

This result is equivalent to the well known fact, that a Gaussian distribution is completely characterized by the first two moments.

Suppose one is interested in a stochastic process represented by the variable ω, observed at times t_1 , t_2 , ..., t_n . The process is Gaussian if the probability distribution of the n observed values is:

$$p(\omega_1,t_1;\omega_2,t_2;\ldots;\omega_n,t_n) = C \exp\left(-\frac{1}{2} \sum_{j=1}^{n} \sum_{k=1}^{n} A_{jk}(\omega_j - \mu_j)(\omega_k - \mu_k) \right) \qquad \text{(V-14)}$$

where the average $\mu_j \equiv \langle \omega_j \rangle \equiv \langle \omega(t_j) \rangle$ is the expectation value of $\omega(t)$ at time t_j, and the average is calculated with the distribution function. The constant C is chosen as a normalization constant, such that

$$\int_{-\infty}^{\infty} d\omega_1 \int_{-\infty}^{\infty} d\omega_2 \ldots \int_{-\infty}^{\infty} d\omega_n\, p(\omega_1,t_1;\omega_2,t_2;\ldots;\omega_n,t_n) = 1 \qquad \text{(V-15)}$$

The coefficients A_{jk} are assumed to form a positive definite (and therefore necessarily symmetric) matrix **A**, to ensure that the probability for values very far from the expectation values tends to zero.

The characteristic functional for this case, which is equivalent to the relaxation function of Eq. (IV-48), is

$$\Phi(\zeta_1,\zeta_2,\ldots,\zeta_n) = \int_{-\infty}^{\infty} d\omega_1 \int_{-\infty}^{\infty} d\omega_2 \ldots \int_{-\infty}^{\infty} d\omega_n\, p(\omega_1,t_1;\omega_2,t_2;\ldots;\omega_n,t_n) \exp\left(i\sum_{j=1}^{n}\zeta_j\omega_j\right) \qquad (V\text{-}16)$$

It is convenient to define the vectors $\boldsymbol{\omega} = (\omega_1, \omega_2, \ldots, \omega_n)$, $\boldsymbol{\mu} = (\mu_1, \mu_2,\ldots, \mu_n)$, $\boldsymbol{\zeta} = (\zeta_1, \zeta_2,\ldots, \zeta_n)$ and $\mathbf{y} = \boldsymbol{\omega} - \boldsymbol{\mu}$. No distinction will be made here between row vectors and their transposed row vectors. It is also useful to define another vector: $\mathbf{v} = \mathbf{A}^{-1}\,\boldsymbol{\zeta}$ and then $\mathbf{u} = \mathbf{y} - i\mathbf{v}$. Thus $\mathbf{y} = \mathbf{u} + i\mathbf{v}$ is a combination of two vectors, both of which have in general complex numbers as elements. Using these definitions, the exponential in the integrand of Eq. (V-16) is:

$$\exp\left[-\frac{1}{2}yAy + i\zeta\omega\right] = \exp\left[-\frac{1}{2}yAy + i\zeta y + i\zeta\mu\right] =$$

$$= \exp\left[-\frac{1}{2}uAu - iuAv + \frac{1}{2}vAv + i\zeta u - \zeta v + i\zeta\mu\right] \qquad (V\text{-}17)$$

Now $\boldsymbol{\zeta}\mathbf{v} = \boldsymbol{\zeta}\,\mathbf{A}^{-1}\,\boldsymbol{\zeta}$, $\mathbf{v}\,\mathbf{A}\,\mathbf{v} = \mathbf{v}\,\boldsymbol{\zeta} = \boldsymbol{\zeta}\,\mathbf{v}$ and $\mathbf{u}\,\mathbf{A}\,\mathbf{v} = \mathbf{u}\,\boldsymbol{\zeta} = \boldsymbol{\zeta}\,\mathbf{u}$, so that the exponential is equal to

$$\exp\left[-\frac{1}{2}uAu + i\zeta\mu - \frac{1}{2}\zeta A^{-1}\zeta\right] \qquad (V\text{-}18)$$

The next step is changing the integration variable from $\boldsymbol{\omega}$ to $\mathbf{u} = \boldsymbol{\omega} - \boldsymbol{\mu} - i\mathbf{v}$. Since $\boldsymbol{\mu}$ is a constant vector, and due to the constancy of $\mathbf{A}$ and $\boldsymbol{\zeta}$ (which is a fixed parameter vector in this calculation) also $\mathbf{v}$ is constant, the integration limits are unchanged. The factors independent of $\mathbf{u}$ are taken out of the integral, so the remaining integral is simply

$$\int_{-\infty}^{\infty} du_1 \int_{-\infty}^{\infty} du_2 \ldots \int_{-\infty}^{\infty} du_n\, C \exp\left(-\frac{1}{2}uAu\right) = 1 \qquad (V\text{-}19)$$

due to the normalization of the distribution function. The result is

$$\Phi(\zeta_1,\zeta_2,\ldots,\zeta_n) = \exp\left(i\sum_{j=1}^{n}\mu_j\zeta_j - \frac{1}{2}\sum_{j=1}^{n}\sum_{k=1}^{n}(A^{-1})_{jk}\,\zeta_j\,\zeta_k\right) \qquad (V\text{-}20)$$

Generalized moments and cumulants are defined for this case by Eq. (IV-48). Since the highest order of $\boldsymbol{\zeta}$ which appears in the exponential of Eq. (V-20) is the second order, it is obvious that only the first order and second order cumulants are non-zero, just as for the case of a single variable with a Gaussian distribution (Eq. (V-13)). It is clear from Eqs. (V-20), (IV-48) that the first order cumulants are $\mu_j = \langle\omega(t_j)\rangle$ and the second order cumulants are $(\mathbf{A}^{-1})_{jk}$. On the other hand, from the standard expression for second order cumulants, Eq. (IV-52b):

$$\langle \omega(t_j)\,\omega(t_k)\rangle_c = \langle \omega(t_j)\,\omega(t_k)\rangle - \langle \omega(t_j)\rangle\langle \omega(t_k)\rangle = \langle \omega(t_j)\,\omega(t_k)\rangle - \mu_j\,\mu_k =$$

$$= \langle\, \big(\omega(t_j) - \mu_j\big)\big(\omega(t_k) - \mu_k\big)\,\rangle \tag{V-21}$$

which are just the correlation functions, or elements of the variance matrix of the variables ω_j . Therefore

$$\big(A^{-1}\big)_{j,k} = \langle\, \big(\omega(t_j) - \mu_j\big)\big(\omega(t_k) - \mu_k\big)\,\rangle \tag{V-22}$$

is the variance matrix, and the characteristic function is completely determined by the expectation values (which are the first order cumulants) and the correlation functions (which are the second order cumulants) of the Gaussian random variables ω_j .

In more general cases a given stochastic process may not be strictly Gaussian, but if it is a sum of many independent stochastic variables, the central limit theorem of statistics implies that the sum will behave approximately as a Gaussian. Therefore the above considerations apply both to processes which are modeled directly by a Gaussian distribution, and to processes which result from the sum of many small stochastic perturbations, and are thus Gaussian for all practical purposes.

In the previous chapter the relaxation function was examined as a function of the modulation rate, without specifying the nature of the stochastic process. For the first two cumulants it was shown explicitly that for relatively short correlation times τ_C one obtains a mono-exponential decaying oscillation, which gives a Lorentzian line upon Fourier transformation. The condition for fast modulation was: $t \gg \tau_C$, where t is a typical time for which a measurement is made. The decay constant was $\Delta^2\,\tau_C$, namely the signal decayed as $\exp\{-t(\Delta^2\,\tau_c)\}$, in which Δ is the modulation amplitude (Eq. (IV-14)), or (from Eq. (V-9)): $\Delta^2 = \sigma^2 + \mu^2$. Taking for simplicity the case in which the Gaussian distribution is centered around $\omega = 0$, one has $\mu = 0$ so that Δ is simply the variance of the Gaussian distribution. Thus Δ is a measure of the width of the distribution, and in order to cover this width reasonably one has to measure with a time resolution which is better (shorter time intervals) than approximately $1/\Delta$. Thus the condition for fast motion is: $\Delta\,\tau_C \ll 1$ (Eq. (IV-17)), and when it holds, the decay constant $\Delta^2\,\tau_C$ is very small even compared with Δ. This constant is the inverse of the characteristic time of decay, so a small value of the constant is equivalent to a long relaxation time $T_R \equiv 1/(\Delta^2\,\tau_C)$, or a narrow spectral line (see Eq. (II-7) and Fig. II.2). The line shape in this case is Lorentzian:

$$I(\omega - \omega_0) = \frac{1}{\pi}\,\frac{\dfrac{1}{T_R}}{(\omega - \omega_0)^2 + \left(\dfrac{1}{T_R}\right)^2} \tag{V-23}$$

and $1/T_R = \Delta^2\,\tau_C$ is the half-width of the line. Summarizing this point, a Gaussian stochastic process leads in the fast modulation time to a situation of strong narrowing.

This is the motional narrowing limit, in which also the Bloch-Redfield theory applies.

In the opposite case of relatively long correlation times it was shown in the previous chapter that the line shape in the frequency domain is very similar to the distribution function of the stochastic variable. If this distribution is Gaussian, one thus has a Gaussian line shape for the case of slow modulation, or:

$$I(\omega - \omega_0) = \frac{1}{\sqrt{2\pi\Delta^2}} \exp\left(- \frac{(\omega - \omega_0)^2}{2\Delta^2} \right) \qquad \text{(V-24)}$$

The half width in this case is clearly of the order of the modulation amplitude Δ (more precisely, the half width at half the maximum height is equal here to $\Delta(2\ln(2))^{1/2}$).

It is seen that indeed the line width for fast stochastic motion is much smaller than the width for slow stochastic motion, which justifies the name of "motional narrowing regime" given in NMR for the range of values in which the modulation is relatively fast. A similar phenomenon occurs in EPR, except that there the physical cause is often a strong exchange interaction between ions which are very close to each other. Thus the name "exchange narrowing" is used in that context.

(c) *Cumulant expansion for a Poisson process*

In some physical situations a system is free most of the time, but ocasionally it suffers strong perturbations. This, for example, is the case in radioactive decay, where a Geiger counter will not register anything most of the time, but occasionally a decay event will appear. This is also the case in pressure broadening of spectral lines, where occasional perturbations of a molecule by other molecules in a gas will cause a change in its spectrum. Denote by N the total number of such random events within a given time interval, and by $\tau_1, \tau_2, ..., \tau_N$ the instants in which these events take place. Since N is not determined in advance, one can only make statistical predictions about it. The probability distribution for having no events (even before or after that interval) is denoted as Q_0 ; the probability distribution for having one event at time τ_1 (not necessarily contained in the above-mentioned interval) is denoted as $Q_1(\tau_1)$; the probability distribution for having one event at time τ_1 and another event at time τ_2 is denoted as $Q_2(\tau_1,\tau_2)$, etc. These probability functions are normalized that the (infinite) sum over the integrals of all $Q_i(...)$ (each one being integrated over all its arguments) is equal to 1:

$$Q_0 + \int_{-\infty}^{\infty} d\tau_1\, Q_1(\tau_1) + \int_{-\infty}^{\infty} d\tau_1 \int_{\tau_1}^{\infty} d\tau_2\, Q_2(\tau_1,\tau_2) + \ldots = 1 \qquad \text{(V-25)}$$

This requirement means that if one sums the total probability that either no event will occur, or one event will occur at some time, or two events will occur, etc. then the total probability of having any of these situations is equal to unity. If the random events are independent (and are all of the same kind), one has for any value of k:

$$Q_k = Q_0\, q(\tau_1)\, q(\tau_2) \ldots q(\tau_k) \tag{V-26}$$

In such a case the process is called a **Poisson process.** In the integrations in Eq. (V-25), the integration region for Q_k is 1/k! of the full k-dimensional space (cf. Eq. (IV-55)). With Q_k factorized as in Eq. (V-26), the normalization condition is equivalent to

$$1 = Q_0 \left\{ 1 + A + \frac{1}{2!}A^2 + \ldots + \frac{1}{k!}A^k + \ldots \right\} = Q_0 \exp(A) \tag{V-27}$$

where $A \equiv {}_{-\infty}\int^{\infty} q(\tau)\, d\tau$. Therefore $Q_0 = \exp(-A)$. Now suppose the total number of events taking place at any time ($-\infty < \tau_j < \infty$ for all j) is: k. Then the number N of events in the interval $a \le t \le b$ is given by

$$N_k = \sum_{j=1}^{k} \chi(\tau_j) \tag{V-28}$$

where $\chi(\tau_j) = 1$ if: $a < \tau_j < b$, and $\chi(\tau_j) = 0$ otherwise. The average value of N for the interval $a \le t \le b$ is calculated by averaging over all possible cases:

$$\langle N \rangle = \sum_{k=1}^{\infty} \langle N_k \rangle = \sum_{k=1}^{\infty} \frac{1}{k!} \int_{-\infty}^{\infty} d\tau_1 \int_{-\infty}^{\infty} d\tau_2 \ldots \int_{-\infty}^{\infty} d\tau_k \sum_{j=1}^{k} \chi(\tau_j)\, Q_k(\tau_1, \tau_2, \ldots, \tau_k) \tag{V-29}$$

In this expression all the τ_j appear on an equal footing, so

$$\langle \sum_{j=1}^{k} \chi(\tau_j) \rangle = k \langle \chi(\tau_1) \rangle \tag{V-30}$$

Consequently

$$\langle N \rangle = \sum_{k=1}^{\infty} \frac{1}{(k-1)!} \int_a^b d\tau_1\, \chi(\tau_1) \int_{-\infty}^{\infty} d\tau_2 \ldots \int_{-\infty}^{\infty} d\tau_k\, Q_k(\tau_1, \tau_2, \ldots, \tau_k) = \int_a^b q(\tau)\, d\tau \tag{V-31}$$

The last equality follows from the normalization condition (the dummy index $k' \equiv k-1$ here corresponds to k in Eq. (V-29)). It is also possible to calculate the characteristic function in this case, from which one finds both the cumulants and the probability distribution of N. The characteristic function, which in this case is the same as the relaxation function, is

$$\langle e^{i\xi N}\rangle = \langle \exp\left(i\xi\sum_{j=1}^{k}\chi(\tau_j)\right)\rangle = e^{-A}\sum_{k=0}^{\infty}\frac{1}{k!}\int_{-\infty}^{\infty}d\tau_1 e^{i\xi\chi(\tau_1)}q(\tau_1)\dots\int_{-\infty}^{\infty}d\tau_k e^{i\xi\chi(\tau_k)} =$$

$$= \exp\left(-\int_{-\infty}^{\infty}q(\tau)\,d\tau\right)\sum_{k=0}^{\infty}\frac{1}{k!}\left(\int_{-\infty}^{\infty}e^{i\xi\chi(\tau)}q(\tau)\,d\tau\right)^k = \exp\left(\int_{-\infty}^{\infty}\left(e^{i\xi\chi(\tau)}-1\right)q(\tau)\,d\tau\right) =$$

$$= \exp\left(\left(e^{i\xi}-1\right)\int_a^b q(\tau)\,d\tau\right) = \exp\left(\left(e^{i\xi}-1\right)\langle N\rangle\right) \tag{V-32}$$

Therefore:

$$\langle e^{i\xi N}\rangle = e^{-\langle N\rangle}\sum_{N=0}^{\infty}\frac{\langle N\rangle^N}{N!}e^{i\xi N} \tag{V-33}$$

On the other hand, if one defines a probability distribution for N as p(N) then

$$\langle e^{i\xi N}\rangle = \sum_{N=0}^{\infty}p(N)\,e^{i\xi N} \tag{V-34}$$

Thus the distribution of N is equal to

$$p(N) = \frac{\langle N\rangle^N}{N!}e^{-\langle N\rangle} \tag{V-35}$$

According to Eq. (V-32), the cumulants are found from:

$$K(\xi) = \left(e^{i\xi}-1\right)\langle N\rangle = \sum_{m=1}^{\infty}\frac{(i\xi)^m}{m!}\kappa_m \tag{V-36}$$

It is clear that

$$\kappa_m = \langle N\rangle \qquad (m = 1,2,\dots) \tag{V-37}$$

for all m.

A physical realization of this model exists, for example, in the pressure broadening problem. If N molecules of an ideal gas are contained in a volume V, and a molecule perturbs a particular "oscillator" if it is found within a small volume v around the location of that oscillator, then an event in the sense defined above would occur if one of the N molecules is found inside v. The average number of perturbations is

$$\langle N \rangle = \frac{Nv}{V} \tag{V-38}$$

which is simply the average number of molecules found in the volume v. This number, like the ratio N/V, is proportional to the pressure of the gas. The calculations made above then apply to this simplified picture of pressure broadening. In order to get more substantial results one needs to assume a model for the local oscillator, and for the relative motion of the perturbers and the oscillator. Such models make it possible to calculate a relaxation function, from which a measured spectrum can be calculated.

V.2 Discretization of the Stochastic Parameter

(a) A finite difference approach to a diffusion-type SLE

In the previous section a general analytical method of solution, the cumulant method, was discussed. Here a very different type of approach will be considered, that of discretizing the continuous stochastic parameter in the course of solving the SLE numerically. The method of cumulants was meant not just to supply a procedure for obtaining a solution, but to do so in a manner which would give physical insight into the problem. Each term in the expansion, and in particular the two lowest order terms, have some physical significance. Thus the rate of convergence of the expansion would be directly related to characteristics of the stochastic process. In the discretization approach one regards the SLE simply as a partial differential equation which has to be solved numerically to a reasonable accuracy. There are various techniques for simplifying the numerical treatment of such equations, but the finite difference technique has been applied in the context of the SLE more than other methods. The purpose of this technique is to simplify the equation so that it can be solved numerically more easily for any stochastic process. The general idea is presented here first, and examples for its application in the context of EPR are given in the next subsections.

As found above, the SLE (Eqs. (V-1), (V-4)) involves a first derivative with respect to time, and some operator representing the effect of the stochastic process. In many practical cases this is a differential operator. For example, in magnetic resonance the process is very often rotational tumbling of molecules, represented by a second order differential operator with respect to angular variables. In CIDEP the important interaction is usually orientation independent, and is a function of the radial distance between two radicals. In such cases the relevant part of the diffusion process is the radial process, involving a second order differential operator with respect to the radial variable. It is therefore possible to assume that in practical applications the SLE is usually a diffusion equation, with a first derivative with respect to time and second derivatives with respect to one or more spatial variables. The essence of the finite difference method is representing an analytical function y(x) by its values for a set of points $x = x_1, x_2, \ldots$ and approximating derivatives of the form: dy/dx by finite differences of the form: $\Delta y/\Delta x \equiv (y(x_j) - y(x_k))/(x_j - x_k)$. Then a single differential equation for the function y(x) is converted to a set of linear algebraic equations, relating the values of y for the chosen set of values of the argument x. In the case of the SLE, the "function" y is the density matrix $\boldsymbol{\rho}$.

The time variable may be treated in one of two ways. On the one hand, it is possible to treat it analytically, and then the derivative with respect to time may be eliminated by a Fourier or Laplace transform. On the other hand, it may be regarded as a numerical variable, for which one chooses a sequence of values, and then the equations are solved for these values. In the numerical approach, the scale of the time variable is essentially fixed by the parameters of the problem. The duration of the experiment is limited in practice by the characteristic time in which the magnetic resonance signal effectively decays. The time resolution is determined by the type of experiment, as will be dicussed in Chapter VIII. From the well known rules of Fourier transform, sufficient coverage of the line shape in the frequency domain is related to a sufficiently good time resolution, and sufficiently good frequency resolution is related to a sufficiently long period in which the signal is measured. The total extent of the line shape in the frequency domain is of the order of the largest interaction parameters in the system, and the needed frequency resolution depends on the relative magnitude of the differences between transition frequencies. Thus if one solves the SLE numerically using the finite difference technique for all variables, the discretization of the time variable has to be carried out so as to satisfy the requirements of sufficient coverage and resolution, which are determined in turn by the parameters of the system.

For the spatial variables the situation is more complicated. If one deals with angular variables, they usually span the full range in which they are defined. In the case of a distance variable (the radial variable), one only needs to consider distances which are not too large, because the interaction obviously decreases with distance. In both cases, however, there is no simple relation between resolution in the stochastic variable and spectral resolution. Moreover, the differential operator may involve not only ordinary derivatives but also mixed derivatives, and combinations of several derivatives, each of which is pre-multiplied by some function of the stochastic variables. Because of the second problem, there is more than one possible choice for the variable to be discretized. Due to the first problem, the useful resolution in the discretization has to be found mainly on a trial-and-error basis.

In the simplest case a diffusion-type SLE is written as:

$$\frac{\partial}{\partial t}\rho(x,t) = L_0(x)\,\rho(x,t) + \frac{\partial^2}{\partial x^2}\rho(x,t) \qquad \text{(V-39)}$$

where x is the stochastic variable. The first term on the right hand side of the equation represents the effect of the ordinary Hamiltonian, including both time independent interactions and the effects of coherent irradiation, if present. The operator $\mathbf{L_0}$ is assumed here to be constant in time. This means that either there is no coherently time dependent interaction, or it has been transformed away by going to the "rotating frame". The second term represents the stochastic operator for this case, which is also chosen as time independent. The first term may be a function of the stochastic variable, but does not operate on it (except by simple multiplication). Choosing a set of N+1 discrete values for x:

$$x\ :\ \{x_0\,;\,x_1 = x_0 + \Delta x\,;\,...\,;\,x_k = x_0 + (k-1)\,\Delta x\,;\,...\,;\,x_N = x_0 + N\,\Delta x\} \qquad \text{(V-40)}$$

one can directly find $\mathbf{L_0}$(x) for each of these values, and operate with it on $\boldsymbol{\rho}$. Derivatives with respect to x have to be replaced by their finite difference equivalents. The first derivative becomes:

$$\left(\frac{\partial}{\partial x}f(x)\right)_k \rightarrow \frac{f(x_{k+1})-f(x_k)}{\Delta x} \equiv \frac{(\Delta f)_k}{\Delta x} \qquad \text{(V-41)}$$

and the second derivative can be taken as:

$$\left(\frac{\partial^2}{\partial x^2}f(x)\right)_k \rightarrow \frac{\dfrac{f(x_{k+1})-f(x_k)}{\Delta x}-\dfrac{f(x_k)-f(x_{k-1})}{\Delta x}}{\Delta x} = \frac{(\Delta f)_k-(\Delta f)_{k-1}}{(\Delta x)^2} =$$

$$= \frac{f(x_{k+1})-2f(x_k)+f(x_{k-1})}{(\Delta x)^2} \qquad \text{(V-42)}$$

The first derivative may be represented with somewhat greater accuracy by:

$$\left(\frac{\partial}{\partial x}f(x)\right)_k \rightarrow \frac{1}{2}\left(\frac{(\Delta f)_k}{\Delta x}+\frac{(\Delta f)_{k-1}}{\Delta x}\right) = \frac{f(x_{k+1})-f(x_{k-1})}{2\,\Delta x} \equiv \frac{\Delta^{(2)}f_k}{2\,\Delta x} \qquad \text{(V-43)}$$

The second derivative could be modified in a similar manner to:

$$\left(\frac{\partial^2}{\partial x^2}f(x)\right)_k \rightarrow \frac{(\Delta^{(2)}f)_{k+1}-(\Delta^{(2)}f)_{k-1}}{(2\,\Delta x)^2} = \frac{f(x_{k+2})-2f(x_k)+f(x_{k-2})}{4(\Delta x)^2} \qquad \text{(V-44)}$$

but this is less accurate than Eq. (V-42).

Example V.2: Comparing approximations for derivatives

In order to estimate the accuracy of the approximations given here for first and second derivatives, it is useful to examine the result of employing these approximations in specific cases. This is done here with a very coarse grid, namely, the domain under consideration is divided into a number of intervals which is too small, thus magnifying the inaccuracies of the different approximations. In the two tables below, the first and second analytically calculated derivatives (f' and f'' , respectively) of two functions are compared with the approximate derivatives calculated from Eqs. (V-41) - (V-44). In Table V.1 the analytically calculated derivatives of the gaussian

$$f(x) = \frac{1}{\sqrt{\pi\alpha}}\exp(-\alpha x^2) \qquad \text{(E-1)}$$

(with $\alpha = 0.125$) are compared with the approximate derivatives calculated from Eqs. (V-41) - (V-44). It is clear that for the first derivative Eq. (V-43) is an improvement over Eq. (V-41), whereas for the second derivative Eq. (V-44) is not as good as Eq. (V-42). The reason is that Eq. (V-44) averages out important changes in the function. A similar comparison is made in Table V.2 for the function

$$f(t) = \cos(\omega t)\exp(-\gamma t) \tag{E-2}$$

(with $\omega = 1.5$, $\gamma = 0.2$), which has the same form as a simple magnetic resonance signal. Again the same conclusion is reached, that Eqs. (V-43) and (V-42) are the better approximations for the first and second derivatives, respectively.

Table V.1. Discretization of first and second derivatives of a gaussian: $f(x)=(\pi\alpha)^{-1/2} * \exp(-\alpha x^2)$ ($\alpha = 0.125$)

x	f(x)	f '(x)	(V-41)	(V-43)	f "(x)	(V-42)	(V-44)
-5	0.070	0.088	0.146	-----	0.092	-----	-----
-4	0.216	0.216	0.302	0.224	0.162	0.156	-----
-3	0.518	0.389	0.450	0.376	0.162	0.148	0.111
-2	0.968	0.484	0.440	0.445	0.000	-0.009	-0.031
-1	1.408	0.352	0.188	0.314	-0.264	-0.253	-0.223
0	1.596	0.000	-0.188	0.000	-0.399	-0.375	-0.314
-1	1.408	-0.352	-0.440	-0.314	-0.264	-0.253	-0.223
-2	0.968	-0.484	-0.450	-0.445	0.000	-0.009	-0.031
-3	0.518	-0.389	-0.302	-0.376	0.162	0.148	0.111
-4	0.216	-0.216	-0.146	-0.224	0.162	0.156	-----
-5	0.070	-0.088	-----	-----	0.092	-----	-----

Note: columns 2,3 and 6 are calculated analytically.

Table V.2. Discretization of first and second derivatives of: $f(t) = \cos(\omega t) * \exp(-\gamma t)$ ($\omega=1.5$, $\gamma=0.2$)

t	f(t)	f '(t)	(V-41)	(V-43)	f "(t)	(V-42)	(V-44)
0	1.000	-0.200	-0.942	-----	-2.210	-----	-----
1	0.058	-1.237	-0.722	-0.832	0.362	0.221	-----
2	-0.664	-0.009	0.548	-0.087	1.523	1.270	0.690
3	-0.116	0.828	0.547	0.548	-0.066	-0.001	0.104
4	0.431	0.102	-0.304	0.122	-1.029	-0.851	-0.450
5	0.128	-0.543	-0.402	-0.353	-0.075	-0.098	-0.122
6	-0.274	-0.131	0.157	-0.122	0.681	0.559	0.288
7	-0.117	0.349	0.288	0.222	0.129	0.130	0.115
8	0.170	0.128	-0.072	0.108	-0.442	-0.360	-0.180
9	0.098	-0.219	-0.201	-0.137	-0.138	-0.129	-----
10	-0.103	-0.111	-----	-----	0.280	-----	-----

Note: columns 2,3 and 6 are calculated analytically.

****

With these approximations for derivatives one can replace Eq. (V-39) by:

$$\frac{\rho(x_k,t_{m+1}) - \rho(x_k,t_{m-1})}{2\Delta t} = L_0(x_k,t_m)\,\rho(x_k,t_m) + \frac{\rho(x_{k+1},t_m) - 2\rho(x_k,t_m) + \rho(x_{k-1},t_m)}{(\Delta x)^2} \qquad (V\text{-}45)$$

if one wishes to discretize all variables in the problem. This is a set of linear algebraic equations in the variables $\rho(x_k,t_m)$, where (x_k,t_m) are points defined on a two dimensional grid. After choosing a certain grid and solving the equations, one has to decide whether the result is accurate enough. Usually, however, it is much more convenient to discretize only the spatial variables. The equation is then:

$$\frac{\partial}{\partial t}\rho(x_k,t) = L_0(x_k)\,\rho(x_k,t) + \frac{\rho(x_{k+1},t) - 2\rho(x_k,t) + \rho(x_{k-1},t)}{(\Delta x)^2} \qquad \text{(V-46)}$$

Fourier transform with respect to t will result in

$$i\omega\,\rho(x_k,\omega) = L_0(x_k)\,\rho(x_k,\omega) + \frac{\rho(x_{k+1},\omega) - 2\rho(x_k,\omega) + \rho(x_{k-1},\omega)}{(\Delta x)^2} \qquad \text{(V-47)}$$

since the Liouville - von Neumann and relaxation operators are both time independent, and therefore commute with integrations over the time variable. One now has a set of linear algebraic equations in the variables $\rho(x_k,\omega)$, which have to be solved on a one dimensional grid of points (x_k) and as a function of the continuous variable ω. In practice the Liouville - von Neumann operator always includes a Larmor term $\omega_0\mathbf{I}$, where **I** is the identity operator. This is modified by transferring the $i\omega\rho$ term to the right hand side, to an off resonance term $-\Delta\omega\mathbf{I} = -(\omega - \omega_0)\mathbf{I}$, . There is no other term in this operator or in the relaxation operator which depends on ω, so the equation can be written schematically as:

$$0 = -i\Delta\omega\,\rho(x_k,\omega) + L_0(x_k)\,\rho(x_k,\omega) + \sum_{j=0}^{j=N} R_{kj}\,\rho(x_j,\omega) \qquad \text{(V-48)}$$

The third term on the right hand side is in the form of a matrix **R** multiplying a vector $\rho(x)$, where the elements of the matrix are

$$R_{kj} = \frac{\delta_{j,k-1} - 2\delta_{j,k} + \delta_{j,k+1}}{(\Delta x)^2} \qquad \text{(V-49)}$$

The first term on the right hand side of Eq. (V-48) is in the same form, but its matrix is diagonal and proportional to the unit matrix. The second term will not be worked out in detail here, but as will be shown in the next section, it can be written as a sum of a constant vector and a matrix multiplying the vector $\rho(x)$. The constant vector includes all the information on the initial value of the density matrix. Thus the equation reduces to

$$\rho_0(x_k) = \sum_{j=0}^{j=N} \left(-i\Delta\omega\,\delta_{jk} + G_{k,j}\right)\rho(x_j,\omega) \qquad \text{(V-50)}$$

Here **G** is a sum of **R** and the matrix resulting from the $\mathbf{L}_0$ term, and ρ_0 is the constant vector resulting from the $\mathbf{L}_0$ term. This equation can be solved by diagonalizing **G**, and then the transformation which diagonalizes **G** also diagonalizes $-i\Delta\omega\mathbf{I} + \mathbf{G}$ for all values of ω. Thus the frequency dependence is trivial, and the main calculation is diagonalization of a single matrix, the dimensions of which are $N \times N$.

Finally, after solving for the density matrix one may calculate the line shape using the standard expression, which is the trace over the product of ρ and a transverse magnetization operator. This expression has to be summed or averaged over all values of the stochastic variable, because the measured magnetization comes from a whole ensemble of molecules, with random values of the stochastic variable. The summation may be done with some weight factor, expressing the probability for a molecule to have a particular value for the stochastic variable (this will be explained in detail in the next Section).

(b) Application to rotational diffusion

If the stochastic process is rotational diffusion, the relaxation operator is the Laplacian, which is equal in polar spherical coordinates to:

$$\nabla^2 = \frac{1}{r^2}\frac{\partial}{\partial r}\left(r^2\frac{\partial}{\partial r}\right) + \frac{1}{r^2\sin(\theta)}\frac{\partial}{\partial\theta}\left(\sin(\theta)\frac{\partial}{\partial\theta}\right) + \frac{1}{r^2\sin^2(\theta)}\frac{\partial^2}{\partial\phi^2} \qquad (V\text{-}51)$$

In the case of intra-molecular interactions, the distance r is fixed, and only the orientation relative to the laboratory frame of reference changes randomly due to the diffusion. Thus the constant r can be ignored, leading to the simpler expression for the Laplacian on the surface of a sphere with unit radius:

$$\left(\nabla^2\right)_{r=1} = \frac{1}{\sin(\theta)}\frac{\partial}{\partial\theta}\left(\sin(\theta)\frac{\partial}{\partial\theta}\right) + \frac{1}{\sin^2(\theta)}\frac{\partial^2}{\partial\phi^2} \qquad \text{(V-52)}$$

It is this simplified operator which is used in magnetic resonance for most rotational diffusion effects. The exceptional cases in which inter-molecular interactions are important, so the distance is not fixed, are treated in the next subsection. The equation of motion of the density matrix with rotational diffusion is thus:

$$\frac{\partial}{\partial t}\rho(\Omega,t) = L_0(\Omega)\,\rho(\Omega,t) + \left(\nabla^2\right)_\Omega\rho(\Omega,t) \qquad \text{(V-53)}$$

where $\Omega \equiv (\theta,\phi)$ specifies the orientation of some intra-molecular axis with respect to an external frame of reference. For example, if the diffusion refers to the relative positions of two spins which interact through a dipolar interaction, the axis is that which joins the locations of the two spins. The external frame of reference is usually chosen according to the directions of the applied magnetic fields. The direction of the large constant field is normally chosen as the **z** axis and the direction of the alternating field inducing transitions is normally chosen as the **x** axis.

One possible approach to the equation is to divide the angular range into small intervals, and then the angular variables are defined on a grid:

$$\theta : \{0\,;\pi/N\,;2\pi/N\,;...\,;k\pi/N\,;...\,;\pi\} \qquad (\Delta\theta = \pi/N) \tag{V-54a}$$

$$\phi : \{0\,;2\pi/N'\,;4\pi/N'\,;...\,;2k\pi/N'\,;...\,;2\pi\} \qquad (\Delta\phi = 2\pi/N') \tag{V-54b}$$

The division in the two angular variables does not have to form the same length of intervals, so one does *not* need to assume N' = 2N or any other special relation between N and N'. In practice their values are determined by the dependence of the equation on these variables. The relevant range of values of (θ,ϕ) is in principle the full range, but in many cases symmetry arguments show that the spectrum calculated for some sector in the sphere is identical to that calculated for another section (or sections) of the sphere. For example, the dependence on θ may be symmetrical about $\theta = \pi/2$, and the dependence on ϕ may be symmetrical about $\phi = \pi$, or there is even a higher symmetry. In such cases it is sufficient to calculate for one sector, and thus the effective angular range is reduced, with a corresponding reduction in the amount of computations.

It often occurs that the relevant interaction depends only on the polar angle θ and not on the azimuthal angle ϕ. In that case only the first term in Eq. (V-52) is needed in the relaxation operator, and the equation of motion (V-53) can be integrated immediately over ϕ. The integration may include a weight factor which depends on ϕ if there are different probabilities for having different values of ϕ.

Discretizing the angular variables the original equation, Eq. (V-53) becomes a set of differential equations with respect to time, and difference equations with respect to the angular variables. For the common case in which there is no dependence on ϕ, one only needs the θ term in the Laplacian, which can be written as

$$(\nabla^2)_\theta = \frac{1}{\sin(\theta)}\frac{\partial}{\partial\theta}\left(\sin(\theta)\frac{\partial}{\partial\theta}\right) = \frac{\cos(\theta)}{\sin(\theta)}\frac{\partial}{\partial\theta} + \frac{\partial^2}{\partial\theta^2} \tag{V-55}$$

Discretizing the derivatives as in Eq. (V-42),(V-43) the Laplacian is replaced by a difference operator:

$$\left((\nabla^2)_\theta f(\theta)\right)_k \rightarrow \frac{\cos(\theta_k)}{\sin(\theta_k)}\frac{f(\theta_{k+1}) - f(\theta_{k-1})}{2\Delta\theta} + \frac{f(\theta_{k+1}) - 2f(\theta_k) + f(\theta_{k-1})}{(\Delta\theta)^2} =$$

$$= \frac{1}{(\Delta\theta)^2}\left((1+\epsilon(\theta))f(\theta_{k+1}) + (1-\epsilon(\theta))f(\theta_{k-1}) - 2f(\theta_k)\right) \tag{V-56}$$

with the definition $\epsilon(\theta) \equiv \Delta\theta/(2\tan(\theta))$. All non-zero angles are equal to or larger than

$\Delta\theta$, and thus $\tan(\theta) \approx \theta$ (for small angles) is not smaller than $\Delta\theta$ even for the smallest angles, and much larger than $\Delta\theta$ for larger angles. Therefore $\epsilon(\theta) \ll 1$ for almost all angles, so that it is only of secondary importance in the difference operator. The discretized equation of motion is:

$$\frac{\partial}{\partial t}\rho(\theta_k,t) = L_0(\theta_k)\,\rho(\theta_k,t)$$

$$+\frac{1}{(\Delta\theta)^2}\Big((1+\epsilon(\theta_k))\,\rho(\theta_{k+1},t) + (1-\epsilon(\theta_k))\,\rho(\theta_{k-1},t) - 2\rho(\theta_k,t)\Big) \qquad \text{(V-57)}$$

which can be written schematically as

$$\frac{\partial}{\partial t}\rho(\theta_k,t) = \sum_{j=0}^{j=N}\Big(\big(L_0(\theta_k)\big)_{k,j} + R_{k,j}\Big)\,\rho(\theta_j,t)) \qquad \text{(V-58)}$$

with the "exchange" operator

$$R_{k,j} = \frac{1}{(\Delta\theta)^2}\Big((1+\epsilon(\theta_k))\,\delta_{j,k+1} + (1-\epsilon(\theta_k))\,\delta_{j,k-1} - 2\delta_{j,k}\Big) \qquad \text{(V-59)}$$

If one Fourier tranforms the time variable the result is a set of algebraic equations which are not linear, due to the $\sin(\theta)$ factors in the Laplacian. If no transformation is carried out, the equations are very similar to the density matrix equations for chemical exchange. The main difference is that instead of an exchange between a small number of chemically distinct sites, or between molecules in distinct configurations, there is an exchange between many different molecular orientations relative to the laboratory frame of reference. The final line shape is then summed over all values of the stochastic variables. For polar spherical coordinates, the volume element for integration is: $(dr)\,(r\,d\theta)\,(r\sin(\theta)\,d\phi) = r^2\,dr\,\sin(\theta)\,d\theta\,d\phi$. As usual, a weight factor may be included in this calculation.

An alternative approach is to work with the variable $x = \cos(\theta)$. This may be convenient because in the integration element one has: $\sin(\theta)\,d\theta = -d(\cos(\theta))$. Then

$$\sin(\theta) = \sqrt{1-x^2} \quad ; \quad \frac{\partial}{\partial\theta} = \frac{\partial x}{\partial\theta}\frac{\partial}{\partial x} = -\sin(\theta)\frac{\partial}{\partial x} \qquad \text{(V-60)}$$

Therefore the θ-dependent Laplacian can be written as:

$$\frac{1}{\sin(\theta)}\frac{\partial}{\partial\theta}\left(\sin(\theta)\frac{\partial}{\partial\theta}\right) = \frac{1}{\sin(\theta)}\frac{\partial}{\partial\theta}\left(-\sin^2(\theta)\frac{\partial}{\partial x}\right) =$$

$$= \frac{-\sin(\theta)}{\sin(\theta)}\frac{\partial}{\partial x}\left((x^2-1)\frac{\partial}{\partial x}\right) = -\left(2x\frac{\partial}{\partial x} + (x^2-1)\frac{\partial^2}{\partial x^2}\right) \tag{V-61}$$

In this approach the range of the stochastic variable is from x = -1 (for $\theta = \pi$) to x = 1 (for $\theta = 0$), and then one divides the range of x into equal intervals:

$$x : \{-1\,;\,-1+2/N\,;\,...\,;\,-1+2k/N\,;\,...\,;\,-1+2N/N=1\} \tag{V-62}$$

Thus it is $\cos(\theta)$ which changes in equal amounts, rather than θ itself. The main term in the dipolar interaction depends on θ through $\cos^2(\theta)$, so working with $\cos(\theta)$ may seem more natural than working with θ. Moreover, this is a way for eliminating the trigonometric functions from the calculation, which may be convenient. In practice, however, this method gives effectively the same results as the method which works directly with θ.

Example V.3: Rotational diffusion of triplet state molecules

In triplet EPR one deals with a molecule in the triplet state, in which the spins of two unpaired electrons couple to a total spin of S = 1. In a molecular principal axis system the Hamiltonian is given, for the case of isotropic g-tensor, by

$$H = \omega_0 S_z + D\left[S_z^2 - \frac{1}{3}S^2\right] + E\left(S_x^2 - S_y^2\right) \tag{E-1}$$

Forming irreducible spherical tensors from spin operators, this can be written (see Appendix D) as

$$H = \omega_0 T^1{}_0 + D\sqrt{2/3}\,T^2{}_0 + E\left(T^2{}_2 + T^2{}_{-2}\right) \tag{E-2}$$

Using the standard notation, this implies (see Appendix D, Example D.1):

$$F^1{}_0 = 0 \quad ; \quad F^2{}_0 = D\sqrt{2/3} \quad ; \quad F^2{}_2 = F^2{}_{-2} = E \tag{E-3}$$

Therefore, in the laboratory frame of reference the Hamiltonian is equal to

$$H = \omega_0 T^1{}_0\cos(\theta) + \sum_{m=-2}^{2} T^2{}_m\left\{D\sqrt{2/3}\,D^2{}_{0,m}(\Omega) + E\left(D^2{}_{2,m}(\Omega) + D^2{}_{-2,m}(\Omega)\right)\right\} \tag{E-4}$$

To discretize the equation of motion for the density matrix one would need to putthe values of (θ,ϕ) on a lattice. Assume for simplicity that E = 0, and also that $\omega_0 \gg$ D,E so that only terms which are diagonal in $\mathbf{S}_z$ are important. This leaves, except for the

S_z term, only $T^2_0\, D(2/3)^{1/2}\, D^2_{0,0}(\Omega)$, which is independent of ϕ. Consequently, only one variable - θ - has to be discretized. Now assume the rotational diffusion proceeeds in small steps only, from each "site" (value of θ, which represents a particular molecular orientation), to its nearest neighbours only. Also assume first order kinetics, i.e., the number of molecules leaving a "site" (per second) is proportional to the number of molecules in that "site". One then gets for one matrix element of the density matrix an equation of motion of the type (V-58), with the exchange operator:

$$R_{i,j} = k_{i+1,i}\,\delta_{j,i+1} + k_{i-1,i}\,\delta_{j,i-1} - \left(k_{i,i+1} + k_{i,i-1} + \frac{1}{T_2}\right)\delta_{j,i} \tag{E-5}$$

The assumption of detailed balance makes it possible to calculate all rate constants $k_{i,j}$ from one rate constant, $k \equiv k_{1,2}$ in terms of the site populations. Thus, for example

$$k_{2,1} = k\frac{n_1}{n_2} \quad ; \quad k_{2,3} = k\frac{n_2 - n_1}{n_2} \quad ; \quad k_{3,2} = k\frac{n_2 - n_1}{n_3} \quad \ldots \tag{E-6}$$

For a steady state experiment one can set the time derivative equal to zero, and the part of the equation involving the non-stochastic Hamiltonian includes a term due to the coherent CW irradiation. This term is time independent, as mentioned above, because of the transformation to the rotating frame. One thus has a set of algebraic equationsfor $\rho(\theta_i)$, and from the solution of these equations it is straightforward to calculate the observed magnetization.

****

(c) *Application to CIDEP*

Using the approach explained in the last subsection it is also possible to treat the equations for CIDEP (Chemically Induced Dynamic Electron Polarization). In CIDEP a pair of radicals, formed in a chemical reaction, is generated with non-equilibrium spin polarization. Common mechanisms for CIDEP are the radical pair mechanism (RPM) and the triplet mechanism (TM). In RPM, a photochemical reaction involves a transient radical pair, which evolves in time due to a difference in the Larmor frequencies of the two unpaired electrons. Their spins may combine to a triplet (S = 1) or a singlet (S = 0), but these states may be mixed if the exchange interaction J(r) between them is weak, which occurs when the two radicals are relatively far from each other. If they have a reencounter, the mixed states cause a significant polarization of the spin, and this has a strong effect on the EPR signal. In TM, a photoexcited triplet state acts as a precursor to a pair of radicals (each in the doublet state), which are formed by the reaction of that precursor with some other molecule. Since the photoexcitation creates the triplet state by a spin selective intersystem crossing mechanism, also the final radicals have a strong electron spin polarization. Here we shall consider the RPM mechanism, as treated by the SLE.

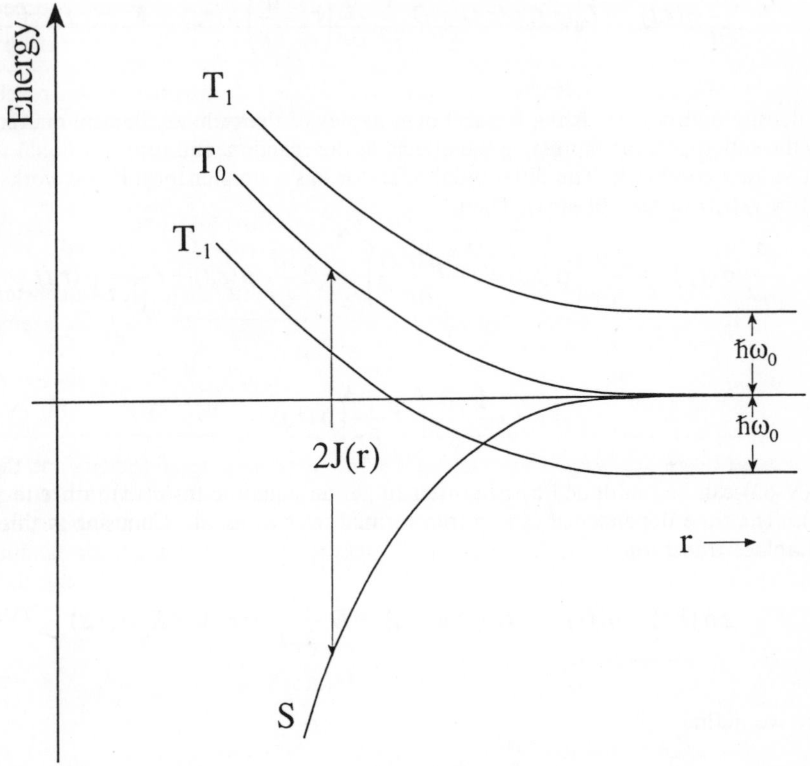

Fig. V.2. *The triplet and singlet energy levels as a function of the inter-radical distance r. The magnitude of the interaction constant J(r) is indicated in the figure.*

The equation for the density matrix with the relative diffusion of the two radicals is similar to Eq. (V-53), but with two differences. First, the distance independent $(\nabla^2)_\Omega$ operator is replaced by the full Laplacian operator of Eq. (V-51). Second, there is a term for the reaction rate constants of the radical pair. Anisotropic interactions are assumed in this case to be averaged over, if one deals with liquid solutions. The exchange interaction can be assumed to depend only on the distance between the two radicals, and not on their relative orientation. Thus there is no orientation dependence, and the density matrix may be trivially integrated over the angles θ and ϕ. The result is a density matrix which depends only on r (the distance) and t , with the equation:

$$\frac{\partial}{\partial t}\rho(r,t) = L_0(r,t)\,\rho(r,t) + D\frac{1}{r^2}\frac{\partial}{\partial r}\left(r^2\frac{\partial}{\partial r}\right)\rho(r,t) + K_r\,\rho(r,t) \qquad \text{(V-63)}$$

The density matrix $\rho(r,t)$ is the integral over angles of the ordinary density matrix, and K_r is the superoperator expressing the effects of the reaction, and is proportional to the reaction rate constants. The differential operator has a simpler form if one works with $\sigma(r,t) \equiv r\rho(r,t)$ instead of $\rho(r,t)$. Then:

$$\frac{\partial^2}{\partial r^2}\sigma(r,t) = \frac{\partial}{\partial r}\left(\rho(r,t) + r\frac{\partial\rho(r,t)}{\partial r}\right) = 2\frac{\partial}{\partial r}\rho(r,t) + r\frac{\partial^2}{\partial r^2}\rho(r,t) =$$

$$= \frac{1}{r}\frac{\partial}{\partial r}\left(r^2\frac{\partial}{\partial r}\right)\rho(r,t) \qquad \text{(V-64)}$$

Eq. (V-63) can be multiplied by r in order to get an equation for $\sigma(r,t)$ rather than for $\rho(r,t)$. The time dependence can be transformed over as usual. Choosing in this case the Laplace transform:

$$s\,\sigma(r,s) - \sigma_0(r) = L_0(r)\,\sigma(r,s) + D\frac{\partial^2}{\partial r^2}\sigma(r,s) + K_r\,\sigma(r,s) \qquad \text{(V-65)}$$

where we define

$$\sigma(r,s) \equiv \int_0^\infty e^{-st}\,\sigma(r,t)\,dt \equiv \int_0^\infty e^{-st}\,r\,\rho(r,t)\,dt \qquad \text{(V-66)}$$

and $\sigma_0(r) = r\,\rho(r,0)$ is the initial condition.

In order to solve Eq. (V-65) it is convenient to employ the finite difference approach, using for the radial diffusion term the approximate expression

$$\frac{\partial^2}{\partial r^2}\sigma(r,s) = \frac{\sigma(r+\Delta r,s) - 2\sigma(r,s) + \sigma(r-\Delta r,s)}{(\Delta r)^2} \qquad \text{(V-67)}$$

The exchange interaction J(r), included in the L_0 term, has a stronger dependence on r than all other expressions in the equation. Therefore the size of the increment Δr has to be chosen in accordance with the rate of change of J(r).

One may integrate the density matrix also over the radial variable r (with the integration factor of r^2 for the spherical coordinates), so as to leave only the time dependence:

$$\rho(t) \equiv \int_d^\infty r^2 \rho(r,t)\, dr \equiv \int_d^\infty r\, \sigma(r,t)\, dr \tag{V-68}$$

The parameter d is equal to the distance of closest approach between the two radicals, where their interaction is maximized. It is the density matrix $\rho(t)$ which is used for calculating expectation values of spin operators, which correspond to the quantities measured experimentally.

In the references given at the end of this Chapter, various possible solutions of these equations are discussed, and their physical meaning is considered. Here we shall deal only with the technical aspect of solving the equations by means of the finite difference technique. Using the approximation of Eq.(V-67) in Eq. (V-65), the continuous diffusion equation becomes a discrete master equation. The stochastic variables, regarded as elements of a vector, are the functions $\rho(r_j, t)$ or $\sigma(r_j, t)$ with $r_j \equiv d + j\Delta r$ ($j = 0, 1, 2, ..., N$). The transition probability matrix **W** is defined by

$$D \frac{\partial^2}{\partial r^2} \sigma(r,s) \quad \rightarrow \quad \mathbf{W} \sigma \tag{V-69}$$

Denoting $\sigma(r_j, s)$ by σ_j, the matrix element $\mathbf{W}_{jk}$ is the probability of going to σ_j from σ_k. The value of N should be chosen so that: $r_N = d + N\Delta r \gg d$, in order to get negligible exchange interaction at the maximal distance: $J(r_N) \ll J(d) \equiv J_0$ and also in order to have negligible reencounter probability at that distance. Assuming there is no net accumulation of radicals at the minimum distance d is formally equivalent to a boundary condition of a reflecting wall:

$$\frac{\partial}{\partial r} \rho(r,t) \Big|_{r=d} = 0 \quad \Leftrightarrow \quad \frac{\partial}{\partial r} \sigma(r,s) \Big|_{r=d} = \frac{\sigma(d,s)}{d} \tag{V-70}$$

Approximating the derivative by its finite difference equivalent, this boundary condition becomes:

$$\frac{\sigma(d+\Delta r, s) - \sigma(d-\Delta r, s)}{2\,\Delta r} - \frac{\sigma(d,s)}{d} = 0 \tag{V-71}$$

from which it follows that

$$\sigma(d-\Delta r, s) = \sigma(d+\Delta r, s) - \frac{(2\,\Delta r)\, \sigma(d,s)}{d} \tag{V-72}$$

This can be substituted into the finite difference equivalent of the second derivative to give:

$$D \frac{\partial^2}{\partial r^2} \sigma(r,s) \Big|_{r=d} = \frac{D}{\Delta r^2} \left(2\, \sigma(d+\Delta r, s) - 2\left(1 + \frac{\Delta r}{d}\right) \sigma(d,s) \right) \tag{V-73}$$

as the boundary condition for Eq. (V-65) at the distance of closest approach. At the maximum distance the probability for going between the next-to-last position and the last position, and the probability of staying there are equated to zero:

$$W_{N-1,N} = W_{N,N} = 0 \tag{V-74}$$

Notice that $\mathbf{W}_{N-1,N} = 0$ is approximately equal to the expression that would be obtained by the assumption of a reflecting wall, if $\Delta r/r_N \ll 1$. Using the boundary conditions at the minimum and maximum distances one may solve the diffusion equation (V-65) as a set of linear algebraic equations.

The integral in Eq. (V-68) is approximated as

$$\int_d^\infty r\,\sigma(r,s)\,dr = \sum_{j=0}^{N} V_j\,\sigma(r_j,s) \tag{V-75}$$

where V_j is the weight factor for r_j, which is equal to:

$$V_j = r_j \Delta r \;\; (1 \le j \le N-1) \quad ; \quad V_0 = \frac{d\,\Delta r}{2} \quad ; \quad V_N = \frac{r_N \Delta r}{2} \tag{V-76}$$

This choice of weights means the range between $r = d$ and $r = r_N$ is divided into intervals which are centered about the points r_j. Thus ordinary points have a weight of 1 (multiplied by r because of the integrand in Eq. (V-75)), whereas the two endpoints have a weight of ½ only (multiplied by r). Since the total probability density is given by $\mathrm{Tr}(\rho(t))$, conservation of probability requires

$$\sum_{j=0}^{N} V_j\,W_{j,k} = 0 \qquad (for\ j = 0,1,\ldots,N) \tag{V-77}$$

Finally, an important reduction in the amount of calculations required can be achieved by the following observation. The exchange interaction is effective only at very short range, and within that range it changes greatly even with small changes in distance. The Brownian diffusion process, however, has a significant effect only over significantly larger distances, and acts over all distances. Thus J(r) is already equal (approximately) to zero even when r is too small for a final separation of the radicals (r_N must be large enough that both conditions would be fulfilled). Therefore the domain $d \le r \le r_N$ can be divided conveniently into two distinct regions. In the region: $d \le r \le r_M$ J(r) is non-zero, and Δr must be small enough to follow the changes in J as a function of r. In the region: $r_M < r \le r_N$ J(r) is equal to zero, and there the intervals can be chosen as $\Delta' r \equiv f\,\Delta r$. Actual calculations have shown that the factor f can be of the order of 10 - 100. Then the weight factors of Eq. (V-76) are multiplied in the second region by f, and the elements of **W** in that region are divided by f^2.

V.3 The Method of Eigenfunctions

(a) *Solution of the SLE in the general case*

Two approaches to the SLE have been considered in the previous sections. On the one hand the theoretically oriented method of cumulants, and on the other hand the computationally oriented finite difference technique. A method which combines to some degree both characteristics will now be presented, the method of eigenfunctions. The main limitation of this method is its restriction to cases in which one can find eigenfunctions of the stochastic operator, and where these eigenfunctions are convenient to work with (the meaning of "convenient" in this context will be defined below). It thus differs from the two previous methods, which are in principle of general applicability. However, for very common practical cases the method is applicable, and has been shown to be very useful.

The starting point for this approach is Eq. (V-4), in which the difficult part is working with the relaxation superoperator $\mathbf{\Gamma}$. The idea is to expand all operators in the equation in eigenfunctions of $\mathbf{\Gamma}$, so that the most difficult part of the equation will become its simplest part. The commutator term will then become somewhat more complicated, but will not be too difficult to handle. It is useful to rewrite the equation as

$$\frac{\partial}{\partial t}\rho(\Omega,t) = \frac{1}{i\hbar}[H(\Omega,t),\rho(\Omega,t)] - \mathbf{\Gamma}_\Omega\big(\rho(\Omega,t) - \rho_0(\Omega,t)\big) \qquad \text{(V-78)}$$

in order to emphasize that both the Hamiltonian and the density matrix depend on the stochastic parameter Ω. The relaxation operator $\mathbf{\Gamma}_\Omega$ has been redefined with an opposite sign, in order to emphasize that it represents a decay process. The operator ρ_0 is the equilibrium density matrix, which is included in the equation in order to ensure a decay to the correct finite temperature equilibrium. As will be seen later on, in most practical cases this term may be omitted with no significant error.

Suppose the eigenfunctions of $\mathbf{\Gamma}_\Omega$ are known:

$$\mathbf{\Gamma}_\Omega G_j(\Omega) = \gamma_j G_j(\Omega) \qquad \text{(V-79)}$$

The eigenvalues may be degenerate, i.e., some different functions may be related to the same eigenvalue: $\gamma_i = \gamma_j$ ($i \neq j$). Suppose further that these eigenfunctions form a complete set, i.e., any function of Ω can be expressed in terms of these functions. Then

$$\rho(\Omega,t) = \sum_k r_k(t)\, G_k(\Omega) \qquad \text{(V-80a)}$$

$$H(\Omega,t) = \sum_k h_k(t)\, G_k(\Omega) \qquad \text{(V-80b)}$$

The expansion coefficients $\mathbf{r}_k(t)$, $\mathbf{h}_k(t)$ are independent of the stochastic parameter Ω, but are not necessarily simple functions. In fact, in the applications to be considered later on they will be spin operators, containing the spin dependence of the density matrix and the Hamiltonian, respectively. The $G_j(\Omega)$, however, will be simple functions of Ω, which represents spatial variables (angles) in those applications. The important point here is the separation between spin operators with a coherent time dependence on the one hand, and parameters with a stochastic time dependence on the other hand. This separation leads to the following form of the equation:

$$\sum_i \left(\frac{\partial}{\partial t} r_i(t)\right) G_i(\Omega) = \frac{1}{i\hbar}\sum_{j,k} [h_j(t), r_k(t)]\, G_j(\Omega) G_k(\Omega) - \sum_l \left(r_l(t) - r^{(0)}{}_l\right) \gamma_l G_l(\Omega) \quad (V\text{-}81)$$

The functions $G_j(\Omega)$ are assumed to be square integrable. Assume further that these functions are orthogonal to each other. This is natural for square integrable eigenfunctions related to different eigenvalues of an operator, and if it does not hold for degenerate eigenfunctions, it can be obtained by taking appropriate linear combinations. The functions are also assumed to be normalized, so that

$$\int G_j^*(\Omega)\, G_k(\Omega)\, d\Omega = \delta_{j,k} \quad \text{(V-82)}$$

It will now be assumed that also the integral of any product of three of the $G_j(\Omega)$ functions converges, and we shall use the notation

$$I(i,j,k) = \int G_i^*(\Omega)\, G_j(\Omega)\, G_k(\Omega)\, d\Omega \quad \text{(V-83)}$$

Now multiply Eq. (V-81) from the left by the complex conjugate of a particular function, $G_m(\Omega)$ and then integrate over Ω. The result is:

$$\frac{\partial}{\partial t} r_m(t) = \frac{1}{i\hbar}\sum_{j,k} [h_j(t), r_k(t)]\, I(m,j,k) - \left(r_m(t) - r^{(0)}{}_m\right)\gamma_m \quad \text{(V-84)}$$

If the $r_k(t)$ operators are a complete set, then the commutators appearing in this equation can be expressed as linear combinations of the operators $r_k(t)$:

$$[h_j(t), r_k(t)] = \sum_n a_{j,k,n}\, r_n(t) \quad \text{(V-85)}$$

The coefficients in this expansion are defined by (see Section I.6):

$$a_{j,k,n} \equiv \frac{1}{N_n} Tr\left\{r^\dagger{}_n(t)\, [h_j(t), r_k(t)]\right\} \quad \text{(V-86)}$$

with the assumption of orthogonality:

$$Tr\left\{r^\dagger{}_i(t)\, r_j(t)\right\} = N_i\, \delta_{i,j} \quad \text{(V-87)}$$

These assumptions hold, for example, for sets of irreducible spherical tensors, which will be used below (see Appendix D). In such a case the equation (taken for all possible values of m) leads to a set of coupled first order linear differential equations, which can be solved by standard methods. Using a Laplace or a Fourier transform, these are transformed to a set of linear algebraic equations.

From the above development it is clear that, provided certain conditions are satisfied, the SLE may be transformed to a set of standard algebraic equations, which may be solved in a straightforward manner. The conditions are basically two, and can be presented in the following simplified form:

(i) The stochastic superoperator $\mathbf{\Gamma}_\Omega$ has eigenfunctions which are a complete set in the stochastic parameter Ω, and which are square integrable and "triple-integrable" (Eqs. (V-82) and (V-83)).

(ii) The operator "coefficients" $\mathbf{r}_k(t)$ (in the expansion of $\boldsymbol{\rho}$ in terms of those eigenfunctions) form a complete set of spin operators.

This formulation of the conditions is actually too restrictive. What is needed in condition (i) is simply that the density matrix (which is taken as unknown at the beginning of the procedure) and the Hamiltonian can be expanded in terms of the eigenfunctions. In condition (ii) it is only needed that the expansion coefficients $\mathbf{h}_k(t)$ of the Hamiltonian, as well as the commutators appearing in Eq. (V-81) or Eq. (V-84), can all be expressed in terms of the expansion coefficients $\mathbf{r}_k(t)$ of $\boldsymbol{\rho}$. The restrictive formulation of these conditions is only made in order to make them simpler to apply in practice. Nevertheless, it should be born in mind that there may be some special cases in which the simpler conditions do not hold, and yet the method can be used.

If these conditions are fulfilled then each term in the calculation has a physical meaning, being related to specific eigenvalues of the stochastic superoperator. The calculation is also relatively simple, because the original integro-differential equation has been converted to a more manageable form. In the next subsection this procedure will be followed for a case of great practical importance, that of rotational diffusion, for which the formalism was originally developed.

(b) Solution for rotational diffusion

In every molecular system, even at quite low temperatures, thermal energy is sufficient for molecular rotation, especially if the molecule is not in a solid matrix. Molecular collisions hinder the rotation, so that in practice the rotation usually proceeds in many angular steps, each step having in general a different magnitude and a different orientation. Thus molecular rotation in liquids (and also in some solids) is actually a process of random rotational steps, known as rotational diffusion. The orientation Ω of the molecule relative to a fixed frame of reference is the stochastic variable in this case. A common model for such motion is that of Brownian rotational diffusion, in which the molecule is typically observed at relatively long time intervals, between which it performs many random steps. The diffusion may be isotropic, i.e., with the same speed of rotation about each principal axis of the molecule, or anisotropic, with different rates of rotation about different axes. In the most general case, the rotational diffusion operator is completely asymmetric, namely, it has three different rates for diffusion about the three principal axes. These axes may be different from the principal axes of other operators (e.g., dipolar interaction tensor) which characterize the molecule. The diffusion operator is needed in the equation (IV-79) for the probability distribution $p(\Omega,t)$ to find the molecule in a particular orientation Ω at time t, provided it was known to have an orientation Ω_0 at time t = 0. Consequently, this is also the operator in the SLE (Eq. (V-78)). In the principal axis system (**x**, **y**, **z**) of the diffusion tensor R, the rotational diffusion operator is:

$$\Gamma_\Omega = R_{xx}L_x^2 + R_{yy}L_y^2 + R_{zz}L_z^2 = R_+L^2 + R_-\left(L_x^2 - L_y^2\right) + \left(R_{zz} - R_+\right)L_z^2 \qquad \text{(V-88)}$$

Here **L** is the angular momentum operator, and $R_\pm \equiv \frac{1}{2}(R_{xx} \pm R_{yy})$ where $R_{\alpha\alpha}$ are the rotational diffusion rates. This diffusion operator is mathematically identical to the Hamiltonian of a free rigid rotor, if one makes the identification: $R_{\alpha\alpha} = 1/(2I_{\alpha\alpha})$ (α = x, y, z), where $I_{\alpha\alpha}$ is the α principal moment of inertia. Thus Eq. (IV-79) is almost identical mathematically to the Schrödinger equation of a free rigid rotor, except for the factor of $i\hbar$ which appears on the left hand side of the Schrödinger equation. As a result, the eigenfunctions of the rigid rotor are also eigenfunctions of the rotational diffusion operator, and the eigenvalues are the same up to a common factor.

The components of the rotational diffusion tensor can be related to data about the viscosity of the medium and characteristics of the diffusing molecule. These relations are known for the general case, but are not necessary for the present discussion. For the case of spherical symmetry, assuming the molecule is a sphere of radius r rotating in a medium of viscosity η, the Stokes-Einstein relation gives the following value for the diffusion rate R (= $R_{\alpha\alpha}$ for any α):

$$R = \frac{k_B T}{8\pi r^3 \eta} \qquad \text{(V-89)}$$

where T is the temperature and k_B is Boltzmann's constant.

The discussion here will be restricted to an axially symmetric diffusion tensor (which includes isotropic diffusion as a special case). The completely asymmetric case can be treated in a similar manner, but the calculation is much more complicated, and it does not occur very often in actual applications. For the case of axial symmetry about the **z** axis, $R_\parallel \equiv R_{zz}$ is the rate of rotation about the symmetry axis, and $R_\perp \equiv R_{xx} = R_{yy}$ is the rate of rotation about the two perpendicular axes. The axially symmetric diffusion operator can be written explicitly as (see Eqs. (A-8), (A-9)):

$$\Gamma_\Omega = R_\perp\left(\frac{1}{\sin(\theta)}\frac{\partial}{\partial\theta}\left(\sin(\theta)\frac{\partial}{\partial\theta}\right) + \frac{1}{\sin^2(\theta)}\frac{\partial^2}{\partial\phi^2}\right) + (R_\parallel - R_\perp)\frac{\partial^2}{\partial\phi^2} \qquad \text{(V-90)}$$

Its eigenfunctions are Wigner's rotation functions, with the eigenvalues $\gamma_{L,M,K} = R_\perp L(L+1) + (R_\parallel - R_\perp) K^2$:

$$\Gamma_\Omega D^L{}_{MK}(\Omega) = \left(R_\perp L(L+1) + \left(R_\parallel - R_\perp\right)K^2\right) D^L{}_{MK}(\Omega) \qquad \text{(V-91)}$$

In the isotropic case ($R_\parallel = R_\perp \equiv R$), the diffusion operator reduces to the Laplacian on a sphere of unit radius (see Eq. (V-52)), and its eigenvalues reduce to: $\gamma_{L,M,K} = R\,L(L+1)$. Notice that for $D^0{}_{00}(\Omega)$ the eigenvalue is equal to zero, with or without spherical symmetry. When the medium is isotropic, the equilibrium distribution is also expected to be isotropic, and therefore described by the eigenfunction $D^0{}_{00}(\Omega)$, which is independent of Ω. In such a case:

$$\Gamma_{\Omega}\, \rho_0(\Omega) = 0 \tag{V-92}$$

The rotation functions are indeed orthogonal to each other. In order to work with normalized functions, we shall define

$$\langle \Omega \,|\, LMK \rangle = \Psi^{L}{}_{MK}(\Omega) \equiv \left(\frac{2L+1}{8\pi^2} \right)^{1/2} D^{L}{}_{MK}(\Omega) \tag{V-93}$$

Then the orthonormality condition is satisfied:

$$\int \left(\Psi^{L}{}_{MK}(\Omega) \right)^{*} \Psi^{L'}{}_{M'K'}(\Omega)\, d\Omega = \delta_{L,L'}\, \delta_{M,M'}\, \delta_{K,K'} \tag{V-94}$$

and the triple integral, defined schematically in Eq. (V-83) is equal to:

$$I(L,M,K;L',M',K';L'',M'',K'') \equiv \int \left(\Psi^{L}{}_{MK}(\Omega) \right)^{*} \Psi^{L'}{}_{M'K'}(\Omega)\, \Psi^{L''}{}_{M''K''}(\Omega)\, d\Omega =$$

$$= (-1)^{M-K} \sqrt{(2L+1)(2L''+1)} \begin{bmatrix} L & L' & L'' \\ -M & M' & M'' \end{bmatrix} \begin{bmatrix} L & L' & L'' \\ -K & K' & K'' \end{bmatrix} \tag{V-95}$$

Using these integrals it is possible to re-derive Eq. (V-84) where each index (m,j and k) is replaced by a set of three numbers (L,M,K). In order to proceed further it is necessary to specify the exact form of the Hamiltonian, so that $\mathbf{h}_j(t)$ will be known. It is then possible to make some assumptions about the form of the $\mathbf{r}_k(t)$ and obtain an equation which involves only numerical functions (including matrix elements), without any spin operators.

The Hamiltonian is assumed to consist of three terms:

$$H = H_0 + H_1(\Omega) + H_2(t) \tag{V-96}$$

The first term, $\mathbf{H}_0$, is independent of both the stochastic variable and the time. It includes the Zeeman term of the electron or electrons if one deals with EPR, or the Zeeman tern of the nucleus or nuclei in NMR. It also includes the isotropic terms of various interactions, such as the hyperfine (electron-nuclear dipolar) interaction in EPR, or the (nucleus-nucleus) dipolar interaction in NMR. The second term, which is also time independent, depends on the stochastic variable, and is of the form

$$H_1(\Omega) = \sum_{\xi} \sum_{m,m'} D^{2}{}_{m,m'}(\Omega)\, F_{\xi}^{(2,-m')} A_{\xi}^{(2,m)} \tag{V-97}$$

where ξ is an index of the interaction (several different interaction tensors may be present in the same system), $\mathbf{A}$ are spin operators in the laboratory frame of reference and F are tensors of interaction constants in a molecular frame of reference. The rotation functions appear because the F tensors are transformed from the molecular frame to the laboratory frame (see Appendix D). The interaction tensors were taken to be of second rank, because in magnetic resonance the interactions almost always appear as zero or second rank tensors. Zero rank tensors may be included in $\mathbf{H_0}$. It has been assumed, for simplicity of presentation, that all interaction tensors have the same principal axes. It has also been assumed here that the molecule is in an isotropic medium, so that one does not have to consider an extra frame of reference characterizing that medium. Finally, the third term is the time dependent CW irradiation term:

$$H_2(t) \;=\; \omega_1 S_x \cos(\omega t) \tag{V-98}$$

which induces transitions in the system.

In many cases the constant Hamiltonian $\mathbf{H_0}$ has matrix elements which are very large compared with those of $\mathbf{H_1}(\Omega)$, and in most cases in EPR $\mathbf{H_2}(t)$ has matrix elements which are small even compared with those of $\mathbf{H_1}(\Omega)$. The equilibrium density matrix $\boldsymbol{\rho_0}$ is equal in general to (see Eq. (IV-99)):

$$\rho_0 \;=\; \frac{\exp\left\{-\left(H_0 + H_1(\Omega)\right)/k_B T\right\}}{Tr\left(\exp\left\{-\left(H_0 + H_1(\Omega)\right)/k_B T\right\}\right)} \;\approx\; p_0(\Omega)\left\{ I - \frac{H_0 + H_1(\Omega)}{k_B T} + \ldots \right\} \tag{V-99}$$

The high temperature approximation has been used on the right hand side of the equation. The identity operator has a trivial effect on the equation of motion, so it is the second term on the right hand side of Eq. (V-99) which approximates $\boldsymbol{\rho_0}$ in the actual calculations. It will be assumed here that $\mathbf{H_1}(\Omega)$ is very small compared with $\mathbf{H_0}$, so that the $\boldsymbol{\rho_0}$ is independent of Ω to a good approximation. Then Eq. (V-92) holds even if the molecule is found in a non-isotropic medium.

If the irradiation is relatively weak, as in ordinary CW experiments, the system is not far from equilibrium throughout the experiment. The only exceptions are those cases in which the system is prepared in a state which is very far from the thermal equilibrium state. This occurs in CIDEP and CIDNP, and in EPR of photoexcited triplets, as well as in pulse experiments. Such cases will be treated in the following chapters, and here the discussion will focus on the more usual situation in EPR, where the system is always close to equilibrium. It is then convenient to work with the deviation from equilibrium, defined by

$$\chi(t) \;\equiv\; \rho(t) - \rho_0 \tag{V-100}$$

In most cases to be treated below this deviation is indeed small, but this will not be taken as a general assumption, because the same formalism will also be used to deal with saturation (strong irradiation). Since $\boldsymbol{\rho_0}$ is time independent, its time derivative can be subtracted from the left hand side of Eq. (V-78) without changing it, so the equation can be written as:

$$\frac{\partial}{\partial t}\chi(\Omega,t) = \frac{1}{i\hbar}[H(\Omega,t),\chi(\Omega,t) + \rho_0] - \Gamma_\Omega \chi(\Omega,t) \qquad \text{(V-101)}$$

Since the Hamiltonian depends on time, it is convenient to decompose the Hamiltonian and the density matrix into Fourier components, in addition to the expansion in rotation functions. The density matrix is decomposed as

$$\chi(t) = \sum_{n=-\infty}^{\infty} e^{in\omega t} Z^{(n)} \qquad \text{(V-102)}$$

in terms of the irradiation frequency ω. Each of the Fourier coefficients $\mathbf{Z}^{(n)}$ is a time independent spin operator, which has in general complex matrix elements. These operators contain the Ω-dependence of the density matrix. Due to the stochastic variable, one only observes averages of these operators over the distribution function of Ω. If the medium is in equilibrium with respect to this distribution, and the medium is isotropic, the average is given by

$$\langle Z^{(n)} \rangle = \int P_0(\Omega)\, Z^{(n)}(\Omega)\, d\Omega \qquad \text{(V-103)}$$

The n = 1 term is the one that rotates close to resonance, so this is the only one measured in practice for CW with weak irradiation. In fact, if one writes

$$Z^{(n)} = Z^{(n)\prime} + iZ^{(n)\prime\prime} \qquad \text{(V-104)}$$

where $\mathbf{Z}^{(n)\prime}$ and $\mathbf{Z}^{(n)\prime\prime}$ have only real matrix elements, then the power absorbed in the measurement is proportional to $\omega_1 \langle \mathbf{Z}^{(1)\prime\prime} \rangle$, because the observed signal is proportional to:

$$S(t) = \langle S_+ \rangle(t) = Tr\{S_+ \rho(t)\} = Tr\{S_+ \chi(t)\} \qquad \text{(V-105)}$$

Nevertheless, the equation of motion of the density matrix couples different $\mathbf{Z}^{(n)}$, so one has to solve in principle for all $\mathbf{Z}^{(n)}$ before calculating the signal from $\mathbf{Z}^{(1)}$.

The Hamiltonian has a very simple Fourier decomposition, since two of its three parts are constant in time and thus constitute its n = 0 term, and $\mathbf{H}_2(t)$ is a sum of an n = 1 term and an n = -1 term. Substituting the Fourier series into Eq. (V-101) one finds:

$$\sum_{n=-\infty}^{\infty} in\omega\, e^{in\omega t} Z^{(n)} = \frac{1}{i\hbar}\sum_{n=-\infty}^{\infty} e^{in\omega t}[H_0 + H_1(\Omega), Z^{(n)}] + \frac{1}{i\hbar}(e^{i\omega t} + e^{-i\omega t})\,\omega_1 [S_x, \rho_0]$$

$$+ \frac{1}{i\hbar}\omega_1 \sum_{n=-\infty}^{\infty} e^{in\omega t}(e^{i\omega t} + e^{-i\omega t})[S_x, Z^{(n)}] - \sum_{n=-\infty}^{\infty} e^{in\omega t}\Gamma_\Omega Z^{(n)} \qquad \text{(V-106)}$$

The commutator of $\mathbf{H}_0 + \mathbf{H}_1(\Omega)$ with ρ_0 does not appear, since it is equal to zero as a consequence of Eq. (V-99).

Multiplying by exp(-ikωt) and integrating over dt from -∞ to ∞ one obtains the k'th component of the equation. Convergence of the integrals can be ensured by including an overall decay factor in the density matrix, representing relaxation due to processes other than rotational diffusion. Alternatively one may take the integrals from -T to T, and then take the limit of the expressions for $T \to \infty$. The result in both cases is the Fourier transform of the previous equation:

$$i k \omega Z^{(k)} = \frac{1}{i\hbar}[H_0 + H_1(\Omega), Z^{(k)}] + (\delta_{k,1} + \delta_{k,-1})\frac{1}{i\hbar}\omega_1 [S_x, \rho_0]$$

$$+ \frac{1}{i\hbar}\omega_1 [S_x, (Z^{(k-1)} + Z^{(k+1)})] - \Gamma_\Omega Z^{(k)} \qquad \text{(V-107)}$$

This may be written somewhat differently, collecting all terms in which the $Z^{(k)}$ appear on one side (and renaming the index k as n):

$$i n \omega Z^{(n)} + \Gamma_\Omega Z^{(n)} - \frac{1}{i\hbar}[H_0 + H_1(\Omega), Z^{(n)}]$$

$$- \frac{1}{i\hbar}\omega_1 [S_x, (Z^{(n-1)} + Z^{(n+1)})] = (\delta_{n,1} + \delta_{n,-1})\frac{1}{i\hbar}\omega_1 [S_x, \rho_0] \qquad \text{(V-108)}$$

If one works in the basis which diagonalizes $\mathbf{H}_0$, then a matrix element of the commutator with $\mathbf{Z}^{(n)}$ is equal to:

$$[H_0, Z^{(n)}]_{i,j} = (H_0)_{i,i}(Z^{(n)})_{i,j} - (Z^{(n)})_{i,j}(H_0)_{j,j} = \hbar\,\omega_{ij}(Z^{(n)})_{i,j} \qquad \text{(V-109)}$$

where ω_{ij} is the transition frequency between energy levels i,j. From Eq. (V-105) it is clear that only density matrix elements χ_{ij} with $m_j - m_i = 1$ will contribute to the signal, so the discussion will be restricted from here on to the equations of motion for these elements only. In the commutator between $\mathbf{S}_x$ and $Z^{(q)}$:

$$[S_x, Z^{(q)}]_{i,j} = \sum_k (S_x)_{i,k}(Z^{(q)})_{k,j} - (Z^{(q)})_{i,k}(S_x)_{k,j} \qquad \text{(V-110)}$$

the relevant elements of $\mathbf{Z}^{(q)}$ will be those with $\Delta m = 0$ or 2. However, if one deals with doublet EPR where the electronic spin levels are only $m = -½$ and $m = ½$, there are no elements of the density matrix with $\Delta m = 2$. Alternatively, in more general EPR problems, elements with $|\Delta m| = 2$ will not be excited if the irradiation is weak. Starting from the equilibrium condition (Eq. (V-99)) in which only the diagonal elements of the density matrix are non-zero, and operating with the same type of weak irradiation continuously, one can only excite elements of the density matrix with $|\Delta m| = 1$. Pictorially, the steady weak irradiation can only rotate $\mathbf{S}_z$ (which has only elements with $\Delta m = 0$) to the x-y plane (S_x and S_y have matrix elements with $|\Delta m| = 1$). It cannot create non-zero elements with $|\Delta m| = 2$ (connected with operators like S_x^2, for example) or higher. Therefore in any of the experimental settings assumed here, the relevant elements of $\mathbf{Z}^{(q)}$ in Eq. (V-110) are only those with $\Delta m = 0$. If no degeneracy is assumed, these are diagonal elements, so that

$$[S_x, Z^{(q)}]_{i,j} = (S_x)_{i,j}\left((Z^{(q)})_{j,j} - (Z^{(q)})_{i,i}\right) \tag{V-111}$$

Thus Eq. (V-108) for one of the "relevant" (ij) index pairs becomes:

$$i\,n\,(\omega + \omega_{ij})\,Z^{(n)} + \Gamma_\Omega Z^{(n)} - \frac{1}{i\hbar}[H_1(\Omega), Z^{(n)}]_{ij} - \frac{1}{i\hbar}\omega_1 (S_x)_{i,j} \times$$

$$\times\left((Z^{(n-1)} + Z^{(n+1)})_{j,j} - (Z^{(n-1)} + Z^{(n+1)})_{i,i}\right) = (\delta_{n,1} + \delta_{n,-1})\frac{1}{i\hbar}\omega_1 [S_x, \rho_0]_{i,j} \tag{V-112}$$

Now one may introduce the expansion over eigenfunctions of Γ_Ω:

$$\chi = \sum_{L,M,K} C^L{}_{MK}\, \Psi^L{}_{MK}(\Omega) \tag{V-113}$$

for the density matrix, or equivalently the expansion of each of the $Z^{(n)}$ in the same manner:

$$Z^{(n)} = \sum_{L,M,K} C^{(n)L}{}_{MK}\, \Psi^L{}_{MK}(\Omega) \tag{V-114}$$

into Eq. (V-112). Multiplying from the left by the complex conjugate of a particular eigenfunction, $\Psi^{L^*}{}_{MK}(\Omega)$ and integrating over Ω one obtains the following equation:

$$i\,n\,(\omega + \omega_{ij})\,(C^{(n)L}{}_{MK})_{i,j} + \gamma_{LMK}(C^{(n)L}{}_{MK})_{i,j} - \frac{1}{i\hbar}\sqrt{8\pi^2/5}\int d\Omega\,(\Psi^L{}_{MK}(\Omega))^* \times$$

$$\times \sum_{\xi,m,m'} F_\xi^{(2,-m')}\Psi^2{}_{m,m'}(\Omega) \sum_{L',M',K'} \Psi^{L'}{}_{M'K'}(\Omega)\,[A_\xi^{(2,m)}, C^{(n)L'}{}_{M'K'}]_{i,j} - \frac{1}{i\hbar}\omega_1 (S_x)_{i,j} \times$$

$$\times\left(\left(C^{(n-1)L}{}_{MK} + C^{(n+1)L}{}_{MK}\right)_{j,j} - \left(C^{(n-1)L}{}_{MK} + C^{(n+1)L}{}_{MK}\right)_{i,i}\right) = (\delta_{n,1} + \delta_{n,-1})\frac{1}{i\hbar}\omega_1 [S_x, \rho_0]_{i,j} \tag{V-115}$$

The integral over the three angular functions may be carried out immediately according to Eq. (V-95). Thus Eq. (V-115) is a set of coupled linear algebraic equations for the (i,j) matrix elements of the operators $C^{(n)L}{}_{MK}$. It is convenient to write it as:

$$\sum_\beta X_{\alpha,\beta}\, C_\beta = V_\alpha \tag{V-116}$$

Here **C** is a vector containing, in some order, all elements $(C^{(n)L}{}_{MK})_{ij}$. For example, one may start by arranging the operators $C^{(n)L}{}_{MK}$ according to their (n) values, then arrange the operators within each (n) according to their values of L,M and K. After all operators have been arranged, one may arrange all (i,j) matrix elements of each operator in a certain order. Thus one has a column vector including all such matrix elements, and the index β (or α) stands for the set of indices (n),L,M,K,i,j. The matrix **X** operating on this column vector consists of a diagonal part, multiplying each element of the column vector by a number, and of a potentially non-diagonal part which couples the different elements among them. The column vector **V** includes the elements on the right hand side of Eq. (V-115), arranged in the same order as chosen for the vector **C**.

From Eq. (V-115), the general element of the matrix **X** is equal to:

$$(X)_{nLMKij,n'L'M'K'i'j'} = \delta_{n,n'}\delta_{L,L'}\delta_{M,M'}\delta_{K,K'}\delta_{i,i'}\delta_{j,j'}\left(in(\omega+\omega_{ij})+\gamma_{LMK}\right)$$

$$-\delta_{i,i'}\delta_{j,j'}\frac{1}{i\hbar}\sqrt{8\pi^2/5}\sum_{\xi,m,m'}F_{\xi}^{(2,-m')}\sum_{L',M',K'}(-1)^{M-K}\sqrt{(2L+1)(2L'+1)}\times$$

$$\times\begin{pmatrix} L & 2 & L' \\ -M & m & M' \end{pmatrix}\begin{pmatrix} L & 2 & L' \\ -K & m' & K' \end{pmatrix}[A_{\xi}^{(2,m)}, C^{(n)L'}{}_{M'K'}]_{i,j}$$

$$-\delta_{L,L'}\delta_{M,M'}\delta_{K,K'}\frac{1}{i\hbar}\omega_1(S_x)_{i,j}\left((\delta_{n',n-1}+\delta_{n',n+1})(\delta_{i',j}\delta_{j',j}-\delta_{i',i}\delta_{j',i})\right) \qquad \text{(V-117)}$$

Different values of the Fourier order (n) are coupled among them only through the term proportional to ω_1. In CW experiments the irradiation is usually weak: $\omega_1 \ll \omega_0$, so that this term is very small compared with the other terms in the equations. It is therefore possible to neglect this term in all cases which do not involve saturation (i.e., very strong irradiation). If it is neglected, one only needs to solve for the n = 1 terms, which contribute to the signal, and then the frequency dependent term in **X** becomes the diagonal term $\omega + \omega_{ij} \equiv (\Delta\omega)_{ij}$, the off-resonance value for the (i,j) transition. In many cases, this term is equal to the unit matrix for the (i,j) part of the full matrix, multiplied by $(\Delta\omega)_{ij}$ (see next Chapter).

The elements of the vector **V** are:

$$(V)_{nLMKij} = (\delta_{n,1}+\delta_{n,-1})\frac{1}{i\hbar}\omega_1[S_x, \int d\Omega\,(\Psi^L{}_{MK}(\Omega))^*\rho_0]_{i,j} \qquad \text{(V-118)}$$

If ρ_0 depends on Ω due to relatively large interactions in $\mathbf{H}_1(\Omega)$ then the result of the integration is non-trivial. If the molecule is in a non-isotropic medium, its ρ_0 will be effectively multiplied by an orientation dependent distribution function, as will be seen later on (even if the interactions in $\mathbf{H}_1(\Omega)$ are weak), and again the integral will be non-trivial. However, if neither of these problems exists, the integral is simply equal to

$$\int d\Omega\,(\Psi^{L}_{MK}(\Omega))^{*}\,\rho_0 \;=\; \delta_{L,0}\,\delta_{M,0}\,\delta_{K,0}\,\frac{1}{\sqrt{8\pi^2}}\,\rho_0 \qquad \text{(V-119)}$$

This is the situation in doublet EPR experiments in isotropic media. In triplet EPR $\mathbf{H}_1(\Omega)$ is relatively large, and the integral has some additional terms. Only in very high magnetic fields (with Larmor frequencies of the order of 10^2 GHz) the Zeeman terms would be very small compared with $\mathbf{H}_1(\Omega)$ of a triplet, so Eq. (V-118) would apply to it in isotropic media.

In order to continue from this point on, more information is needed about the Hamiltonian. In the next Chapter several different cases will be examined, including a case of saturation. Here, only one important example will be presented. Assume an EPR experiment with the following constant Hamiltonian:

$$H_0 \;=\; \omega_0 S_z - \sum_j \omega_{n(j)} I_{z_j} - \gamma_e \sum_j a_j S_z I_{z_j} \qquad \text{(V-120)}$$

The first term is the Zeeman term for a single electron, with $\omega_0 \equiv g_s\,\beta_e\,B_0$ (g_s is the average g-value of the electron and B_0 is the magnitude of the constant magnetic field). The second term is a sum over the Zeeman terms of some nuclei, with $\omega_{n(j)} \equiv \gamma_j\,B_0$ being the Larmor (angular) frequency for the j'th nucleus. The third term is a sum over the isotropic part of the hyperfine interaction between the electron and the nuclei. $\mathbf{H}_1(\Omega)$ is given by Eq. (V-97), where the interaction tensors are the g-tensor and the hyperfine tensor.

Therefore, in the high temperature approximation and neglecting the contribution of $\mathbf{H}_1(\Omega)$ to ρ_0:

$$[S_x, \rho_0] \;=\; -iq\,S_y\Big\{\omega_0 - \gamma_e \sum_j a_j I_{z_j}\Big\} \qquad \text{(V-121)}$$

where q is the normalization factor of the equilibrium density matrix, divided by $k_B T$. This can be substituted directly into Eq. (V-118) for the case of a small $\mathbf{H}_1(\Omega)$. Then $\mathbf{V}$ is proportional to: ($\mathbf{S}_+$ - $\mathbf{S}_-$), which usually leads to a simplification of the line shape calculation, as will be explained in the next section. Further details about the solution of problems of this type appear in the discussion of specific cases in the next Chapter.

V.4 Numerical Methods of Solution in the Eigenfunctions Method

(a) The general computational scheme

The method of eigenfunctions presented in the previous Section leads, as shown there, to a set of linear algebraic equations, from which the line shape may be calculated. We shall summarize here the procedure which is applied in practice in this method, emphasizing the computational implications of the various steps. At the end of this summary some possible numerical difficulties will become evident, and methods for overcoming them are described in the next two subsections.

Given a problem with a known (or assumed) Hamiltonian and a known (or assumed) stochastic process, one has to construct for it the stochastic Liouville equation in the form (V-4). Having done this, one has to find eigenfunctions of the stochastic superoperator $\mathbf{\Gamma}_\Omega$. If such eigenfunctions are found, and they form a complete set of functions for the set of parameters Ω, one may continue to the next step. If the set of functions is not complete, it may still be possible to continue in some cases, but then the validity of each step has to be checked carefully. Granted that the eigenfunctions and the corresponding eigenvalues are known, one expands the density matrix and the Hamiltonian in terms of the eigenfunctions. The expansion coefficients of the density matrix and the Hamiltonian have to expressed in the same form, so that the commutators between them can be calculated. It is also necessary that these commutators are expressed in the same form, which will usually occur if they are all expanded in terms of a complete set of spin operators. Then the matrix elements of the commutators appearing on the left hand side of Eq. (V-115) are replaced by matrix elements of operators belonging to that set.

One then has to solve Eq. (V-116), which is a set of linear algebraic equations in the unknowns $\mathbf{C}_\beta$, and then the line shape is calculated from $\mathbf{Z}^{(1)}$ through Eqs. (V-103), (V-114). The line shape of Eq. (V-105), Fourier transformed as in Eq. (V-107), may therefore be written as:

$$S(\omega) = Tr\left\{S_+ X^{-1} V\right\} = \sum_{i,j} \left(S_+\right)_{i,j} \left(X^{-1} V\right)_{j,i} \qquad \text{(V-122)}$$

In many cases a simplification of this formula is possible, because $\mathbf{V}$ is proportional to $\mathbf{S}_x$ or $\mathbf{S}_y$ as seen above, and is thus a sum of terms proportional to $\mathbf{S}_+$ and $\mathbf{S}_-$. If the electron-nucleus interaction (i.e., hyperfine interaction) and electron-electron interactions (e.g., the ZFS interactions in a triplet) are not too strong, $\mathbf{X}$ does not connect states with different values of $\mathbf{S}_z$. In such a case , also $\mathbf{X}^{-1}$ will not connect states with different values of $\mathbf{S}_z$. The non-zero contribution to Eq. (V-122) will then come only from the term in $\mathbf{V}$ proportional to $\mathbf{S}_-$, so the equation may be rewritten as:

$$S(\omega) = Tr\left\{V^\dagger X^{-1} V\right\} = \langle V \,|\, X^{-1} \,|\, V\rangle \qquad \text{(V-123)}$$

This is the usual form of the equation in doublet EPR. In triplet EPR the equation has to be modified, as will be discussed in the next Chapter. At this point Eq. (V-123) will serve as the basis for the computational considerations. We shall also make the assumption that the stochastic superoperator represents a process of rotational diffusion of one type or another (Brownian, jumps, etc. - examples will be discussed in the next Chapter).

In actual applications of the above expressions, the first point to be considered is the dimensions of the matrix $\mathbf{X}$ and the vector $\mathbf{V}$. Even when the coupling between different (n) values is neglected, as done here, the dimensions of the problem are infinite. This is because L = 0,1,2,... (to infinity) and for each L one has 2L+1 values of M, and for each of these there are 2L+1 values of K. Finally, for each set of values of L, M and K there are several possible index pairs (i,j) connected to quantum mechanical transitions or elements of the density matrix (the diagonal elements (ii) are also included, in general). It is obvious that a numerical solution can be found only if one may approximate the infinite matrix by a finite dimensional one. Doing such an approximation requires knowledge of the structure of the matrix. This depends on the specific problem under considration, but some general rules can be stated for all relevant cases in magnetic resonance. First, it follows from Eq. (V-117) that a particular value of L can only be coupled to values L' which do not differ from it by more than 2: $|L - L'| \leq 2$. This results from the fact that interactions in magnetic resonance can be described by zero or second rank tensors. Second, the only off-diagonal matrix elements in **X** are those proportional to the interaction constants $F_\xi^{(2,-m')}$, whereas the diagonal contains both elements proportional to $F_\xi^{(2,-m')}$ and elements proportional to the off-resonance term $\omega + \omega_{ij}$ and the eigenvalues γ_{LMK} of the stochastic superoperator.

In practice one only needs those components of the density matrix which are proportional to those rotation functions included in the equilibrium distribution $p_0(\Omega)$, because only they will contribute to the average result in Eq. (V-103). If the medium is isotropic, only the L=0, M=0, K=0 component is needed. In principle, since all L values are coupled indirectly by steps of 1 or 2 ($|L - L'| = 1$ or 2), one has to solve the full infinite matrix in order to get a correct answer. In practice, however, the coupling between different L values is only important if the off-diagonal terms are not small compared with the diagonal terms. The interaction constants are independent of L, M and K, and so is the off-resonance term. Moreover, the off-resonance term is of the same order of magnitude as those interaction terms. However, $\gamma_{LMK} \approx R L^2$ for large L values (R is the rotational diffusion rate constant) so the diagonal elements increase with the square of L. Thus the off-diagonal elements become unimportant when L is large enough so that

$$RL^2 \gg F \tag{V-124}$$

Here F is the largest interaction constant among the $F_\xi^{(2,-m')}$. Therefore the infinite set of equations can be truncated at some maximum L value L_{max} such that $(L_{max})^2 \gg F/R$. This is only a rough criterion for truncation, and more accurate criteria for practical purposes can be found in each problem by careful analysis of the problem, as will be mentioned below. However, this is a general criterion and is based on analytical considerations, whereas the more accurate criteria are not of general validity, and are based on various empirical estimates. In practical cases one can often restrict the M and K values which are used with a given L, but this depends on the details of the specific case. Ignoring such possible restrictions for the moment, the matrix **X** is of dimension $d_{trun.} \times d_{trun.}$, and the vector **V** is of dimension $d_{trun.}$, where $d_{trun.}$ is equal to the number N of quantum mechanical transitions (number of (ij) pairs) multiplied by the number of rotation functions up to $L_{max.}$:

$$d_{trun.} = N\sum_L (2L+1)^2 = N\left\{\left[\frac{2}{3}L_{max}(2L_{max}+1) + 2L_{max} + 1\right](L_{max}+1)\right\} \tag{V-125}$$

It is of great practical importance that the matrix is very sparse, i.e., it has a relatively small number of non-zero elements. This can be visualized as follows. Arrange all $\Psi^L_{MK}(\Omega)$ according to L values, and inside each L value, by the M, K and (ij) values. All functions with the same L value are regarded as comprising one distinct subset of the full basis. Each such set of basis functions is labeled by its L value. Then the matrix consists of blocks labeled as (L,L'), connecting the (L) subset with the (L') subset. Part of the blocks are on the diagonal (connecting each L value with itself) and part of them connect different L values. As mentioned already, the off-diagonal blocks cannot connect L values which differ by more than 2. In a particular "row" with index L the non-zero blocks occur in the columns with index L' = L-2, L-1, L, L+1 and L+2 (unless one of these numbers is negative, because then that column does not exist). Thus if L_{max} is large, the matrix consists formally of $(L_{max} + 1)^2$ blocks (of L-dependent sizes !), but only about $5(L_{max} + 1)$ of them contain non-zero elements.

From the truncation criterion it is clear that the magnitude of the required L_{max} will increase when R decreases, i.e., when motion becomes slower. This is the same situation as found above in the perturbation expansion and in the cumulant expansion: faster motion leads to a smaller number of terms, and a simple approximation. Slower motion leads to more terms, and a more complicated calculation. However, as in the cumulant expansion, the calculation can be carried out effectively also for slow motions, and - unlike the cumulant method - the analytical form of the solution does not become more complicated with the slowing down of the motion. It is only the dimensions of the calculation which become larger, so that one only needs an effective way of dealing with a large numerical problem of this kind. If, for example, $F/R \approx 10^3$ then one needs $L_{max} \approx 10^2$ (it may be smaller or larger than this number by, say, a factor of 2). Then the total dimension is $d_{trun.} \approx 10^6$ or, if L_{max} can be decreased or increased by a factor of 2, then $d_{trun.}$ is between 10^5 and 10^7. In actual cases the dimension can be significantly decreased because of the above mentioned restrictions on M and K. Nevertheless it is clear that fairly large dimensions are involved, and therefore it is highly important to find numerical methods which do the computations in the most effective way, taking advantage of the great sparsity of the matrix.

(b) Using the Lanczos algorithm for diagonalization

There are two different ways of solving Eq. (V-116). One is to solve it directly by inverting the matrix **X**. This has to be done for every value of $\omega - \omega_0 \equiv \Delta\omega$ (for any (ij), $(\Delta\omega)_{ij}$ is equal to $\Delta\omega$ plus a relatively small term which usually depends only on (ij), and not on ω and ω_0). The number of values of $\Delta\omega$ is the number of points in the line shape, which is typically a few hundreds. The other possibility is to subtract from **X** the matrix $\Delta\omega\mathbf{I}$ and diagonalize the resulting matrix. This is a somewhat longer computation, but once done it avoids the need to repeat the whole calculation for each value of $\Delta\omega$. Both problems are standard problems in linear algebra, but they are characterized here by the possibly large size of the matrix and by its sparseness.

The Lanczos algorithm (LA) is a very effective technique for converting a large and sparse matrix to tridiagonal form. A tridiagonal matrix can be diagonalized very efficiently, so bringing the original matrix to tridiagonal form is the major step in the whole calculation (if the second alternative for solving (V-116) is chosen). A brief review of the theory and use of the method will now be given, and then its application and its practical advantages for the SLE will be explained. The overall approach of the LA is to perform recursive steps in order to create successively larger tridiagonal matrix approximations to the original matrix. These steps form projections to subspaces (in the present context - quantum mechanical subspaces), which approximate the exact solution subspace. The procedure is continued until one reaches convergence, namely, until the result is sufficiently accurate. In all practical cases the number of steps needed for convergence is much smaller than the dimension of the matrix. This reduction in size is especially dramatic when the dimension of the matrix is very large.

The original algorithm is constructed for real symmetric matrices as follows. Suppose one is given a real vector **v** of length N, and a real symmetric N × N matrix **A**, and one is interested in a simple transformation of **A** which will lead to a tridiagonal form in an appropriate basis. In order to achieve this, one first creates a sequence of vectors: **v**, **A v**, $\mathbf{A}^2$ **v**,..., $\mathbf{A}^{n-1}$ **v** (known as a Krylov sequence). The tridiagonal form approximating **A** is constructed in the n-dimensional subspace spanned by these Krylov vectors, in the following way. The Gram-Schmidt orthogonalization procedure is applied on this sequence of vectors, generating from them a set of mutually orthogonal vectors. The procedure creates an orthonormal basis $\{\mathbf{q}_1, \mathbf{q}_2, \ldots, \mathbf{q}_n\}$ for the subspace, and at the same time also generates the transformed matrix **A**.

The first few steps of the procedure will now be described explicitly, and then the general formulas of the algorithm will be derived by induction. The first step is to define the first basis vector by

$$\beta_0 q_1 = v \tag{V-126}$$

and require that $\mathbf{q}_1$ be normalized: $(\mathbf{q}_1)^{tr}\, \mathbf{q}_1 = 1$, where the row vector $(\mathbf{q}_1)^{tr}$ is the transpose of the column vector $\mathbf{q}_1$. This implies that $\beta_0 = (\mathbf{v}^{tr}\, \mathbf{v})^{1/2}$. The next step is to define a linear combination of **v** and **A v**, imposing the requirement that this combination is both normalized and orthogonal to $\mathbf{q}_1$. The combination is defined by

$$\beta_1 q_2 = \left(A - \alpha_1 I\right) q_1 \tag{V-127}$$

in which **I** is the unit matrix. Multiplying Eq. (V-127) from the left by $(\mathbf{q}_1)^{tr}$ leads to:

$$0 = (q_1)^{tr} A\, q_1 - \alpha_1 \quad \rightarrow \quad \alpha_1 = (q_1)^{tr} A\, q_1 \tag{V-128}$$

Multiplying Eq. (V-127) from the left by $(\mathbf{q}_2)^{tr}$ leads to:

$$\beta_1 = (q_2)^{tr} A\, q_1 \tag{V-129}$$

Since symmetry of a matrix is preserved in a similarity transformation, the symmetry of **A** in the original basis implies its symmetry in the new basis, so that

$$(q_2)^{tr} A\, q_1 = (q_1)^{tr} A\, q_2 \tag{V-130}$$

In the third step $\mathbf{q}_3$ is generated from **v**, **A v**, and $\mathbf{A}^2$ **v**:

$$\beta_2 q_3 = \left(A - \alpha_2 I\right) q_2 - \gamma_1 q_1 \tag{V-131}$$

Again, the conditions of orthonormality are imposed. Multiplying the equation from the left by $(\mathbf{q}_1)^{tr}$ results in

$$\gamma_1 = (q_1)^{tr} A\, q_2 = \beta_1 \tag{V-132}$$

Multiplying by $(\mathbf{q}_2)^{tr}$ leads to:

$$\alpha_2 = (q_2)^{tr} A\, q_2 \tag{V-133}$$

and multiplication by $(q_3)^{tr}$ gives

$$\beta_2 = (q_3)^{tr} A\, q_2 \tag{V-134}$$

The significant change in this process comes in the fourth step, defined through

$$\beta_3\, q_4 = (A - \alpha_3 I)\, q_3 - \gamma_2 q_2 - \gamma_1\, q_1 \tag{V-135}$$

If this equation is multiplied from the left by $(q_1)^{tr}$ then, using Eq. (V-127):

$$\gamma_1 = (q_1)^{tr} A\, q_3 = (A\, q_1)^{tr} q_3 = (\beta_1 q_2 + \alpha_1 q_1)^{tr} q_3 = 0 \tag{V-136}$$

Therefore $\mathbf{q}_4$ depends only on three out of the four vectors **v**, ..., $\mathbf{A}^3$ **v**, and **A** does not connect $\mathbf{q}_1$ with $\mathbf{q}_3$. Multiplying Eq. (V-135) by $(q_2)^{tr}$, $(q_3)^{tr}$ and $(q_4)^{tr}$ respectively one obtains the following equations:

$$\gamma_2 = (q_2)^{tr} A\, q_3 = \beta_2 \tag{V-137}$$

$$\alpha_3 = (q_3)^{tr} A\, q_3 \tag{V-138}$$

$$\beta_3 = (q_4)^{tr} A\, q_3 \tag{V-139}$$

Thus for m > 1 the general formula for a basis vector emerges as

$$\beta_m\, q_{m+1} = (A - \alpha_m I)\, q_m - \beta_{m-1}\, q_{m-1} \tag{V-140}$$

with the coefficients

$$\alpha_m = (q_m)^{tr} A\, q_m \tag{V-141}$$

$$\beta_m = (q_{m+1})^{tr} A\, q_m \tag{V-142}$$

Equations (V-140) - (V-142) will now be proved by induction. Assume they are correct for $m \le M$. Then for $m = M + 1$ one may define:

$$\beta_{M+1}\, q_{M+2} = (A - \alpha_{M+1} I)\, q_{M+1} - \gamma\, q_M - \sum_{j=1}^{M-1} \gamma_j\, q_j \tag{V-143}$$

and it remains to prove that $\gamma = \beta_M$ and that $\gamma_j = 0$. Multiplying the equation from the left by any $(q_k)^{tr}$ such that $k \le M - 1$:

$$\gamma_k = (q_k)^{tr} A\, q_{M+1} = (A\, q_k)^{tr} q_{M+1} = (\beta_k q_{k+1} + \alpha_k q_k + \beta_{k-1} q_{k-1})^{tr} q_{M+1} = 0 \tag{V-144}$$

so that all the γ_j vanish. Multiplying by $(\mathbf{q}_M)^{tr}$ results in

$$\gamma = (q_M)^{tr} A\, q_{M+1} = \beta_M \tag{V-145}$$

by which the validity of the induction process has been proved.

The vectors of the new basis are the columns in the transformation matrix $\mathbf{Q}_{(n)}$ (the subscript indicates the number of steps performed), taking **A** or any other matrix from the original basis to the new basis. The transformed matrix is:

$$T_{(n)} = (Q_{(n)})^{tr} A\, Q_{(n)} \tag{V-146}$$

and its matrix elements have already been calculated in the process. The matrix **A** is therefore transformed to

$$T_{(n)} = \begin{pmatrix} \alpha_1 & \beta_1 & 0 & 0 & \dots & 0 & 0 & 0 \\ \beta_1 & \alpha_2 & \beta_2 & 0 & \dots & 0 & 0 & 0 \\ 0 & \beta_2 & \alpha_3 & \beta_3 & \dots & 0 & 0 & 0 \\ \dots & \dots & \dots & \dots & \dots & \dots & \dots & \dots \\ 0 & 0 & 0 & 0 & \dots & \beta_{n-2} & \alpha_{n-1} & \beta_{n-1} \\ 0 & 0 & 0 & 0 & \dots & 0 & \beta_{n-1} & \alpha_n \end{pmatrix} \tag{V-147}$$

This is indeed a tridiagonal matrix, which can be handled easily using standard methods.

The whole process can be reformulated in the language of quantum mechanical Hilbert space as follows. Instead of the arbitrary vector **v** one starts from the Hilbert space "ket" vector $|\mathbf{v}\rangle$, and instead of the matrix **A** one has an operator **A**, represented by a matrix according to the usual rules. At this stage it is still assumed that the vectors and the matrix are real, and that the matrix is symmetric. The "starting vector" is defined again by

$$\beta_0 |\Phi_1\rangle = |v\rangle \tag{V-148}$$

In the context of the SLE, **v** is simply the vector **V** defined in Eqs. (V-116) - (V-118). Now define for any k ($1 \le k \le n$) the following sum of projection operators:

$$P_k = \sum_{j=1}^{k} |\Phi_j\rangle\langle\Phi_j| \tag{V-149}$$

Then a set of basis vectors $\{|\Phi_1\rangle, |\Phi_2\rangle, \ldots, |\Phi_n\rangle\}$ for a subspace of Hilbert space is generated from the set of vectors $\{|\mathbf{v}\rangle, \mathbf{A}\,|\mathbf{v}\rangle, \ldots, \mathbf{A}^{n-1}\,|\mathbf{v}\rangle\}$ through the expression:

$$\beta_k \,|\Phi_{k+1}\rangle = \{I - P_k\} A \,|\Phi_k\rangle \tag{V-150}$$

where **I** is the unit operator. For k = 1 this simplifies to

$$\beta_1 \,|\Phi_2\rangle = \{I - |\Phi_1\rangle\langle\Phi_1|\} A \,|\Phi_1\rangle \tag{V-151}$$

Operating from the left with the "bra" vector $\langle\Phi_1|$ on Eq. (V-151), it is immediately clear that the constant α_1 defined above appears here as:

$$\alpha_1 = \langle\Phi_1|A|\Phi_1\rangle \tag{V-152}$$

Operating with the "bra" vector $\langle\Phi_2|$ on Eq. (V-151) one obtains

$$\beta_1 = \langle\Phi_2|A|\Phi_1\rangle \tag{V-153}$$

which is analogous to the expressions derived above. For k = 2, Eq. (V-150) gives

$$\beta_2 \,|\Phi_3\rangle = \{I - |\Phi_1\rangle\langle\Phi_1| - |\Phi_2\rangle\langle\Phi_2|\} A \,|\Phi_2\rangle = \{I - \langle\Phi_2|A|\Phi_2\rangle\}\,|\Phi_2\rangle$$

$$- \{\langle\Phi_1|A|\Phi_2\rangle\}\,|\Phi_1\rangle = \{I - \alpha_2 I\}\,|\Phi_2\rangle - \beta_1\,|\Phi_1\rangle \tag{V-154}$$

again in complete analogy to the expressions obtained previously. For higher values of m a similar formula can be written, using the same method as above to prove that **A** does not have non-zero elements between basis vectors $|\Phi_j\rangle$ and $|\Phi_k\rangle$ for which: $|j - k| \geq 2$.

So far it was assumed throughout the calculation that only real numbers are involved. It is not difficult to generalize all this to a case in which the matrix **A** is complex hermitian, i.e., $(\mathbf{A})^* = (\mathbf{A})^{tr}$. However, in our applications the matrix **X** is complex but not hermitian. From Eq. (V-78) or (V-101) it may be expressed as

$$X = \frac{1}{i\hbar}[H, \ldots] - \Gamma \tag{V-155}$$

The Liouville - von Neumann commutator is a hermitian matrix in Liouville space (see Section I.6), so denoting it by **B** it satisfies: $(\mathbf{B})^* = (\mathbf{B})^{tr}$. Being multiplied by $(i\hbar)^{-1}$ it becomes anti-hermitian ,i.e., $\hbar^{-1}(-i\mathbf{B})^* = \hbar^{-1}(i\mathbf{B})^{tr}$. This is still not a serious problem, because one could diagonalize the hermitian **B** and obtain trivially the diagonalization of i**B**. Formally this means one would transform the non-hermitian matrix to a hermitian one - in this case, simply by multiplying it by a scalar. The difficulty is that the stochastic operator term is a real term which is normally symmetric. The total matrix is therefore not hermitian (even not necessarily symmetric), and cannot be made hermitian by a simple transformation, because

$$\frac{1}{\hbar}(-i B)^* - (\Gamma)^* = \frac{1}{\hbar}(i B)^{tr} - \Gamma \tag{V-156}$$

Nevertheless, in most practically relevant cases there is a simple basis transformation which converts **X** to a complex symmetric form, i.e., $(\mathbf{X})^{tr} = \mathbf{X}$. The transformation will be described in the next Chapter, in connection with cases in which it can be carried out. Here we shall only note an important feature relevant to the application of LA in those cases.

It can be shown that the main difficulty in applying the LA to complex symmetric matrices is caused by the use of the standard scalar product, defined by $\langle\Phi|\Psi\rangle = (\Phi^*)^{tr}\,\Psi$. This scalar product occurs every time an equation is multiplied from the left by a "bra" vector $\langle\Phi|$. It has been proved that the problem is solved by using a modified scalar product, in which the complex conjugation is omitted:

$$\langle\Phi|\Psi\rangle \equiv (\Phi)^{tr}\,\Psi \tag{V-157}$$

so that the condition of orthonormality, $\langle\Phi_j|\Phi_k\rangle = \delta_{j,k}$ is modified to:

$$(\Phi_j)^{tr}\,\Phi_k = \delta_{j,k} \tag{V-158}$$

With this modification the LA can be applied to the problem of tridiagonalizing the complex symmetric **X**.

Finally, a few words should be said about the implementation of the LA on a computer. One of the great advantages of the algorithm is that only non-zero elements of the matrix have to be stored, and these elements are not altered throughout the calculation. No new matrix elements are generated during the calculation. The algorithm is very simple, so it can be coded in a very short computer program. It converges relatively fast to the correct solution subspace, and thus allows one to approximate a high-dimensional solution with a small number of components of this solution.

Numerical applications of the LA to problems in EPR have shown that the method may not generate very accurately specific eigenvalues or eigenvectors, but it reproduces very well the total subspace spanned by all those eigenvectors which are of greatest importance. These are the eigenvectors corresponding to eigenvalues with small real parts, i.e., with slow decay rates. It turns out that most eigenvalues represent fast decaying components of the signal, so that only a relatively small number of components is needed to generate the signal to a good accuracy. The great success of the method results from its ability to find the relatively small number of highly significant eigenvectors, possibly in some linear combinations, out of the very large number of eigenvectors of **A**.

An additional advantage of the method is the close relation between the LA solution and expressions obtained by certain projection methods important in statistical mechanics. Thus the solution has a clear physical meaning, besides serving as a useful numerical approximation.

There are nevertheless some disadvantages to the method. The first is that numerical roundoff errors accumulate to spoil the ortogonality of the vectors $\mathbf{q}_j$ created by the algorithm. If one chooses to work to a value of n which is much larger than necessary, the quality of the spectrum will start deteriorating instead of improving continuously with increasing n. The method also has no clear criterion to determine the degree of inaccuracy in each stage in the calculation. Moreover, the LA gives no indication as to what is the minimal value of N (the original dimension of the problem) which will lead to a sufficiently accurate solution. One therefore needs to try several different basis sets in order to decide which of them is optimal for the given case. All these problems can be eliminated by the related method of conjugate gradients, which will now be discussed.

(c) *Using the conjugate gradients technique for inversion*

An alternative technique to deal with the original problem is the conjugate gradients method (CGM) which takes a different viewpoint, but is equivalent to the Lanczos method. Rather than trying to diagonalize the matrix **X** in Eq. (V-116), one tries in the CGM to approximate the solution vector **C** to a sufficiently good accuracy. This is equivalent to inverting **X** and calculating $\mathbf{C} = \mathbf{X}^{-1}\,\mathbf{V}$. More generally, a matrix **A** of dimension N×N and a real vector **v** of dimension N are given. The problem is to find a vector **u** which solves the equation:

$$A u = v \tag{V-159}$$

which is of the same form as Eq. (V-116). At this stage it is assumed that the matrix is real symmetric: $\mathbf{A}^{tr} = \mathbf{A}$ where $\mathbf{A}^{tr}$ is the transpose of **A**. The matrix is also assumed to be positive definite, i.e., for any non-zero N-dimensional vector **x**: $\mathbf{x}^{tr}\,\mathbf{A}\,\mathbf{x} > 0$. Consequently the matrix must be non-singular. The CGM approximates Eq. (V-159) by defining a residual vector

$$r_k \equiv v - A u_k \tag{V-160}$$

which would be equal to zero if the approximate vector $\mathbf{u}_k$ equals **u**. The equation is solved iteratively. The vector $\mathbf{u}_k$ is the k'th order approximation to **u**, and $\mathbf{r}_k$ is the residual, or the error, at the k'th stage. In order to minimize the error one first defines the functional:

$$f[y] \equiv \frac{1}{2} y^{tr} A y - y^{tr} v \tag{V-161}$$

where **y** is any real N-dimensional column vector, and the row vector $\mathbf{y}^{tr}$ is its transpose. In particular, if **y** = **u** where **u** is the solution of Eq. (V-159), then

$$f[u] \equiv \frac{1}{2} u^{tr} A u - u^{tr} v = -\frac{1}{2} u^{tr} A u = -\frac{1}{2}\left(A^{-1} v\right)^{tr} v = -\frac{1}{2} v^{tr} A^{-1} v \tag{V-162}$$

This is the minimum value of the functional, as will now be proved. Suppose one adds to **u** an arbitary real vector **z**. Then

$$f[u+z] \equiv \frac{1}{2}(u+z)^{tr}A(u+z) - (u+z)^{tr}v = f[u] + \frac{1}{2}u^{tr}Az + \frac{1}{2}z^{tr}Au$$

$$+ \frac{1}{2}z^{tr}Az - z^{tr}Au = f[u] + \frac{1}{2}z^{tr}Az \tag{V-163}$$

The symmetry of **A** has been used in the final equality. Since **A** is assumed to be positive definite, $\mathbf{z}^{tr}\,\mathbf{A}\,\mathbf{z} \geq 0$ so that $f[\mathbf{u}+\mathbf{z}] \geq f[\mathbf{u}]$, which proves that the minimum of f is attained for the argument $\mathbf{y} = \mathbf{u}$.

It is now clear that finding a vector which minimizes the functional is equivalent to solving the set of linear algebraic equations in Eq. (V-159). A well known mathematical procedure for minimization is the method of steepest descent, which in the present context would try to minimize at each step along the direction of the residual vector $\mathbf{r}_j$ of that stage. However, it is possible to show that such a calculation may lead in some cases to a situation in which a new step spoils part of the achievement of previous steps (See Ref. 22 below). In order to avoid this problem, a different method was developed. The difference between two successive residual vectors is defined as:

$$r_j - r_{j-1} = -A(u_j - u_{j-1}) \equiv -a_j A p_j \tag{V-164}$$

This equation defines the **conjugate direction vectors** $\mathbf{p}_j$ (the reason for this name will become apparent later on). A sufficient condition for Eq. (V-164) is:

$$a_j p_j = u_j - u_{j-1} \tag{V-165}$$

If the matrix **A** is non-singular (as it is assumed to be), this is also a necessary condition for Eq. (V-164). Suppose $\mathbf{p}_j$ is given by some formula. What value of the parameter a_j will make $\mathbf{u}_j$ the best approximation to **u** ? Obviously, the required value is that which minimizes $f[\mathbf{u}_j]$, since this will bring $\mathbf{u}_j$ as close as possible to the correct solution **u**. Therefore the derivative of $f[\mathbf{u}_j]$ according to a_j will be zero at this value of a_j. Using the symmetry of **A**:

$$f[u_j] \equiv \frac{1}{2}(u_{j-1} + a_j p_j)^{tr} A (u_{j-1} + a_j p_j) - (u_{j-1} + a_j p_j)^{tr} v =$$

$$= f[u_{j-1}] + a_j (p_j)^{tr} A u_{j-1} - a_j (p_j)^{tr} v + \frac{1}{2}(a_j)^2 (p_j)^{tr} A p_j \tag{V-166}$$

The condition $\partial f[\mathbf{u}_j]/\partial a_j = 0$ implies:

$$(p_j)^{tr} A u_{j-1} - (p_j)^{tr} v + a_j (p_j)^{tr} A p_j = 0 \tag{V-167}$$

which, using Eq. (V-160) becomes:

$$a_j = \frac{(p_j)^{tr} r_{j-1}}{(p_j)^{tr} A p_j} \tag{V-168}$$

With this value for the parameter, Eq. (V-166) becomes:

$$f[u_j] = f[u_{j-1}] - a_j (p_j)^{tr} r_{j-1} + \frac{1}{2}(a_j)^2 (p_j)^{tr} A p_j =$$

$$= f[u_{j-1}] - \frac{1}{2} \frac{\left((p_j)^{tr} r_{j-1}\right)^2}{(p_j)^{tr} A p_j} \tag{V-169}$$

It is clear that a non-zero product $(\mathbf{p}_j)^{tr}\, \mathbf{r}_{j-1}$ lowers the value of the functional, bringing it closer to the desrired minimum point. It follows also that p_j is orthogonal to $\mathbf{r}_j$, because multiplying Eq. (V-164) from the left by $(\mathbf{p}_j)^{tr}$ one obtains:

$$(p_j)^{tr} (r_j - r_{j-1}) = -a_j (p_j)^{tr} A p_j = -(p_j)^{tr} r_{j-1} \quad \rightarrow \quad (p_j)^{tr} r_j = 0 \tag{V-170}$$

Now that the implications of the j'th step are known, it remains to set up a systematic algorithm for calculating (V-160) iteratively. Before writing down such an algorithm we should summarize what we already know about the result of doing these iterations. Eq. (V-165) indicates that, having done j-1 steps, the j'th iteration can be calculated using $\mathbf{p}_j$, so $\mathbf{u}_j$ is given recursively by

$$u_j = u_0 + a_1 p_1 + a_2 p_2 + \dots + a_{j-1} p_{j-1} + a_j p_j \tag{V-171}$$

It is reasonable to demand that all the $\mathbf{p}_k$'s are linearly independent of each other. Moreover, these vectors can be cosen so as to satisfy the requirement:

$$(p_j)^{tr} A p_k = 0 \qquad (for\ k < j) \tag{V-172}$$

which is a generalized orthogonality relation, and is the reason for the name "conjugate directions" given to the $\mathbf{p}_j$'s. Then Eq. (V-166) becomes (using Eq. (V-171)):

$$f[u_j] = f[u_{j-1}] + a_j (p_j)^{tr} A u_0 - a_j (p_j)^{tr} v + \frac{1}{2}(a_j)^2 (p_j)^{tr} A p_j \tag{V-173}$$

Choosing $\mathbf{u}_0 = \mathbf{0}$ (the zero vector), the expression for f[$\mathbf{u}_j$] has now been seprated into two distinct parts. The first part, f[$\mathbf{u}_{j-1}$] depends only on the vectors $\mathbf{p}_1$, $\mathbf{p}_2$, ..., $\mathbf{p}_{j-1}$ whereas the second part depends only on $\mathbf{p}_j$. Thus the global minimum of the sum is obtained by minimizing separately each of the two terms. Assuming that minimization has already been done in the previous step, it is only necessary to minimize the second term, which is done by a correct choice of the parameter a_j. Thus Eq. (V-168) gives much more than the relative minimum of of f[$\mathbf{u}_j$] in the restricted one-dimensional space in which only a_j is allowed to vary. It actually gives us a method for obtaining the global minimum of the functional in the j-dimensional space, spanned by all the $\mathbf{p}_k$'s with $1 \le k \le j$.

The only missing link in the process is an explicit formula for $\mathbf{p}_j$. From Eqs. (V-166),(V-167) it is clear that $\mathbf{p}_j$ has a component parallel to $\mathbf{r}_{j-1}$, in addition to a component orthogonal to it. For j=1 one may choose $\mathbf{p}_1 = \mathbf{r}_0$, and since $\mathbf{u}_0 = \mathbf{0}$, $\mathbf{r}_0 = \mathbf{v}$. A more thorough treatment of the issue leads to the formula (for j>1)

$$p_j = r_{j-1} + b_j p_{j-1} \tag{V-174}$$

to ensure optimal efficiency of the process. Therefore (using Eq. (V-168))

$$(p_j)^{tr} r_{j-1} = (r_{j-1})^{tr} r_{j-1} + b_j (p_{j-1})^{tr} r_{j-1} = (r_{j-1})^{tr} r_{j-1} \tag{V-175}$$

and, from Eq. (V-172):

$$0 = (r_{j-1} + b_j p_{j-1})^{tr} A p_{j-1} \qquad \rightarrow \qquad b_j = -\frac{(r_{j-1})^{tr} A p_{j-1}}{(p_{j-1})^{tr} A p_{j-1}} \tag{V-176}$$

It is now possible to express both a_j and b_j in simpler forms. From Eqs. (V-168) and (V-175) it follows immediately that:

$$a_j = \frac{(r_{j-1})^{tr} r_{j-1}}{(p_j)^{tr} A p_j} \tag{V-177}$$

From Eq. (V-164):

$$A p_{j-1} = \frac{1}{a_{j-1}} (r_{j-2} - r_{j-1}) \tag{V-178}$$

which, assuming ortogonality of the different $\mathbf{r}_j$'s , yields

$$-(r_{j-1})^{tr} A p_{j-1} = \frac{1}{a_{j-1}} (r_{j-1})^{tr} r_{j-1} \tag{V-179}$$

Orthogonality of the $\mathbf{r}_j$'s can be proved by induction. First, from Eq. (V-170):

$$(r_0)^{tr} r_1 = (p_1)^{tr} r_1 = 0 \tag{V-180}$$

Then assume $\mathbf{r}_0, \mathbf{r}_1, \ldots, \mathbf{r}_k$ are orthogonal to each other ($k \ge 1$). The next residual vector is orthogonal to all previous ones, because for any value of j ($j \le k$):

$$(r_j)^{tr} r_{k+1} = (r_j)^{tr} r_k - a_{k+1} (r_j)^{tr} A p_{k+1} = \delta_{j,k} (r_k)^{tr} r_k - a_{k+1} \times$$

$$\times (p_{j+1} - b_{j+1} p_j)^{tr} A p_{k+1} = \delta_{j,k} (r_k)^{tr} r_k - \delta_{j,k} a_{k+1} (p_{k+1})^{tr} A p_{k+1} = 0 \tag{V-181}$$

where Eqs. (V-164), (V-172), (V-174) and (V-177) have been used. Combining Eqs.

(V-176), (V-177) and (V-179) results in

$$b_j = \frac{(r_{j-1})^{tr} r_{j-1}}{(r_{j-2})^{tr} r_{j-2}} \tag{V-182}$$

On the basis of Eqs. (V-174), (V-182) one may construct $\mathbf{p}_j$ as:

$$p_j = r_{j-1} + b_j\left(r_{j-2} + b_{j-1}\left(r_{j-3} + \ldots\right)\right) = (r_{j-1})^{tr} r_{j-1} \sum_{k=0}^{j-1} \frac{r_k}{(r_k)^{tr} r_k} \tag{V-183}$$

using the assumption: $\mathbf{p}_1 = \mathbf{r}_0 = \mathbf{v}$. Thus there is a complete procedure for constructing the conjugate direction vectors, from which the approximate solution vectors $\mathbf{u}_j$ can be calculated using Eq. (V-171).

The method described so far thus allows one to solve a set of linear algebraic equations by systematic iterations. Moreover, at each stage one may estimate the magnitude of the error by computing the norm (the square of the "length") of the residual vector:

$$e_k = (r_k)^{tr} r_k \tag{V-184}$$

As in the case of the Lanczos algorithm, the whole process can be reformulated in terms of quantum mechanical vectors. If every column vector $\mathbf{q}$ is replaced by a Hilbert space "ket" vector $|\mathbf{q}\rangle$, and its transpose $\mathbf{q}^{tr}$ is replaced by the "bra" vector $\langle\mathbf{q}|$ then all formulas given above become relevant to the discussion of the EPR problem. In practice there is still one difficulty, namely, that in magnetic resonance applications one deals with a complex symmetric matrix, and not with a real symmetric (or complex hermitian) matrix. However, one may show that the method is applicable to such problems provided the scalar product is modified as in Eq. (V-157) above. The only exception to this rule is that the magnitude of the error is calculated in the conventional way , namely

$$e_k = \langle r_k | r_k \rangle = \sum_m (r_k)_m^{\,*} (r_k)_m \tag{V-185}$$

Here $(\mathbf{r}_k)_m$ is the m'th element in the vector $\mathbf{r}_k$.

It is possible to show that the conjugate gradients technique for solving a set of linear equations is equivalent at every stage to the Lanczos method for transforming a matrix to tridiagonal form (see Refs. 22, 19, 24 and 25). Thus at every stage of the iteration one may, in principle, "translate" a result in the CGM to the corresponding result in the LA (the actual "translation" process is a certain algebraic transformation). All the virtues of the LA as a fast and efficient method are therefore common also to the CGM. However, there is the advantage of estimating at each stage the magnitude of the error, so one can estimate the degree of convergence without having to calculate the line shape. In LA one could only achieve this by transforming from LA to CGM, but this would be very cumbersome and would rob the LA of its great efficiency. It is also possible to show that using CGM one can estimate the contribution of each basis vector to the solution (see Refs. 24, 19). By doing this for just a few points in the spectral range one can find a reasonable size of the truncated basis (see discussion at end of Sec. **V.4.(a)** above), which will give a sufficiently accurate solution with the smallest possible basis.

Suggested References

* On the use of cumulant expansions to solve the SLE:

1. R. Kubo, *J. Phys. Soc. Japan* **17**, 1100 (1962).
2. R. Kubo in *"Fluctuation, Relaxation and Resonance in Magnetic Systems"*, edited by D. ter Haar (Oliver and Boyd, Edinburgh, 1962), p. 23.
3. R. Kubo, *J. Math. Phys.* **4**, 174 (1962).
4. R. Kubo in *"Stochastic Processes in Chemical Physics" - Adv. Chem. Phys.* vol. **XV**, edited by K.E. Shuler (1969), p. 101.
5. R. Kubo, *J. Phys. Soc. Japan*, vol. **26,** Supplement (1969), p. 1.
6. R. Kubo, M. Toda and N. Hashitsume, *"Statistical Physics II"*, second edition (Springer, Berlin, 1991).
7. J.H. Freed, *J. Chem. Phys.* **49**, 376 (1968).

* On discretization of the stochastic parameter:

8. J.R. Norris and S.I. Weissman, *J. Phys. Chem.* **73**, 3119 (1969).
9. H. Sillescu, *J. Chem. Phys.* **54,** 2110 (1971); K. Hensen, W.-O. Riede, H. Sillescu and A. v. Wittgenstein, *J. Chem. Phys.* **61**, 4365 (1974).
10. R.G. Gordon and T. Messenger in *"Electron Spin Relaxation in Liquids"*, edited by L.T. Muus and P.W. Atkins (Plenum Press, New York, 1972), Ch. XIII.
11. J.H. Freed and J.B. Pedersen, *Adv. Mag. Reson.* **8**, 1 (1976).
12. G.P. Zientara and J.H. Freed, *J. Chem. Phys.* **70**, 2587 (1979); A.E. Stillman, G.P. Zientara and J.H. Freed, *J. Chem. Phys.* **71**, 113 (1979); G.P. Zientara and J.H. Freed, *J. Chem. Phys.* **71**, 744 (1979).
13. G. Kothe, *Mol. Phys.* **33**, 147 (1977).
14. K.-H. Wassmer, E. Ohmes, M. Portugall, H. Ringsdorf and G. Kothe, *J. Am. Chem. Soc.* **107**, 1511 (1985).

* On the solution with the method of eigenfunctions:

15. L.D. Favro, *Phys. Rev.* **119**, 53 (1960); L.D. Favro in *"Fluctuation Phenomena in Solids"*, edited by R.E. Burgess (Academic Press, New York, 1965), p. 79.
16. J.H. Freed, G.V. Bruno and C.F. Polnaszek, *J. Phys. Chem.* **75**, 3385 (1971).
17. J.H. Freed in *Spin Labeling: Theory and Applications*, Vol. I, L. Berliner, editor (Academic Press, New York, 1976), Ch. 3.
18. E. Meirovitch, D. Igner, E. Igner, G. Moro and J.H. Freed, *J. Chem. Phys.* **77**, 3915 (1982).
19. D.J. Schneider and J.H. Freed in *Lasers, Molecules and Methods*, edited by J.O.

Hirschfelder, R.E. Wyatt and R.O. Coalson, Vol. 73 in *Adv. Chem. Phys.* (1989), p. 387.

* On the Lanczos and conjugate gradients techniques:

20. C. Lanczos, *J. Res. Natl. Bur. Stand.* **45**, 255 (1950); **49**, 33 (1952).
21. M.R. Hestenes and E. Stiefel, *J. Res. Natl. Bur. Stand.* **49**, 409 (1952).
22. G.H. Golub and C.F. van Loan, *"Matrix Computations"* (Johns Hopkins University, Baltimore, MD 1983).

* On applying the Lanczos and conjugate gradients techniques in EPR line shape simulations:

Ref. 19 and

23. G. Moro and J.H. Freed, *J. Chem. Phys.* **74**, 3757 (1981).
24. K.V. Vasavada, D.J. Schneider and J.H. Freed, *J. Chem. Phys.* **86**, 647 (1987).
25. D.J. Schneider and J.H. Freed in *Spin Labeling*, Vol. VIII, edited by L. Berliner and J. Reuben (Academic Press, New York, 1989), p.1.

CHAPTER VI

APPLICATIONS TO CW MAGNETIC RESONANCE

In the previous chapters the Stochastic Liouville Equation was developed, and methods for its solution were discussed. A particular emphasis was laid on the solution with eigenfunctions of the relaxation superoperator, and the equation was presented in some detail in the context of EPR. In the present Chapter some applications of the SLE to CW magnetic resonance will be discussed, mostly for EPR but also for NMR. The examples chosen demonstrate the wide range of applicability of the formalism in magnetic resonance. Applications to more elaborate experimental techniques such as pulsed magnetic resonance, double resonance and ENDOR will be discussed in the next Chapter.

VI.1 Doublet EPR - isotropic rotational diffusion in isotropic solvents

The most basic application of the method in magnetic resonance is to the case of a two-level system, i.e. a spin S = ½ system, with simple interactions, undergoing the simplest kind of rotational diffusion. In the context of EPR a two-level system, commonly known as a doublet, exists for a radical with a single unpaired electron. Simple interactions are those with a simple orientation dependence, and also simple rotational diffusion is that which has the simplest orientation dependence. Suppose first that the Hamiltonian has only a constant term as in Eq. (V-120), the orientation-dependent term of Eq. (V-97) being equal to zero. It is then clear from Eq. (V-115) or (V-117) that, assuming no saturation is present, the set of equations is completely uncoupled. Each element of the type $(\mathbf{C}^{(n)L}{}_{MK})_{ij}$ is uncoupled from all other elements of this type, and therefore it relaxes with the decay constant γ_{LMK}. In particular, for L = 0 the decay constant is equal to zero, so for the case of an isotropic environment the signal would not decay at all due to the rotational diffusion process. In practice there would be other relaxation processes, leading to a finite relaxation time, but this does not concern us here. We conclude that in order to have the simplest example of relaxation due to rotational diffusion there must be a non-zero orientation dependent part in the Hamiltonian. This is not surprising, since the SLE was initially developed from the assumption that the Hamiltonian depends on a stochastic parameter Ω which is responsible for relaxation, and in the present case Ω stands for the molecular orientaion. In this Section some important cases of simple orientation dependent Hamiltonians will be treated, with the simplest possible rotational diffusion process. This means the diffusion itself is isotropic, with the corresponding eigenvalues

$$\gamma_{LMK} = -RL(L+1) \tag{VI-1}$$

(see Eq. (V-88); the negative sign results from the negative sign of the relaxation superoperator in Eq. (V-78)). The relaxation process will depend in each case on the specific interactions existing in the system.

(a) Axially symmetric g-tensor - non-saturated lines

In this example and in most other examples in this Chapter, the experiment is assumed to be a CW experiment with weak irradiation, so that no saturation occurs. Then the term in Eq. (V-117) coupling different Fourier orders can be neglected. It is only the commutators in that equation which couple different elements of the supervector **C** in Eq. (V-116). These commutators can be expanded, in general, in the relevant set of operators, coupling different pairs of (i,j) indices, and their coefficients couple different sets of (LMK) indices. The commutators depend on the Hamiltonian of the molecule (or the radical) under consideration. The molecule is assumed to be dissolved in a liquid or a glassy solvent, so that all molecular orientations with respect to the external magnetic field are possible. The constant part of the Hamiltonian is assumed here to consist only of the Zeeman term: $\mathbf{H_0} = \omega_0\, \mathbf{S_z}$, so its eigenstates are $|-\frac{1}{2}\rangle$ and $|\frac{1}{2}\rangle$, where the index of each state is the corresponding eigenvalue of $\mathbf{S_z}$ (divided by $\hbar$). The orientation dependent interaction is:

$$H_1 = F D^2{}_{0,0}(\Omega)\, S_z \tag{VI-2}$$

This is the simplest example for the Hamiltonian of Eq. (V-97). The interaction constant F between the spin and the external magnetic field is equal to

$$F = \frac{2}{3}\frac{\beta_e B_0}{\hbar}(g_\parallel - g_\perp) \tag{VI-3}$$

where β_e is the Bohr magneton and B_0 is the constant magnetic field. The constants $g_\parallel = g_{zz}$ and $g_\perp = g_{xx} = g_{yy}$ are the components of the g-tensor parallel to the molecular z-axis and perpendicular to it, respectively. This interaction commutes with the constant Hamiltonian $\mathbf{H_0}$, and therefore cannot induce transitions between different quantum states belonging to it. The influence of this interaction is only to add to the energy an orientation dependent term, and the magnitude of this term is modulated by the rotational diffusion. Thus the inclusion of this interaction amounts to including a particular mechanism for modulating the energies, or the transition frequency, in this two level system. The system thus formally resembles an oscillator with a single frequency, modulated by a stochastic process, which is the most basic example for which the SLE was developed in Chapter IV.

The commutators can be calculated by expanding the operators $\mathbf{C}^L{}_{MK}$ belonging to the Fourier order n = 1 (which, for the sake of simplicity, is not written explicitly in the superscript) in the set of angular momentum operators used in Eq. (VI-2). For the case of S = ½, which does not interact with other spins, a complete set of angular momentum operators is the following: $\{\mathbf{I}, \mathbf{S_z}, \mathbf{S_+} = \mathbf{S_x}+i\mathbf{S_y}$ and $\mathbf{S_-} = \mathbf{S_x}-i\mathbf{S_y}\}$, where **I** is the unit operator. Therefore the most general operator can be expanded as:

$$Q = q_1 I + q_z S_z + q_+ S_+ + q_- S_- \tag{VI-4}$$

where the numbers q_1 , q_z , q_+ and q_- are the expansion coefficients which characterize the specific operator. The only commutator occuring in Eq. (V-117), applied to the present case, is:

$$[S_z, C] = c_+ S_+ - c_- S_- \tag{VI-5}$$

where the standard commutation relations of angular momentum operators have been employed (see Appendix A). Here **C** denotes some $\mathbf{C}^L_{MK}$, and c_α are its corresponding coefficients in the expansion (VI-4). From Eq. (V-105) it is clear that the only matrix elements (i,j) of the density matrix which contribute to the line shape are those with $m_i - m_j = -1$ (m_k is the eigenvalue of S_z for state k). In the present problem this fact implies: $|i\rangle = |-½\rangle$ and $|j\rangle = |½\rangle$. It is therefore sufficient to solve Eq. (V-116) only for this set of values of (i,j), unless it is coupled to other sets of (i,j) values. For any operator **C**:

$$\langle -½ | C | ½ \rangle = c_- \langle -½ | S_- | ½ \rangle = c_- \tag{VI-6}$$

and therefore

$$\langle -½ | [S_z, C] | ½ \rangle = -c_- \langle -½ | S_- | ½ \rangle = -c_- \tag{VI-7}$$

Thus, for the relevant (i,j) pair, one may simply write

$$[S_z, C]_{i,j} = -C_{i,j} \tag{VI-8}$$

Then $\mathbf{C}_{i,j}$ is not coupled by the commutator to any $\mathbf{C}_{k,l}$ with (k,l) ≠ (i,j), which means that one can ignore the equations for all other sets of (k,l), and also that the commutator can be replaced by its equivalent from Eq. (VI-8). In the coefficient (i.e., in the rotation function which is the coefficient in the expansion of $\mathbf{H}_1$ in terms of angular momentum operators) one has m = m' = 0, so that K = K', M = M' (see Eqs. (V-97) and (V-117)). The matrix element of the supermatrix **X** is therefore:

$$(X)_{LMKij,L'M'K'i'j'} = \delta_{L,L'}\delta_{M,M'}\delta_{K,K'}\delta_{i,i'}\delta_{j,j'}\big(i(\omega+\omega_{ij}) + \gamma_{LMK}\big) + \delta_{i,i'}\delta_{j,j'} \times$$

$$\times \frac{1}{i\hbar}\sqrt{8\pi^2/5}\, F\, (-1)^{M-K}\sqrt{(2L+1)(2L'+1)} \begin{pmatrix} L & 2 & L' \\ -M & 0 & M \end{pmatrix} \begin{pmatrix} L & 2 & L' \\ -K & 0 & K \end{pmatrix} \tag{VI-9}$$

This can be simplified by using: $\omega_{-½,½} = -\omega_0$, and also by using the isotropy of the rotational diffusion. As mentioned above, from Eq. (V-103) it follows that the only elements which contribute directly to the line shape for the case of an isotropic solution are those with L = M = K = 0. In Eq. (VI-9) the M = 0 states are only coupled to M = 0 states, and K = 0 states are only coupled to K = 0 states. Thus for the relevant subset of equations: M = M' = 0, K = K' = 0. The relevant elements of **X** are

$$(X)_{L,L'} = \big(i(\omega-\omega_0) - RL(L+1)\big) + \frac{1}{i\hbar}\sqrt{8\pi^2/5}\, F\sqrt{(2L+1)(2L'+1)} \begin{pmatrix} L & 2 & L' \\ 0 & 0 & 0 \end{pmatrix}^2 \tag{VI-10}$$

and the only index which can take more than one value is L (or L'). From the general properties of the 3-j coefficients it follows that the second and third column can be interchanged in the 3-j appearing in Eq. (VI-10), without changing its value. Moreover, this 3-j is non-zero only if L, L' have the same parity, i.e., either both are even or both are odd. Since for the line shape only the even value L = 0 is needed, the values of L, L' in (VI-10) must be even in order to contribute to the solution. From standard tables or from the general formula for 3-j's it is known that:

$$\begin{pmatrix} L & L & 2 \\ 0 & 0 & 0 \end{pmatrix} = -\left(\frac{L(L+1)}{(2L+3)(2L+1)(2L-1)}\right)^{1/2} \qquad \text{(VI-11)}$$

$$\begin{pmatrix} L & L\pm 2 & 2 \\ 0 & 0 & 0 \end{pmatrix} = \sqrt{\frac{3}{2}}\left(\frac{(L+1)(L+2)}{(2L+5)(2L+3)(2L+1)}\right)^{1/2} \qquad \text{(VI-12)}$$

For all values of L' such that $|L' - L| > 2$ the 3-j in Eq. (VI-10) vanishes. Since only one index is involved, it is easy to estimate the results of truncation of the infinite set of equations implied in (VI-10). For large L:

$$\begin{pmatrix} L & L & 2 \\ 0 & 0 & 0 \end{pmatrix} \xrightarrow[L\to\infty]{} -\frac{1}{2\sqrt{2L}} \qquad \text{(VI-13)}$$

$$\begin{pmatrix} L & L\pm 2 & 2 \\ 0 & 0 & 0 \end{pmatrix} \xrightarrow[L\to\infty]{} \sqrt{\frac{3}{8}}\,\frac{1}{\sqrt{2L}} \qquad \text{(VI-14)}$$

Thus the off-diagonal elements in **X**, which couple different L values, are approximately proportional for large L to $(2L+1)F/2L \approx F$. Since the diagonal elements include the term $RL(L + 1)$, or approximately $R L^2$ for large L, the off-diagonal terms can be neglected if

$$F \ll RL^2 \quad \Leftrightarrow \quad F/R \ll L^2 \qquad \text{(VI-15)}$$

This condition allows one to truncate the equations for L values which are large compared with $(F/R)^{1/2}$, which is exactly the criterion obtained from more general considerations in Eq. (V-124). An additional simplification is the restriction to M = 0, K = 0 states, which reduces drastically the dimensions of the computational problem.

Example VI.1: Rotational diffusion of planar S_5^+

An experiment for which the above theory is relevant was conducted for a 10^{-2} molar solution of sulfur in 60% oleum (Ref. 8). Based on previous experimental EPR work with the same material, the paramagnetic species was identified as a planar S_5^+ ring, which exists in the solution in a concentration of less than 10^{-4} molar, as estimated from the intensity of the EPR lines. In this species, the only orientation dependent interaction is an axially symmetric g-tensor interaction. The line shape was measured as a function of temperature from 170 K to 290 K.

The line shape showed a marked temperature dependence, changing from narrow lines below 200 K to broad lines at about 240 K and higher (see Fig. VI.1). The motional correlation time became shorter by almost an order of magnitude over a temperature range of 50 degrees in this critical region. It is thus an important temperature range from the experimental viewpoint. Also from a theoretical point of view the interesting temperature range is the slow motional region, in which motional narrowing theories like the Bloch-Redfield theory are not applicable, and on the other hand the effects of motion are significant.

The rotational diffusion theory of the SLE was used here in an attempt to fit the experimental spectra, with two different motional models. One is the model of Brownian rotational diffusion, which was described in subsection V.3.(b), and for which the formulas given in the present section are appropriate. The other is a model of random rotational large angle jumps. This model is implicitly included in the theory discussed above, the only difference being that the angular jumps are large rather than small, and consequently the eigenvalues of the relaxation superoperator change from: RL(L + 1) to a more complicated expression (this is the "jump diffusion" model in Refs. 2,4).

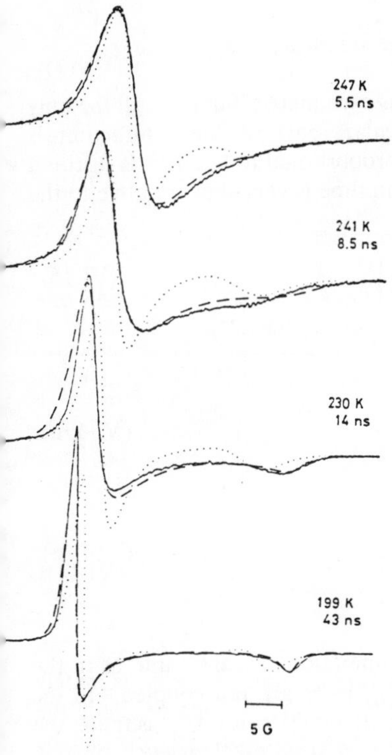

Fig. VI.1. *EPR line shapes of S_5^+ undergoing rotational diffusion. In this species, the only orientation dependent interaction is an axially symmetric g-tensor. The solid lines are the experimental spectra, the dotted ones - simulated with the Brownian diffusion model, and the dashed ones - with the model of random jumps.*

As seen in the Figure, the experimental spectra were not fit by the model of Brownian rotational diffusion, but were fit in a satisfactory manner by the model of random rotational jumps. The conclusion is that the paramagnetic ion presumably undergoes rotational diffusion, proceeding by large angle random rotational jumps. It is interesting to note that in this work the simulations were done both by the method of eigenfunctions and by a discretization of the angular variables. The two techniques gave identical spectra, but the eigenfunctions method resulted in a faster calculation.

****

(b) Asymmetric secular g-tensor - non-saturated lines

In a somewhat more general case the g-tensor is asymmetric, but it is still the only orientation dependent interaction. Only the "secular" part of the interaction is considered, namely, only that part of the interaction proportional to $\mathbf{S}_z$. This is justified if the constant magnetic field is high, or the correlation time is very short relative to the inverse Larmor frequency. Now the Hamiltonian is:

$$H_1 = \left\{ F^{(2,0)} D^2_{0,0}(\Omega) + F^{(2,2)}\left(D^2_{-2,0}(\Omega) + D^2_{2,0}(\Omega) \right) \right\} S_z \tag{VI-16}$$

with the definitions

$$F^{(2,0)} = \frac{2}{3}\frac{\beta_e B_0}{\hbar}\left[g_{zz} - \frac{1}{2}\left(g_{xx}+g_{yy}\right)\right] \tag{VI-17a}$$

$$F^{(2,2)} = F^{(2,-2)} = \frac{1}{\sqrt{6}}\frac{\beta_e B_0}{\hbar}\left(g_{xx}-g_{yy}\right) \tag{VI-17b}$$

Since the only operator in $\mathbf{H}_1$ is $\mathbf{S}_z$, the non-zero commutators are the same as in the previous subsection (see Eq. (VI-8)), and different (i,j) pairs are not coupled. In the rotation functions m' = 0, so that each K value is coupled only to itself. Assuming the medium is isotropic, one only needs the equations for K = 0 states. However, since m = 0,2 or -2 there is a coupling between different M values, including for a given L all even values of M which satisfy: $|M| \le L$. The matrix element of X is:

$$(X)_{LM0ij,L'M'0ij} = \delta_{L,L'}\delta_{M,M'}\left(i(\omega-\omega_0)+\gamma_{LM0}\right) + \frac{1}{i\hbar}\sqrt{8\pi^2/5}\,\times$$

$$\times \sum_{L',m} \delta_{M',M+m} F^{(2,m')} \sqrt{(2L+1)(2L'+1)} \begin{pmatrix} L & 2 & L' \\ -M & m & M+m \end{pmatrix}\begin{pmatrix} L & 2 & L' \\ 0 & m' & 0 \end{pmatrix} \tag{VI-18}$$

As in the previous case, the appearance of a 3-j in which the lower row consists only of zeroes implies that L, L' must have the same parity, and therefore the relevant values for calculating the line shape are only the even values of L, L'. Thus the equations couple all even L values, and for each of them - all possible M values which are even are included, in principle. The dimensions of the problem are larger than in the case of the axially symmetric g-tensor, and the structure of the supermatrix X is a little more complicated. The general features of the problem, however, are the same in both cases. This includes, for example, the convergence properties of the calculation, which are fairly similar, although a detailed analysis of this question leads to more involved criteria for truncation, in which the relative magnitudes of $F^{(2,0)}$ and $F^{(2,2)}$ are taken into account.

(c) *Axially symmetric g-tensor and hyperfine tensor - non-saturated lines*

A more difficult situation occurs when one deals with a system having more than two levels. In the context of EPR this happens, for example, when there is still only a single unpaired electron, but it is coupled to one or more nuclei via the hyperfine interaction. The expression for $\mathbf{H_0}$ is given in Eq. (V-120). Part of this interaction is secular, i.e. it is diagonal in the electronic spin operator $\mathbf{S_z}$ and in the nuclear spin operators $\mathbf{J_{zj}}$ (j is the index of the nucleus). Such an interaction commutes with the Zeeman term in $\mathbf{H_0}$ and therefore does not couple different quantum states of the Zeeman term, just as in the case of g-tensor interaction.

If the hyperfine interaction is very strong one must consider the possibility that the electronic spin will have transitions due to this interaction (its non-secular terms). In cases of practical interest it is usually sufficient to include pseudo-secular terms, involving nuclear spin flips without any change in the electronic spin state. These terms are proportional to $\mathbf{S_z\, J_{+j}}$ and to $\mathbf{S_z\, J_{-j}}$, and therefore do not commute with $\mathbf{H_0}$. This non-commutativity is the formal manifestation of the fact that these terms can induce transitions in the system.

In order to illustrate the main characteristics of such a situation the simplest case is taken, that of a single electronic S = ½ spin interacting with a single nuclear J = ½ spin. It is assumed here for simplicity that both the g-tensor and the hyperfine tensor are axially symmetric, and that their principal axes systems coincide. The orientation dependent interaction is then

$$H_1 = \left\{F_g^{(2,0)} + F_h^{(2,0)} J_z\right\} D^2_{0,0}(\Omega)\, S_z - \left[\frac{3}{8}\right]^{1/2} F_h^{(2,0)} \left\{D^2_{0,1}(\Omega)\, J_+ + D^2_{0,-1}(\Omega)\, J_-\right\} S_z \qquad \text{(VI-19)}$$

Here $F_g^{(2,m)}$ are components of the g-tensor, and $F_h^{(2,m)}$ are components of the hyperfine interaction. $F_g^{(2,0)}$ is defined as in Eqs. (VI-3) or (VI-17), and the hyperfine constant is defined as:

$$F_h^{(2,0)} = \frac{2}{3}\left(A_\parallel - A_\perp\right) \qquad \text{(VI-20)}$$

From the general formula (Eq. ((V-97) or Appendix D) it is clear that the two hyperfine terms must have the same interaction constant, as given here. The fact that one of them is multiplied by a numerical factor of $-(3/8)^{1/2}$ results from the use of unnormalized spin operators. The properly normalized irreducible spherical operators, constructed from the appropriate spin operators, would be here:

$$A^{(2,0)} = \left(\frac{2}{3}\right)^{1/2} \left\{ S_z J_z - \frac{1}{4}\left(S_+ J_- + S_- J_+\right) \right\} \quad \text{(VI-21)}$$

and

$$A^{(2,\pm 1)} = \mp \frac{1}{2}\left\{ S_\pm J_z + S_z J_\pm \right\} \quad \text{(VI-22)}$$

It is the difference in normalization constants of these operators which is responsible for the appearance of the numerical factor. Differently stated, this factor ensures proper normalization of both hyperfine terms.

The medium is assumed again to be isotropic, so that the line shape requires only the L = 0, M = 0, K = 0 state and all states which are coupled to it. Now m = 0, so that only M = M' = 0 is needed. However, m' = 0, ±1 so that different K values are coupled. As for the commutators between $\mathbf{H}_1$ and the components of the density matrix, the system now has four levels, and the density matrix or any other general operator have to be expanded in terms of sixteen operators:

$$\begin{aligned} Q &= q_{1,1} I_S I_J + q_{1,z} I_S J_z + q_{1,+} I_S J_+ + q_{1,-} I_S J_- \\ &+ q_{z,1} S_z I_J + q_{z,z} S_z J_z + q_{z,+} S_z J_+ + q_{z,-} S_z J_- + q_{+,1} S_+ I_J + q_{+,z} S_+ J_z \\ &+ q_{+,+} S_+ J_+ + q_{+,-} S_+ J_- + q_{-,1} S_- I_J + q_{-,z} S_- J_z + q_{-,+} S_- J_+ + q_{-,-} S_- J_- \end{aligned} \quad \text{(VI-23)}$$

In this formula $\mathbf{I_S}$ is the unit operator in the Hilbert space of the electronic spin, and $\mathbf{I_J}$ is the corresponding unit operator in the Hilbert space of the nuclear spin. In ordinary calculations they would often be omitted, but here it is important to write them explicitly in order to get the correct expression for the general case. The needed commutators can be calculated using the general formula for any four operators:

$$[AB, CD] = A[B,C]D + AC[B,D] + [A,C]DB + C[A,D]B \quad \text{(VI-24)}$$

which can be verified directly. In the present problem, when the first and third operators ("**A**" and "**C**") in each summand operate on the electronic spin, whereas the second and fourth operator ("**B**" and "**D**") operate on the nuclear spin, there is a special condition:

$$[A,B] = [B,C] = [A,D] = [C,D] = 0 \quad \text{(VI-25)}$$

so the general formula reduces to

$$[A_S B_J, C_S D_J] = A_S C_S [B_J, D_J] + [A_S, C_S] D_J B_J \quad \text{(VI-26)}$$

where the subscripts indicate the subspaces on which the respective operators operate. Using this expression for a general operator **Q** one finds the following commutators:

$$[S_z, Q] = \hbar\Big\{\big(q_{+,1}S_+I_J + q_{+,z}S_+J_z + q_{+,+}S_+J_+ + q_{+,-}S_+J_-\big) - \big(q_{-,1}S_-I_J + q_{-,z}S_-J_z + q_{-,+}S_-J_+ + q_{-,-}S_-J_-\big)\Big\} \tag{VI-27}$$

$$\begin{aligned}[S_zJ_z, Q] = \hbar\Big\{& q_{1,+}S_zJ_+ - q_{1,-}S_zJ_- + q_{z,+}S_z^2J_+ - q_{z,-}S_z^2J_- \\ & + q_{+,1}S_+J_z + q_{+,z}S_+J_z^2 + q_{+,+}(S_zS_+J_+ + S_+J_+J_z) + q_{+,-}(-S_zS_+J_- + S_+J_-J_z) \\ & - q_{-,1}S_-J_z - q_{-,z}S_-J_z^2 + q_{-,+}(S_zS_-J_+ - S_-J_+J_z) - q_{-,-}(S_zS_-J_- + S_-J_-J_z)\Big\}\end{aligned} \tag{VI-28}$$

$$\begin{aligned}[S_zJ_+, Q] = \hbar\Big\{& -q_{1,z}S_zJ_+ + 2q_{1,-}S_zJ_z - q_{z,z}S_z^2J_+ + 2q_{z,-}S_z^2J_z \\ & + q_{+,1}S_+J_+ + q_{+,z}(-S_zS_+J_+ + S_+J_zJ_+) + q_{+,+}S_+J_+^2 + q_{+,-}(2S_zS_+J_z + S_+J_-J_+) \\ & - q_{-,1}S_-J_+ - q_{-,z}(S_zS_-J_+ + S_-J_zJ_+) - q_{-,+}S_-J_+^2 + q_{-,-}(2S_zS_-J_z - S_-J_-J_+)\Big\}\end{aligned} \tag{VI-29}$$

$$\begin{aligned}[S_zJ_-, Q] = \hbar\Big\{& q_{1,z}S_zJ_- - 2q_{1,+}S_zJ_z + q_{z,z}S_z^2J_- - 2q_{z,+}S_z^2J_z \\ & + q_{+,1}S_+J_- + q_{+,z}(S_zS_+J_- + S_+J_zJ_-) + q_{+,+}(-2S_zS_+J_z + S_+J_+J_-) + q_{+,-}S_+J_-^2 \\ & - q_{-,1}S_-J_- + q_{-,z}(S_zS_-J_- - S_-J_zJ_-) - q_{-,+}(2S_zS_-J_z + S_-J_+J_-) - q_{-,-}S_-J_-^2\Big\}\end{aligned} \tag{VI-30}$$

In this system there are two "allowed" transitions, in which only the electronic spin changes its state:

1: $|m_S = -½, m_J = ½\rangle \equiv |\beta\alpha\rangle \rightarrow |m_S = ½, m_J = ½\rangle \equiv |\alpha\alpha\rangle$

2: $|m_S = -½, m_J = -½\rangle \equiv |\beta\beta\rangle \rightarrow |m_S = ½, m_J = -½\rangle \equiv |\alpha\beta\rangle$

and two "forbidden" transitions, in which also the nuclear spin changes its state:

3: $|m_S = -½, m_J = ½\rangle = |\beta\alpha\rangle \rightarrow |m_S = ½, m_J = -½\rangle = |\alpha\beta\rangle$

4: $|m_S = -½, m_J = -½\rangle = |\beta\beta\rangle \rightarrow |m_S = ½, m_J = ½\rangle = |\alpha\alpha\rangle$

The allowed transitions are between eigenstates of $\mathbf{H_0}$, and they are induced by microwave irradiation in an EPR experiment. The forbidden transitions are related to the pseudo-secular terms in $\mathbf{H_1}$, which give rise to a weak mixing of those eigenstates. Consequently their intensity is relatively weak. In each of these transitions, the initial state is denoted by $|i\rangle$ and the final state by $|j\rangle$. Then the frequencies ω_{ij}, which are the energy differences related to $\mathbf{H_0}$ divided by $\hbar$, are

$$\omega_{(1)} = -\left[\omega_0 - \frac{\gamma_e a}{2}\right] \qquad \omega_{(2)} = -\left[\omega_0 + \frac{\gamma_e a}{2}\right] \tag{VI-31a}$$

$$\omega_{(3)} = -\left(\omega_0 + \omega_n\right) \qquad \omega_{(4)} = -\left(\omega_0 - \omega_n\right) \tag{VI-31b}$$

where ω_0 is the electronic Larmor frequency, a is the isotropic hyperfine constant and ω_n is the nuclear Larmor frequency (as in Eq. (V-120)). The subscripts are written as (α) in order to avoid confusion with the frequencies ω_0 and ω_1 which are proportional to the strength of the constant and time dependent magnetic fields. Since for high magnetic fields ω_0 is much larger than the other terms in these expressions, all these frequencies are negative, as in Eq. (VI-10). Only for a field which is weak compared with the hyperfine coupling, the first frequency in (V-31a) will be positive. It is assumed here, as in all other cases, that the field is relatively high.

In order to calculate the line shape, one needs to solve the equations for the matrix elements of $\mathbf{C}^L{}_{MK}$ related to the allowed transitions. It is convenient to use the notation:

$$|1_-\rangle = |3_-\rangle \equiv |\beta\alpha\rangle \qquad |1_+\rangle = |4_+\rangle \equiv |\alpha\alpha\rangle$$

$$|2_-\rangle = |4_-\rangle \equiv |\beta\beta\rangle \qquad |2_+\rangle = |3_+\rangle \equiv |\alpha\beta\rangle \tag{VI-32}$$

This notation is chosen so that transition k is from state $|k_-\rangle$ (its "$|i\rangle$" state) to state $|k_+\rangle$ (its "$|j\rangle$" state). The required matrix elements are thus $\langle 1_-|...|1_+\rangle$, $\langle 2_-|...|2_+\rangle$. Taking these matrix elements for each commutator one finds simple expressions for the secular terms:

$$\langle 1_- | [S_z, Q] | 1_+\rangle = \hbar\langle 1_- | \left\{ \left(q_{+,1} S_+ I_J + q_{+,z} S_+ J_z\right) - \left(q_{-,1} S_- I_J + q_{-,z} S_- J_z\right) \right\} | 1_+\rangle =$$

$$= -\hbar\left\{ q_{-,1}\langle 1_- | S_- I_J | 1_+\rangle + q_{-,z}\langle 1_- | S_- J_z | 1_+\rangle \right\} = -\hbar\left[q_{-,1} + \frac{1}{2}q_{-,z}\right] =$$

$$= -\hbar\langle 1_- | Q | 1_+\rangle = -\hbar\, Q_{(1)} \tag{VI-33}$$

$$\langle 2_- | [S_z, Q] | 2_+ \rangle = -\hbar \{ q_{-,1} \langle 2_- | S_- I_J | 2_+ \rangle + q_{-,z} \langle 2_- | S_- J_z | 2_+ \rangle \} =$$

$$= -\hbar \left[q_{-,1} - \frac{1}{2} q_{-,z} \right] = -\hbar \langle 2_- | Q | 2_+ \rangle = -\hbar\, Q_{(2)} \qquad \text{(VI-34)}$$

$$\langle 1_- | [S_z J_z, Q] | 1_+ \rangle = -\hbar \left\{ q_{-,1} \langle 1_- | S_- J_z | 1_+ \rangle + q_{-,z} \langle 1_- | S_- J_z^2 | 1_+ \rangle \right\} =$$

$$= -\langle 1_- | Q | 1_+ \rangle \langle 1_- | J_z | 1_+ \rangle = -\frac{\hbar}{2} Q_{(1)} \qquad \text{(VI-35)}$$

$$\langle 2_- | [S_z J_z, Q] | 2_+ \rangle = -\hbar \left\{ q_{-,1} \langle 2_- | S_- J_z | 2_+ \rangle + q_{-,z} \langle 2_- | S_- J_z^2 | 2_+ \rangle \right\} =$$

$$= -\langle 2_- | Q | 2_+ \rangle \langle 2_- | J_z | 2_+ \rangle = \frac{\hbar}{2} Q_{(2)} \qquad \text{(VI-36)}$$

and more general results for the pseudo-secular terms:

$$\langle 1_- | [S_z J_+, Q] | 1_+ \rangle = \hbar q_{-,-} \langle 1_- | 2 S_z S_- J_z - S_- J_- J_+ | 1_+ \rangle =$$

$$= -\frac{1}{2} \hbar q_{-,-} = -\frac{1}{2} \hbar \langle 4_- | Q | 4_+ \rangle \equiv -\frac{\hbar}{2} Q_{(4)} \qquad \text{(VI-37)}$$

$$\langle 2_- | [S_z J_+, Q] | 2_+ \rangle = \hbar q_{-,-} \langle 2_- | 2 S_z S_- J_z - S_- J_- J_+ | 2_+ \rangle =$$

$$= -\frac{1}{2} \hbar q_{-,-} = -\frac{\hbar}{2} Q_{(4)} \qquad \text{(VI-38)}$$

$$\langle 1_- | [S_z J_-, Q] | 1_+ \rangle = -\hbar q_{-,+} \langle 1_- | 2 S_z S_- J_z + S_- J_+ J_- | 1_+ \rangle =$$

$$= -\frac{1}{2} \hbar q_{-,+} = -\frac{1}{2} \hbar \langle 3_- | Q | 3_+ \rangle \equiv -\frac{\hbar}{2} Q_{(3)} \qquad \text{(VI-39)}$$

$$\langle 2_- | [S_z J_- , Q] | 2_+ \rangle = -\hbar q_{-,+} \langle 2_- | 2 S_z S_- J_z + S_- J_+ J_- | 2_+ \rangle =$$

$$= -\frac{1}{2}\hbar q_{-,+} = -\frac{\hbar}{2} Q_{(3)} \tag{VI-40}$$

In these commutators the "relevant" matrix elements $\mathbf{Q}_{(1)}$ and $\mathbf{Q}_{(2)}$ of an operator, like the density matrix, are coupled to the "irrelevant" matrix elements $\mathbf{Q}_{(3)}$ and $\mathbf{Q}_{(4)}$. In order to solve Eqs. (V-116), (V-117) one therefore needs to write down also the equations for the "irrelevant" elements. The additional commutators needed for (V-117) are, for the secular terms:

$$\langle 3_- | [S_z , Q] | 3_+ \rangle = -\hbar q_{-,+} \langle 3_- | S_- J_+ | 3_+ \rangle = -\hbar \langle 3_- | Q | 3_+ \rangle = -\hbar Q_{(3)} \tag{VI-41}$$

$$\langle 4_- | [S_z , Q] | 4_+ \rangle = -\hbar q_{-,-} \langle 4_- | S_- J_- | 4_+ \rangle = -\hbar \langle 4_- | Q | 4_+ \rangle = -\hbar Q_{(4)} \tag{VI-42}$$

$$\langle 3_- | [S_z J_z , Q] | 3_+ \rangle = \hbar q_{-,+} \langle 3_- | S_z S_- J_+ - S_- J_+ J_z | 3_+ \rangle =$$

$$= \hbar q_{-,+} \left[-\frac{1}{2} + \frac{1}{2} \right] = 0 \tag{VI-43}$$

$$\langle 4_- | [S_z J_z , Q] | 4_+ \rangle = -\hbar q_{-,-} \langle 4_- | S_z S_- J_- + S_- J_- J_z | 4_+ \rangle =$$

$$= -\hbar q_{-,-} \left[-\frac{1}{2} + \frac{1}{2} \right] = 0 \tag{VI-44}$$

and for the pseudo-secular terms:

$$\langle 3_- | [S_z J_+ , Q] | 3_+ \rangle = -\hbar \langle 3_- | \left\{ q_{-,1} S_- J_+ + q_{-,z} (S_z S_- J_+ + S_- J_z J_+) \right\} | 3_+ \rangle =$$

$$= -\hbar \left[q_{-,1} + q_{-,z} \left[-\frac{1}{2} + \frac{1}{2} \right] \right] = -\frac{\hbar}{2} (Q_{(1)} + Q_{(2)}) \tag{VI-45}$$

$$\langle 4_- | [S_z J_+ , Q] | 4_+ \rangle = 0 \tag{VI-46}$$

$$\langle 3_- | [S_z J_-, Q] | 3_+ \rangle = 0 \tag{VI-47}$$

$$\langle 4_- | [S_z J_-, Q] | 4_+ \rangle = \hbar \langle 4_- | \{ -q_{-,1} S_- J_- + q_{-,z} (S_z S_- J_- - S_- J_z J_-) \} | 4_+ \rangle =$$

$$= \hbar \left[-q_{-,1} + q_{-,z} \left[-\frac{1}{2} + \frac{1}{2} \right] \right] = -\frac{\hbar}{2} (Q_{(1)} + Q_{(2)}) \tag{VI-48}$$

Having calculated these commutators, an explicit expression for the matrix elements of the supermatrix **X** of Eq. (V-117) can be written:

$$(X)_{L0K\alpha, L'0K'\beta} = \delta_{L,L'} \delta_{K,K'} \delta_{\alpha,\beta} \left(i(\omega + \omega_{(\alpha)}) + \gamma_{L0K} \right) + i \sqrt{2\pi^2/5} \sum_{L',m'} \delta_{m',0} \times$$

$$\times \{ F_g^{(2,0)} (2 \delta_{\alpha,\beta}) + F_h^{(2,0)} \{ \delta_{\alpha,\beta} (\delta_{\alpha,1} - \delta_{\alpha,2}) - \sqrt{3/8} (\delta_{m,1} ((\delta_{\alpha,1} + \delta_{\alpha,2}) \delta_{\beta,4}$$

$$+ \delta_{\alpha,3} (\delta_{\beta,1} + \delta_{\beta,2})) + \delta_{m,-1} ((\delta_{\alpha,1} + \delta_{\alpha,2}) \delta_{\beta,3} + \delta_{\alpha,4} (\delta_{\beta,1} + \delta_{\beta,2})))\}\} \times$$

$$\times \delta_{K',K+m'} \sqrt{(2L+1)(2L'+1)} \begin{pmatrix} L & 2 & L' \\ 0 & 0 & 0 \end{pmatrix} \begin{pmatrix} L & 2 & L' \\ -K & m' & K+m' \end{pmatrix} \tag{VI-49}$$

In this equation the indices α, β label the transition (α, β = 1, 2, 3 or 4). The general elements of the starting vector **V** are:

$$V_{LMK\alpha} = -\omega_{(\alpha)} \omega_1 \delta_{L,0} \delta_{M,0} \delta_{K,0} \frac{1}{\sqrt{8\pi^2}} \langle \alpha_- | S_y | \alpha_+ \rangle \tag{VI-50}$$

The matrix element of $\mathbf{S}_y$ is non-zero only for α = 1 or 2. However, since the supermatrix **X** couples these values of α with β = 3 or 4, one needs to solve the equations for all four values of α, using the appropriate matrix elements of **X**. For example, for α = 1 the equation is

$$\sum_{\beta=1}^{4} X_{1,\beta} C_\beta = V_1 \tag{VI-51}$$

where $\mathbf{V}_1$ stands for $\mathbf{V}_{L0K1}$ and $\mathbf{X}_{1,\beta}$ stands for

$$(X)_{LOK1,L'0K'\beta} = \delta_{L,L'}\,\delta_{K,K'}\,\delta_{1,\beta}\left(i(\omega + \omega_{(1)}) + \gamma_{LOK}\right) + i\,\sqrt{2\pi^2/5}\sum_{L',m'}\delta_{m',0}\times$$

$$\times\left\{F_g^{(2,0)}(2\,\delta_{1,\beta}) + F_h^{(2,0)}\left\{\delta_{1,\beta} - \sqrt{3/8}\,(\delta_{m,1}\,\delta_{\beta,4} + \delta_{m,-1}\,\delta_{\beta,3})\right\}\right\}\times$$

$$\times\,\delta_{K',K+m'}\sqrt{(2L+1)(2L'+1)}\begin{pmatrix} L & 2 & L' \\ 0 & 0 & 0 \end{pmatrix}\begin{pmatrix} L & 2 & L' \\ -K & m' & K+m' \end{pmatrix} \tag{VI-52}$$

so that, for example, $\mathbf{X}_{1,3}$ is

$$(X)_{LOK1,L'0K'3} = i\,\sqrt{2\pi^2/5}\sum_{L',m'}\delta_{m',0}\,F_h^{(2,0)}\left\{-\sqrt{3/8}\,\delta_{m,-1}\right\}\times$$

$$\times\,\delta_{K',K+m'}\sqrt{(2L+1)(2L'+1)}\begin{pmatrix} L & 2 & L' \\ 0 & 0 & 0 \end{pmatrix}\begin{pmatrix} L & 2 & L' \\ -K & m' & K+m' \end{pmatrix} \tag{VI-53}$$

Example VI.2: *Motion of spin labels on silicagel surfaces*

Small molecules adsorbed or chemically bound to silicagel surface compounds can undergo, in the general case, random rotational motion at the adsorption or binding site. Such motion has been investigated using nitroxide spin labels, adsorbed or bound to some of the silicagel surface compounds. In the experiments of Ref. [9], several SiO_2 surface compounds were used, with nitroxide spin labels attached to them.

The spectra obtained at a temperature of -150° C have the same shape as the typical EPR rigid limit spectra of rigid NO (N-O group) radicals. Therefore by fitting these low temperature spectra one gets the interaction parameters in the radicals. The systems considered here are characterized by a g-tensor which is not axially symmetric, and by an anisotropic hyperfine interaction which is almost axially symmetric.

The higher temperature spectra were fitted by assuming isotropic rotational Brownian diffusion and using, in addition to the rigid limit interaction parameters, also a motional correlation time τ_C. The correlation time is related in this motional model to the rate constant R by

$$\tau_C = \frac{1}{6R} \tag{E-1}$$

The results of the fitting are shown in Fig. VI.2, in which the experimental temperatures and the calculated correlation times are indicated.

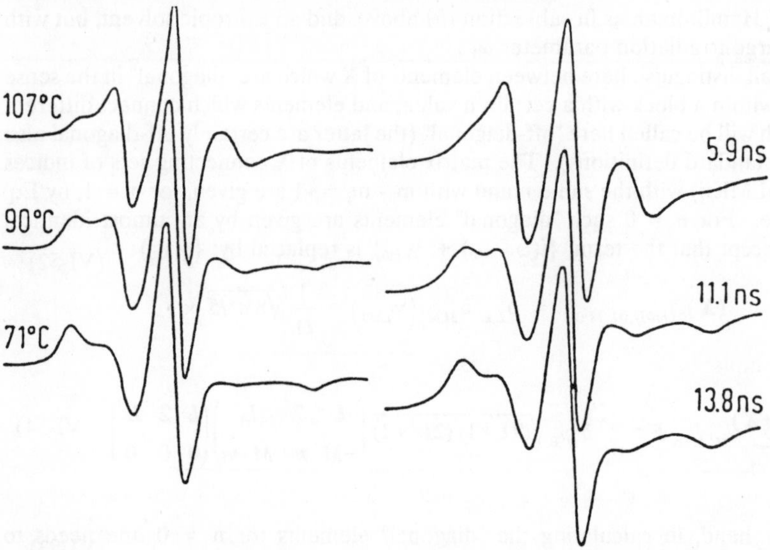

Fig. VI.2. *Experimental (left column) and calculated (right column) EPR spectra of nitroxides bound to silicagel surface compounds (see details in Ref. [9]).*

It should be noted that in this problem the simple fitting procedure did not give satisfactory results, and the experimental spectra were fitted by assuming two different adsorption sites, having different correlation times (but equal magnetic interaction parameters). The superposition of two such calculated spectra gave the result shown in each of the spectra on the right hand side of Fig. VI.2. In principle it is possible that different physical models could also account for the experimental findings, but if this model is taken as realistic it implies that some of the adsorbed molecules have greater motional freedom, and therefore faster rotation rates, than other adsorbed molecules.

****

(d) *Asymmetric g-tensor - saturated line*

In previous calculations in this Chapter it was assumed that irradiation is weak, so that no saturation occurs. If irradiation is relatively strong, the transitions are partially saturated, and then the calculations become more general. The crucial point is that in Eq. (V-116) different Fourier orders (n values) are coupled, which means that the evolution of the n = 1 terms relevant for the spectrum depends on the evolution of other terms as well. Specifically, the n = 1 order which appears in the line shape is coupled to the n = 0 and to the n = 2 orders, and these are coupled in turn to other orders, and so on. In principle one must take into account all Fourier orders, but in practice it is sufficient to consider the n = 1 and the n = 0 orders, as will be shown below. A calculation will now be done for a simple case belonging to this category, namely, for the same internal Hamiltonian as in subsection *(b)* above, and an isotropic solvent, but with a relatively large irradiation parameter ω_1 .

We shall distinguish here between elements of **X** which are "diagonal" in the sense of operating within a block with a certain n value, and elements which connect different n values which will be called here "off-diagonal" (the latter are certainly off-diagonal also according to standard definitions). The matrix elements of **X** connecting sets of indices of the type (nLMKij) with the same n and with $m_i - m_j = -1$ are given, for n = 1, by Eq. (VI-18) above. For n = 0 such "diagonal" elements are given by an almost identical expression, except that the term: $\{i(\omega - \omega_0) + \gamma_{LMO}\}$ is replaced by: $\{\gamma_{LMO}\}$:

$$(X)_{0LM0ij,0L'M'0ij} = \delta_{L,L'}\delta_{M,M'}(\gamma_{LMO}) + \frac{1}{i\hbar}\sqrt{8\pi^2/5}\,\times$$

$$\times\sum_{L',m}\delta_{M',M+m}F^{(2,-m')}\delta_{m',0}\sqrt{(2L+1)(2L'+1)}\begin{pmatrix} L & 2 & L' \\ -M & m & M+m \end{pmatrix}\begin{pmatrix} L & 2 & L' \\ 0 & 0 & 0 \end{pmatrix} \qquad \text{(VI-54)}$$

On the other hand, in calculating the "diagonal" elements for n = 0 one needs to consider only those for which i = j , as will be seen below. One therefore needs to return to Eq. (VI-5) from which it follows that, taking the (i,i) matrix element of the commutator ($|i\rangle = |-½\rangle$ or $|i\rangle = |½\rangle$ for the S = ½ problem), this element is equal to zero. Thus, instead of Eq. (VI-8) which led to the interaction term in Eq. (VI-9) and Eq. (VI-18), there is a zero term here. The corresponding elements of **X** are then equal to:

$$(X)_{0LM0ii,0L'M'0i'j'} = \delta_{i,i'}\delta_{i,j'}\delta_{L,L'}\delta_{M,M'}(\gamma_{LMO}) \qquad \text{(VI-55)}$$

Note that (i,i) is only coupled to itself or to (i'j') with i' ≠ j' , but not to (i'i') with i' ≠ i, because non-secular terms are not present in the Hamiltonian.

The terms coupling the two orders, n = 0 with n = 1, are given for i ≠ j by:

$$(X)_{1LM0ij,n'L'M'0i'j'} = \delta_{1,n'}\delta_{L,L'}\delta_{M,M'}\delta_{i,i'}\delta_{j,j'}(i(\omega-\omega_0)+\gamma_{LMO})$$

$$+\ \delta_{i,i'}\delta_{j,j'}\frac{1}{i\hbar}\sqrt{8\pi^2/5}\sum_{L',m}\delta_{M',M+m}F^{(2,0)}\sqrt{(2L+1)(2L'+1)}\begin{pmatrix} L & 2 & L' \\ -M & m & M+m \end{pmatrix}\begin{pmatrix} L & 2 & L' \\ 0 & 0 & 0 \end{pmatrix}$$

$$-\ \delta_{L,L'}\delta_{M,M'}\frac{1}{i\hbar}\omega_1(S_x)_{ij}\big((\delta_{n',0}+\delta_{n',2})(\delta_{i',j}\delta_{j',j}-\delta_{i',i}\delta_{j',i})\big) \qquad \text{(VI-56)}$$

and for i = j by:

$$(X)_{0LM0ii,n'L'M'0i'j'} = \delta_{0,n'}\,\delta_{L,L'}\,\delta_{M,M'}\,\delta_{i,i'}\delta_{i,j'}\left(\gamma_{LM0}\right)$$

$$+\ \delta_{i,i'}\,\delta_{i,j'}\,\frac{1}{i\hbar}\,\sqrt{8\pi^2/5}\,\sum_{L',m}\delta_{M',M+m}\,F^{(2,0)}\,\sqrt{(2L+1)\,(2L'+1)}\begin{pmatrix} L & 2 & L' \\ -M & m & M+m\end{pmatrix}\begin{pmatrix} L & 2 & L' \\ 0 & 0 & 0\end{pmatrix}$$

$$-\ \delta_{L,L'}\,\delta_{M,M'}\,\frac{1}{i\hbar}\,\omega_1\,(S_x)_{i,i+1}\left(\left(\delta_{n',1}+\delta_{n',-1}\right)\left(\delta_{i',i+1}\,\delta_{j',i}-\delta_{i',i}\,\delta_{j',i+1}\right)\right) \qquad \text{(VI-57)}$$

In Eq. (VI-56) one may take (i,j) to have the values needed for the line shape, i.e., i = -½ and j = ½ for the S = ½ case. As for i' and j', all possible values must be considered. It is clear from the middle term in Eq. (VI-56) that either i' = j' = j (first term in last parentheses in that equation) or i' = j' = i (second term in the same parentheses). This is the reason that for n = 0 (or for n = 2) only elements with i = j are needed. This is also the reason that Eq. (VI-57) was only written for such matrix elements.

Now that n = 0 terms are involved one also needs the non-zero terms of the starting vector. Assuming, as usual, that the interactions within the molecule are small compared with the Zeeman interaction, the starting vector elements are:

$$(V)_{nLM0ij} = \left(\delta_{n,1}+\delta_{n,-1}\right)\frac{1}{i\hbar}\,\frac{1}{\sqrt{8\pi^2}}\,\omega_1\,[\,S_x\,,\,\rho_0\,]_{i,j}\,\delta_{L,0}\,\delta_{M,0} \qquad \text{(VI-58)}$$

The commutator is proportional to $\mathbf{S}_y$ (see Eq. (V-121)), so for the case of i = j its matrix element is equal to zero.

On the basis of all the expressions shown here it is possible to solve for the elements of the vector **C**. Starting with the vector elements having i ≠ j, the relevant equation is:

$$\left(i\,(\omega-\omega_0)+\gamma_{LM0}\right)C_{(1)LM0(i,j)} - i\sum F(L,M;L',M')\,C_{(1)L'M'0(i,j)} + i\,d\,\{C_{(0)LM0(j,j)}$$

$$-\ C_{(0)LM0(i,i)} + C_{(2)LM0(j,j)} - C_{(2)LM0(i,i)}\} = q\,\frac{\omega_0\,\omega_1}{\hbar}\,(S_y)_{i,j}\,\frac{\delta_{L,0}\,\delta_{M,0}}{\sqrt{8\pi^2}} \qquad \text{(VI-59)}$$

using the notation: $d = (\omega_1/\hbar)\,(S_x)_{i,i+1}$ and

$$F(L,M;L',M') = \frac{1}{\hbar}\sqrt{8\pi^2/5}\sum_{L',m}\delta_{M',M+m}F^{(2,0)}\sqrt{(2L+1)(2L'+1)}\times$$

$$\times\begin{pmatrix} L & 2 & L' \\ -M & m & M+m \end{pmatrix}\begin{pmatrix} L & 2 & L' \\ 0 & 0 & 0 \end{pmatrix} \tag{VI-60}$$

The constant q is the normalization factor of the equilibrium density matrix divided by k_BT. Similarly, the equation for elements with $i = j$ is:

$$\gamma_{LM0}C_{(0)LM0(i,j)} + id\left(C_{(1)LM0(i\pm1,j)} - C_{(1)LM0(i,j\pm1)} + C_{(-1)LM0(i\pm1,j)} - C_{(-1)LM0(i,j\pm1)}\right) = 0 \quad (VI\text{-}61)$$

Therefore, if one now chooses $i = -½$, $j = ½$ (which, as seen previously, is the relevant case for an $S = ½$ system) one has:

$$\gamma_{LM0}\left\{C_{(0)LM0(j,j)} - C_{(0)LM0(i,i)}\right\} = 2id\left\{C_{(1)LM0(j,i)} - C_{(1)LM0(i,j)}\right\} = 4d\left\{Im\left(C_{(1)LM0ij}\right)\right\} \quad (VI\text{-}62)$$

where Im(...) is the imaginary part of the complex number (...). The terms with $n = -1$ are omitted here, since they are expected to be very small if irradiation is not very strong. This means that although there is some saturation, namely ω_1/ω_0 is not very small, this ratio is still small compared with 1. Terms with indices less than $-½$ do not exist in this case.

These equations will now be treated for the special case in which the g-tensor is axially symmetric, so that the form of **F** is simplified. It is also assumed that other relaxation mechanisms may be operating, in addition to the rotational diffusion. Therefore in Eq. (VI-59) γ_{LM0} is replaced by $(\gamma_{LM0} + 1/T_2)$ and in Eqs. (VI-61), (VI-62) γ_{LM0} is replaced by $(\gamma_{LM0} + 1/T_1)$, where the relaxation times T_1 and T_2 are rotationally invariant. The terms in Eq. (VI-59) related to $n = 2$ will be neglected, since they are small if irradiation is not very strong. The equation for the relevant elements of **C** is thus:

$$\left(i(\omega-\omega_0) + \gamma_{LM0} + \frac{1}{T_2}\right)C_{(1)LM0(i,j)} - i\sum F(L,M;L',M')\,C_{(1)L'M'0(i,j)}$$

$$+\left(\frac{4d^2}{\gamma_{LM0} + \dfrac{1}{T_1}}\right)i\left\{Im\left(C_{(1)LM0(i,j)}\right)\right\} = iq\,\omega_0\,\omega_1\frac{\delta_{L,0}\,\delta_{M,0}}{\sqrt{8\pi^2}} \tag{VI-63}$$

Since it is assumed here that the g-tensor is axially symmetric, the **F** term does not couple different L and M values (same as in Eq. (VI-10)), so only $M = K = 0$ is needed. The solution of this equation gives the saturated spectrum for this case. This will be discussed further in Sec. VII.1.(d).

VI.2 Doublet EPR - more general rotational diffusion problems

(a) Anisotropic rotational diffusion in isotropic solvents

The discussion of rotational diffusion can be extended to more general situations, both in terms of the rotational mechanism and in terms of the motional constraints created by the solvent. For the rotational mechanism one may be interested in anisotropic rotational diffusion, in which the molecule tends to rotate at different speeds about different axes. This is indeed the usual case when the molecule has a distinctly non-spherical shape. For example, if the molecule is approximately shaped as an elongated ellipsoid, its rotation about its long axis will normally be faster than its rotation about its shorter axes. Thus the diffusion tensor will not be proportional to the unit tensor as in the case of isotropic rotation. If the molecule is approximately axially symmetric, also the diffusion tensor is expected to be approximately axially symmetric.

Dealing with such anisotropy is straightforward, because the diffusion operator still has almost the same form as in the case of spherical symmetry, and its eigenfunctions are still Wigner's rotation functions, only the eigenvalues being different (see Eq. (V-88)). Thus the discussion above for the isotropic case applies, with the necessary changes, also to the cases of anisotropic diffusion. An example of such diffusion is now presented.

Example VI.3: Models of rotational diffusion in liquid and frozen solutions

EPR line shapes of the peroxylamine disulfonate (PADS) radical were measured for several temperatures in 85% glycerol/water solution and in frozen water and D_2O (Ref. 2). This particular radical was chosen because its g-tensor and hyperfine tensor could be determined accurately at the rigid limit, i.e., at sufficiently low temperatures, and because it does not suffer from inhomogeneous broadening due to unresolved intramolecular dipolar interactions. The rigid limit parameters were found by fitting an experimental spectrum at T = - 90° C. The temperature dependence indicated that some kind of motion, i.e., rotational diffusion of the radical occurs both in the liquid and in the frozen solutions. It is thus interesting to find a reasonable model for the motion, as well as the temperature dependence of the motional rate.

Several characteristics of the motion were investigated, both experimentally and theoretically. Among these are the temperature dependence of the line shape and the degree of anisotropy in the diffusion tensor. The main model employed was that of Brownian rotational diffusion. The parameter $N \equiv R_3 / R_1$ was defined as the ratio between the rate constant for rotation about the main (long) axis and rotation about the short axis (axial symmetry is assumed). The parameter $\tau_R \equiv 1/6R$, with $R \equiv (R_3 R_1)^{1/2}$ describes an average correlation time for the motion. In Fig. VI.3 three line fittings are shown for a spectrum measured at T = - 50° C. All three fits were calculated with the same overall τ_R, which was found to give best results, and the three simulated line shapes differ in the anisotropy parameter N. In this case, N = 3 gives better results than the other two values used.

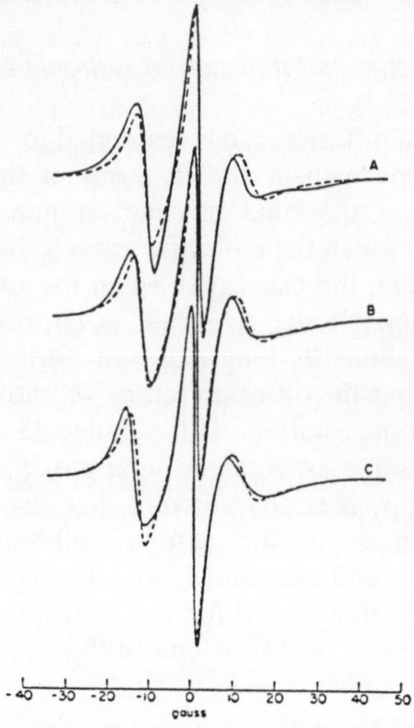

Fig. VI.3. A comparison of simulated and experimental spectra of PADS in frozen D_2O at $T = -50°$ C. The dashed line is the experimental line shape, and the solid line is calculated for Brownian rotational diffusion with an average $\tau_R = 2$ and an anisotropy ratio of: A) $N = 1$, B) $N = 3$ and C) $N = 6$.

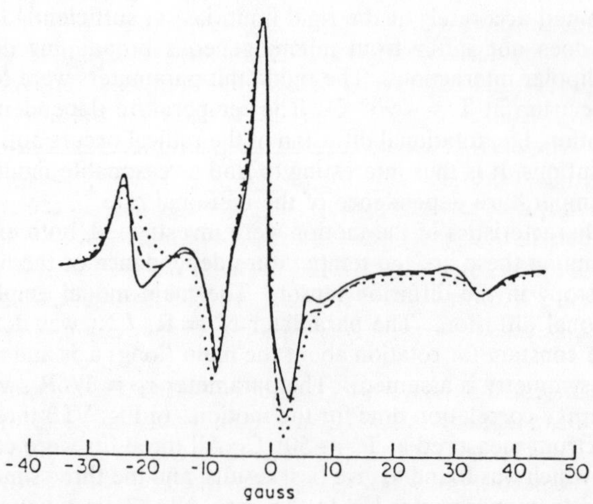

Fig. VI.4. A comparison of line shapes simulated for a nitroxide (with axially symmetric interaction tensors) undergoing Brownian rotational diffusion with $\tau_R = 15$ (solid line), free diffusion with $\tau_R = 10$ (dashed line) and jump diffusion with $\tau_R = 7$ (dotted line).

Several simulations were carried out with different models of the motion, differing in the state of the molecule between collisions and in the nature of the collisions [2]. In Brownian diffusion it is assumed that molecules rotate for some time, then collide and change their orientation slightly, and so on. In jump diffusion the collisions are assumed to be strong, with a major reorientation, but no motion is assumed for the time between collisions. In the free diffusion model it is assumed that collisions are strong, but the molecules rotate freely between collisions. Line shapes calculated with these models for a nitroxide with axially symmetric interaction tensors are presented in Fig. VI.4. From this figure it is apparent that the line shape depends on an interplay of the effect of the model and the effect of the rate parameters, so the distinction between models may not always be easy in practice. Nevertheless, even when the parameters were adjusted to make the graphs similar to each other, one may distinguish between the predictions of different models, through comparisons with experimental results. In the particular case studied here it was found that simulations with free diffusion gave the best fit to experimental results. The Brownian diffusion model, with its sharper features in the line shape, and the jump model, with its broader line shape, were less successful in this case.

****

(b) *Rotational diffusion in anisotropic media*

A more serious problem is including constraints on the motion, due to lack of isotropy in the solvent. The typical example for such constraints is that of a liquid crystalline solvent, which is fluid yet is partly ordered. The order of the liquid crystal molecules imposes restrictions on the rotational motion of the solute molecules. Take, for example, a nematic solvent, which is the simplest type of a liquid crystal. The molecules of such a solvent are usually rod-like in shape, and are arranged (in the appropriate temperature range) so that the rods tend to be roughly parallel to a single axis, the director, characterizing the whole bulk of the liquid crystal. Now assume, to be specific, that the solute molecules rotate about their long axis. The solute molecules can rotate easily if their axis of rotation is parallel to the rods, because then the rotational motion does not cause any collision with the solvent molecules. However, if the axis of rotation is perpendicular to the rods, the motion may be significantly hindered, because then the motion may cause collisions with solvent molecules. In the opposite case, when the solute molecule rotates about a short axis, it can rotate easily if the axis of rotation is perpendicular to the rods, because then the rotational motion is in a plane parallel to the rods. However, if the axis of rotation is parallel to the rods, the motion will be greatly hindered, because then the motion is in a plane which intersects the solvent molecules.

In order to describe these restrictions mathematically one must model the effect of the solvent in an appropriate way, and then solve the SLE with this constraint included. We shall discuss here the calculations necessary both for doublets and for triplets. In the next section some additional consequences of anisotropy, relevant for the EPR of photoexcited triplets, will be explained.

It is possible to model the influence of the solvent by assuming it just leads to an excluded volume effect, so that it allows the solute molecules to rotate only in a limited angular range. A more general formalism, used here, assumes the diffusing molecule is in a cylindrically symmetric restoring potential. Thus any rotation angle is possible, but not all rotations are equally favorable from an energetic point of view. Assume the paramagnetic species is dissolved in a nematic liquid crystal (LC), i.e., a uniaxial LC. The symmetry axis of the nematic LC is called the director, and its direction is one of the parameters influencing the lineshape. The partly ordered solvent exerts some ordering influence on the solute molecules. Specifically, in a nematic liquid crystal it is energetically favorable for the solute molecules to align with their main axis parallel to the director if they are prolate (cigar-like), and perpendicular to the director if they are oblate (disk-like). This is taken into account by assuming a potential energy function which depends on the molecular orientation relative to the director:

$$U(\Omega) = -k_B T \sum_{L=2,4} \left\{ \lambda^L{}_0 D^L{}_{0,0}(\Omega) + \lambda^L{}_2 \left(D^2{}_{0,2}(\Omega) + D^2{}_{0,-2}(\Omega) \right) \right\} \qquad \text{(VI-64)}$$

where $\Omega=(\alpha,\beta,\gamma)$ is the set of Euler angles specifying the molecular orientation relative to the director. The potential is constructed so that a positive value for the parameters $\lambda^L{}_M$ corresponds to a minimum energy configuration with the (approximate) symmetry axis of the molecule parallel to the director ($\beta=0$), and a negative value for these parameters corresponds to minimum energy when the molecular symmetry axis is perpendicular to the director ($\beta=\pi/2$). The first case is natural for prolate molecules, and the second case for oblate molecules, such as porphyrins and related moieties.

This potential changes the equilibrium probability distribution for molecular orientations from an isotropic one to:

$$p_0(\Omega) = \frac{\exp(-U(\Omega)/k_B T)}{\int d\Omega \, \exp(-U(\Omega)/k_B T)} \qquad \text{(VI-65)}$$

and adds to the isotropic reorientation operator $\mathbf{\Gamma}_{isot}$ a contribution depending on the potential. The equilibrium density matrix is proportional to this probability distribution. Equation (IV-79) for the isotropic case can be written, using a dyadic notation, as:

$$\frac{\partial}{\partial t} p(\Omega) = -\mathbf{\Gamma}_{isot} p(\Omega) = -\boldsymbol{J} \cdot \boldsymbol{R} \cdot \boldsymbol{J} p(\Omega) = -\sum_{\alpha,\beta=x,y,z} R_{\alpha\beta} J_\alpha J_\beta \, p(\Omega) \qquad \text{(VI-66)}$$

where **R** is the diffusion tensor and **J** is the angular momentum operator. Note that $\mathbf{\Gamma}$ was defined here with an opposite sign relative to Eq. (IV-79), in order to characterize $\mathbf{\Gamma}$ by a positive rate parameter rather than by a negative parameter. In the principal axis system (PAS) of **R**, which is used in most relevant calculations in the present book, the only non-zero elements of the diffusion tensor are R_{xx} , R_{yy} (both equal to $R_\perp$ in the axial symmetry case) and R_{zz} (denoted by $R_\parallel$ in the axial symmetry case).

In a general approach to the problem of Brownian rotational diffusion, it is possible to show that the torque applied by the potential results in the following modification of the above equation:

$$\frac{\partial}{\partial t}p(\Omega) = -\left(\boldsymbol{\Gamma}_{isot} + \boldsymbol{\Gamma}_{poten}\right)p(\Omega) =$$

$$= -\boldsymbol{J}\cdot\boldsymbol{R}\cdot\boldsymbol{J}\,p(\Omega) - \boldsymbol{J}\cdot\left\{\left[\boldsymbol{R}\cdot\boldsymbol{J}\left[\frac{U(\Omega)}{k_B T}\right]\right]p(\Omega)\right\} \tag{VI-67}$$

which reduces to the form:

$$\frac{\partial}{\partial t}p(\Omega) = -\left\{R_\perp\left(J_x^2 + J_y^2\right) + R_\parallel J_z^2\right\}p(\Omega) - J_x\left\{\left[R_\perp J_x\frac{U(\Omega)}{k_B T}\right]p(\Omega)\right\}$$

$$- J_y\left\{\left[R_\perp J_y\frac{U(\Omega)}{k_B T}\right]p(\Omega)\right\} - J_z\left\{\left[R_\parallel J_z\frac{U(\Omega)}{k_B T}\right]p(\Omega)\right\} \tag{VI-68}$$

in the PAS of **R** in the case of axial symmetry.

It is convenient to transform the operators so as to make the diffusion superoperator symmetric. The density matrix and the distribution function are transformed as

$$\rho^S(\Omega,t) = \frac{1}{\sqrt{p_0(\Omega)}}\rho(\Omega,t) \quad ; \quad p^S(\Omega,t) = \frac{1}{\sqrt{p_0(\Omega)}}p(\Omega,t) \tag{VI-69}$$

and the relaxation superoperator is transformed or symmetrized as:

$$\boldsymbol{\Gamma}_\Omega^S = \frac{1}{\sqrt{p_0(\Omega)}}\boldsymbol{\Gamma}_\Omega\sqrt{p_0(\Omega)} \tag{VI-70}$$

where $\boldsymbol{\Gamma}_\Omega = \boldsymbol{\Gamma}_{isot} + \boldsymbol{\Gamma}_{poten}$, and the transformed superoperator $\boldsymbol{\Gamma}_\Omega^s$ is hermitian. It follows from the definitions that the same equation holds for the transformed quantities:

$$\frac{\partial}{\partial t}p^S(\Omega,t) = -\boldsymbol{\Gamma}_\Omega^S\,p^S(\Omega,t) \tag{VI-71}$$

as for the original ones. The isotropic part $\boldsymbol{\Gamma}_{isot}$ is unchanged by the transformation, but the term due to the potential becomes:

$$\boldsymbol{\Gamma}_{poten}^S = \frac{1}{2k_B T}\left\{\boldsymbol{J}\cdot\boldsymbol{R}\cdot\boldsymbol{J}\,p(\Omega)\right\} + \frac{1}{(2k_B T)^2}\left\{\left(i\boldsymbol{J}U(\Omega)\right)\cdot\boldsymbol{R}\cdot\left(i\boldsymbol{J}U(\Omega)\right)\right\} \tag{VI-72}$$

In the PAS of **R** and assuming axial symmetry, the last expression is equal to:

$$\Gamma_{poten}^{\ S} = \frac{1}{2k_B T}\left\{R_\perp\left(J^2 U(\Omega)\right) + \left(R_\parallel - R_\perp\right)\left(J_z^2 U(\Omega)\right)\right\}$$

$$- \frac{1}{(2k_B T)^2}\left\{R_\perp\left(J_+ U(\Omega)\right)\left(J_- U(\Omega)\right) + R_\parallel\left(J_z U(\Omega)\right)^2\right\} \qquad \text{(VI-73)}$$

When the SLE is written in terms of these transformed quantities, the matrix of Eq. (V-116) is complex symmetric, which enables the use of the numerical solution methods described in Sec. V.4.

For the special case in which the diffusion tensor is spherically symmetric, Eq. (VI-71) thus becomes:

$$\frac{\partial}{\partial t}p(\Omega,t) = -R\nabla_\Omega^2 p(\Omega,t) + \frac{R}{k_B T}\frac{1}{\sin(\beta)}\frac{\partial}{\partial\beta}\left\{\sin(\beta)\left(-\frac{\partial U(\Omega)}{\partial\beta}\right)p(\Omega,t)\right\} \qquad (VI\text{-}74)$$

in which p(Ω,t) is understood to be the transformed distribution (written above as $p^S(\Omega,t)$). The addition of terms depending on the potential results, applying Eq. (V-117), in an additional contribution to the relaxation part in the supermatrix **X**. The detailed form of this contribution (see Refs. 12, 13) will not be given here, but the practical effect of such terms will be demonstrated in the Examples below.

Example VI.4: Slow motion in a liquid crystalline solvent

One of the first experiments in a non-isotropic medium that were interpreted with the SLE formalism was done for cholestane (CSL) spin probes in nematic phase V liquid crystalline solvent (Ref. 6). These probe molecules are relatively large, and therefore their orientation is determined by the anisotropic medium to a greater degree than that of the Tempone nitroxide radical, which was studied earlier (Ref. 4). In this case, due to the strong ordering, it is found that a single parameter, λ^2_0 is sufficient to describe the potential. Since the constant term of -½ in $D^2_{0,0} = ½(3\cos^2(\beta) - 1) = (3/2)\cos^2(\beta) - ½$ contributes only to the overall normalization of the potential, one may define λ by:

$$\frac{U(\beta)}{k_B T} = -\lambda\cos^2(\beta) \qquad \text{(E-1)}$$

instead of the usual definition of λ^2_0 in the expansion of the restoring potential (Eq. (VI-64) above). Using the model of Brownian rotational diffusion in an anisotropic medium, simulations were done to fit the experimental line shapes. The result of such simulations for several temperatures is shown in Fig. VI.5. The fit obtained is generally good, and there is practically no difference between using the model of isotropic diffusion and the model of anisotropic diffusion. This is because correlation times are long, so the spectrum is not sensitive to the details of the motion. Only from the higher temperature spectra (not shown here), corresponding to fast motion, is it possible to determine that the motion is indeed anisotropic.

Table E.1. Parameters for Brownian diffusion in a restoring potential (CSL in phase V)

Temperature (in ° C)	Assuming isotropic diffusion				Assuming anisotropic diffusion				
	τ_R [a]	λ	S	A'[b]	$\tau_{R\perp}$ [a]	N	λ	S	A'[b]
19	9	4.5	0.61	1.45	17	5	4.3	0.61	0.9
3	20	5.3	0.67	1.5					
-6	35	6.0	0.71	1.8	78	5	6.0	0.71	1.3
-16	100	6.9	0.76	2.2					

[a] in ns (10^{-9} s).
[b] in gauss.

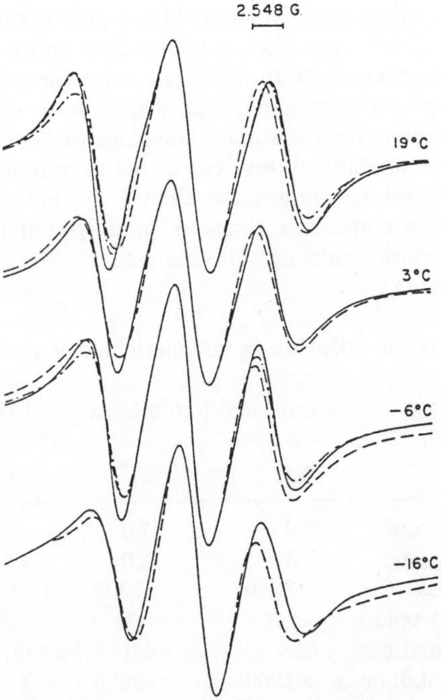

Fig. VI.5. Experimental (dashed line) and simulated spectra, with isotropic Brownian diffusion (solid line) and with anisotropic Brownian diffusion (dot-dashed line), for CSL in nematic phase V solvent. The temperatures are indicated in the Figure; other parameters are given in Table E.1.

Magnetic parameters. i.e., interaction constants, were determined from a low temperature spectrum (T = -152° C) corresponding to the rigid limit. Rate parameters and the corresponding correlation times for the motion were derived from the series of temperature dependent spectra. The fitting procedure yielded information also on the ordering parameters of the liquid crystalline solvent. The parameters used for the simulations shown in Fig. VI.5 are given in Table E.1. In this table, λ is the parameter in the restoring potential, and $S = \langle D^2_{0,0} \rangle$ is the usual order parameter of a nematic liquid crystal, relating here to the solvent. N is the anisotropy ratio $N = R_{\|}/R_{\perp}$ of rotation rates, and A' is a residual line width parameter (see Ref. 6).

****

Example VI.5: *Order and dynamics in a liquid crystal*

Another study involving a non-isotropic medium was done for CSL spin probes incorporated into a liquid crystal (Ref. 10). The main study was done on certain polymer liquid crystals, and part of the work was also done on the monomeric analogue of these polymers, which also has a liquid crystalline phase. The purpose of the work was to examine the motion of CSL within the liquid crystals. In this case the calculations for the SLE were done with a discretized version of the equation solved on a finite grid (see subsection V.2.(b) above), and the effect of the medium was simulated by assuming an orientation dependent distribution for the molecules, and summing over the contributions of all orientations. Fitting experimental line shapes, both order parameters for the medium and rate parameters for the solute molecules were obtained. The line fitting obtained in some of the simulations resulted in the graphs shown in Fig. VI.6. The parameters for these simulations are given in Table E.2. One of the results of these and other simulations was that correlation times in the polymeric liquid crystals are much longer than in the monomeric analogue. This indicates the grteater restriction of motion in the polymeric case.

Table E.2. Parameters for diffusion in polymeric and monomeric liquid crystals[a]

Case	Temperature (K)	phase	motional parameters $\tau_{R\|}$ [b]	$\tau_{R\perp}$ [b]	order parameter S
a	393.5	isotropic	3.0	21.0	-
b	369.0	nematic	7.8	78.0	0.63
c	263.0	glassy	>2000	>2000	0.92
d	370.0	isotropic	0.27	1.35	-
e	360.0	nematic	0.4	4.0	0.48
f	263.0	crystalline	>2000	>2000	0.0

[a] Lines a-c refer to the polymer, and lines d-f to the monomer.
[b] in ns (10^{-9} s).

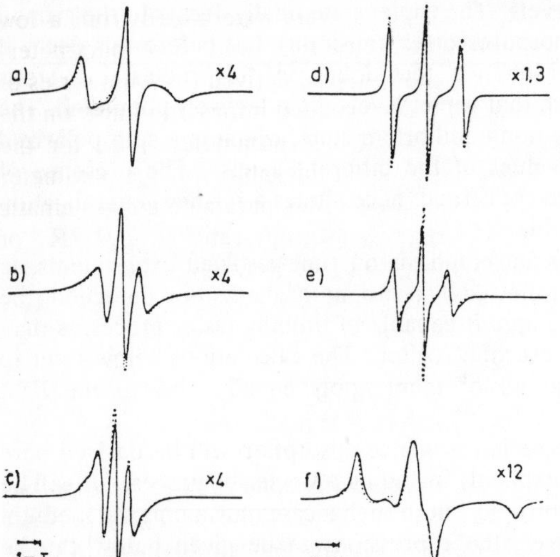

Fig. VI.6. Experimental (solid line) and simulated (dashed line) spectra, for CSL spin probes in macroscopically aligned samples of a polymer (a-c) and of its monomeric analogue (d-f). The parameters are given in Table E.2.

****

VI.3 EPR of photoexcited triplets undergoing rotational diffusion

(a) *Photoexcited triplet states - time dependence*

In most magnetic resonance experiments one works with systems which are initially at thermal equilibrium, and return to thermal equilibrium at the end of the experiment. The ordinary line shape formula depends on the assumption that the initial state is that of thermal equilibrium. This is appropriate for doublet systems, which are generally long lived species (disregarding in the present context CIDEP and similar methods). In such systems the EPR signal is related to a single electron, for which the spin is $S = ½$. There are two S_z eigenstates, having the eigenvalues $s_z = -½$ and $½$, which have different energies, and this is the reason for the name doublet.

In triplet EPR, however, one is concerned with an $S = 1$ state of a molecule, having three energy levels (with $s_z = -1, 0$ and 1). This is usually found only in transient photoexcited systems, because for most systems the ground state is a spin singlet ($S = 0$). Such systems are generated typically by using laser light to excite a molecule to an excited singlet state. Once in that high energy state, the molecule may decay back to the ground state (a fluorescence transition) or, in a small fraction of cases, reach a spin triplet state via a non-radiative intersystem crossing (ISC) mechanism. This mechanism results from spin-orbit interaction and is therefore strongly dependent on the spin quantum numbers of the triplet levels. The triplet state finally decays to the ground state via another ISC transition (a phosporescence transition), but before this decay it may be observed with magnetic resonance techniques in this metastable state. Furthermore, the process of ISC is such that triplet levels are selectively populated. The triplet is thus generated with a highly non-equilibrium state, which is a spin polarized state, due to the different spin (s_z) values of the different levels. The presence of stochastic processes, with which we are concerned, necessitates a further generalization of the treatment of the line shape problem.

In recent years there has been an emphasis on time-resolved experiments, in which the excitation by a laser pulse is followed by transient magnetization detection (the diode detection technique). This technique is capable of probing faster processes than the conventional EPR, in the submicrosecond region. The calculations below refer to experiments of this type, but almost all of them apply equally also to the light modulation technique.

The general expression for magnetic resonance absorption will be derived here with an arbitrary initial condition and explicitly including a stochastic process. It will be assumed that the medium may be anisotropic, but in such a case only a uniaxial medium will be considered explicitly. However, the expressions to be given below can be appropriately generalized for non-isotropic solvents which are not uniaxial. The density matrix evolves according to the following equation:

$$\frac{\partial}{\partial t}\rho(\Omega,t) = \frac{1}{i\hbar}\left[H_0 + H_1(\Omega) + H_2(t), \rho(\Omega,t)\right] - \boldsymbol{\Gamma}_\Omega\left(\rho(\Omega,t) - \rho_{eq}\right) - \boldsymbol{\Gamma}_k\rho(\Omega,t) \qquad \text{(VI-75)}$$

The Hamiltonian consists of three terms:

$$H_0 = \omega_0 S_z \qquad \text{(VI-76a)}$$

$$H_1(\Omega) = \sum_{m,m'} d^2_{m,m'}(\psi)\, D^2_{m',m''}(\Omega)\, F^{(2,-m'')}\, A_{lab}^{(2,m)} \qquad \text{(VI-76b)}$$

$$H_2(t) = 2\,\omega_1 \cos(\omega t) \qquad \text{(VI-76c)}$$

corresponding to the Zeeman interaction with the external magnetic field, intra-molecular interactions and interaction with the microwave irradiation, respectively. Here the stochastic process contributes to relaxation via the operator $\boldsymbol{\Gamma}_\Omega$. The other relaxation operator which appears here explicitly originates in the kinetic rate equations for the triplet levels, responsible for their population and depopulation. Due to this term the triplet state eventually decays to the original ground state. The equilibrium density matrix ρ_{eq} is included in the equation in order to ensure relaxation to a finite temperature thermal equilibrium. The angle ψ specifies the orientation of the symmetry axis (the director) of the nematic liquid crystalline solvent with respect to the direction of the static magnetic field. For isotropic solvents one simply sets: $\psi = 0$. The Euler angles $\Omega = (\alpha,\beta,\gamma)$ give the orientation of the molecular axes relative to the director. For simplicity, the principal axis system (PAS) of the diffusion tensor is assumed identical to the PAS for the ZFS tensor.

The interaction constants $F^{(2,m'')}_{mol.}$ are the components of the ZFS tensor in the molecular axes system:

$$F^{(2,0)} = \sqrt{2/3}\,D \qquad F^{(2,\pm1)} = 0 \qquad F^{(2,\pm2)} = E \tag{VI-77}$$

and $A^{(2,m)}_{lab.}$ are the components of the corresponding second-rank tensor constructed from spin operators in the laboratory system of axes:

$$A_{lab}^{(2,0)} = \frac{1}{\sqrt{6}}\left(3S_z^2 - S^2\right) = \sqrt{3/2}\left[S_z^2 - \frac{1}{3}S^2\right] \tag{VI-78a}$$

$$A_{lab}^{(2,\pm1)} = \mp\frac{1}{2}\left(S_\pm S_z + S_z S_\pm\right) \tag{VI-78b}$$

$$A_{lab}^{(2,\pm2)} = \frac{1}{2}\left(S_\pm\right)^2 \tag{VI-78c}$$

The interaction constants $F^{(2,m'')}$ are small compared with the resonance frequency in high fields (e.g., X-band EPR), but not very small. Therefore, the equilibrium density matrix ρ_{eq}, defined in Eq. (V-99) has a non-negligible orientation dependence, due to the relatively large magnitude of the orientation-dependent part of the triplet Hamiltonian. This is one of the main features which distinguish typical triplet systems from typical doublet systems. This operator will be normalized so that the trace is equal to 1, which means that the total triplet population is normalized to 1.

The initial process of populating the triplet levels by intersystem crossing from the photoexcited singlet state need not be treated directly. However, as will be seen below, the rate constants A_i for populating the triplet strongly influence the line shape, since they determine the initial condition for solving Eq. (VI-75). As for the differential equation itself (Eq. (VI-75)), the kinetic term is taken to be the same term calculated when there is no diffusion. The equations for the triplet levels are, denoting each spin state $|i\rangle$ by its S_z eigenvalue:

$$\Gamma_k \rho_{-1,-1} = -\frac{1}{3}k_{-1} - \left(k_{-1} + 2W\right)\rho_{-1,-1} + W\left(\rho_{0,0} + \rho_{1,1}\right) \tag{VI-79a}$$

$$\Gamma_k \rho_{0,0} = -\frac{1}{3}k_0 - \left(k_0 + 2W\right)\rho_{0,0} + W\left(\rho_{-1,-1} + \rho_{1,1}\right) \tag{VI-79b}$$

$$\Gamma_k \rho_{1,1} = -\frac{1}{3}k_1 - \left(k_1 + 2W\right)\rho_{1,1} + W\left(\rho_{-1,-1} + \rho_{0,0}\right) \tag{VI-79c}$$

Here k_α are the rates of depopulation (due to ISC) for specific levels, and W is a spin-lattice relaxation rate due to other mechanisms. These equations describe the decay of the triplet levels, either to the ground state (via k_α) or to an internal equilibrium (via W).

This decay is slow on the time scale of the EPR measurements discussed here, so it is only of secondary importance. These equations are correct when the magnetic field is along one of the canonical orientations (**x**, **y** and **z**) of the molecular PAS. For non-oriented systems the molecules have arbitrary orientations with respect to the magnetic field. One then needs to express the dependence of the kinetic population (A_α) and depopulation (k_α) constants on orientation.

The population rates A_α and the depopulation rates k_α for the three levels of the triplet originate in the spin-orbit Hamiltonian, which connects the excited singlet state to the triplet state (ISC). Since the ISC mechanism is usually highly level-selective, these rates depend on the orientation of the molecular "α" axis with respect to the external magnetic field. It is simple to calculate the effects of these constants on the inter-level dynamics when the magnetic field is along one of the canonical orientations (**x**, **y** and **z**) of the molecular PAS. On the basis of such calculations it is possible to extract from experimental line shapes the six basic parameters: A_x, A_y, A_z and k_x, k_y, k_z. However, in practice the interesting line shapes are often obtained in samples in which the molecules are not uniformly oriented with respect to the magnetic field. The calculation will be performed first for an isotropic solvent, so that in Eq. (VI-76) the angle Ω is the only relevant angle. More general cases will be considered later on. In the isotropic case there is no director, and one may formally define a "director" as being along the main magnetic field. Thus the orientation $\Xi = (\theta,\phi)$ of the molecule relative to the magnetic field is identical to Ω. The values of the A_α and k_α parameters for each of the canonical orientations are simply related to the basic experimental parameters:

$$F_0 = F_p \qquad F_1 = F_{-1} = \frac{F_q + F_r}{2} \qquad (\textit{orientation along } \boldsymbol{p}) \tag{VI-80}$$

where (**p**, **q**, **r**) is equal to (**x**, **y**, **z**) or its cyclic permutations, and F = A or k. If $\Xi' = (\theta',\phi')$ is the orientation of the magnetic field relative to the molecule (describing the inverse of the rotation by Ξ), it is possible to show that:

$$F_0 = F_x \sin^2(\theta')\cos^2(\phi') + F_y \sin^2(\theta')\sin^2(\phi') + F_z \cos^2(\theta') \tag{VI-81}$$

and

$$F_1 = F_x \frac{\sin^2(\theta')\sin^2(\phi') + \cos^2(\theta')}{2} + F_y \frac{\sin^2(\theta')\cos^2(\phi') + \cos^2(\theta')}{2}$$

$$+ F_z \frac{\sin^2(\theta')}{2} = \frac{F_y + F_z}{2}\sin^2(\theta')\cos^2(\phi') + \frac{F_z + F_x}{2}\sin^2(\theta')\sin^2(\phi')$$

$$+ \frac{F_x + F_y}{2}\cos^2(\theta') \tag{VI-82}$$

For the population constants it is common to define the total population probability: $A_T = A_x + A_y + A_z$. From these expressions one may calculate also the quantity $\alpha(\Xi')$, defined as:

$$\alpha \equiv \frac{A_1 - A_0}{A_T} = \sum_{L,M} \alpha^L{}_M D^L{}_{M,0}(\Xi') \tag{VI-83}$$

which is an important factor influencing the line shape of spin polarized triplets. The second part of this equation defines an expansion of α in rotation functions, for which the non-zero coefficients are:

$$\alpha^2{}_0 = \frac{A_x + A_y}{2} - A_z \qquad \alpha^2{}_{\pm 2} = \sqrt{3/8}\,(A_y - A_x) \tag{VI-84}$$

It is then possible to show that also for the inverse orientation, Ξ, of the molecule relative to the magnetic field one has:

$$\alpha(\Xi) \equiv \sum_{L,M} \alpha^L{}_M D^L{}_{M,0}(\Xi) = \sum_M \alpha^2{}_M D^2{}_{M,0}(\Xi) \tag{VI-85}$$

with exactly the same coefficients $\alpha^L{}_M$ as in Eq. (VI-83). A similar relation can be written for k(Ξ). The special case in which there is no spin polarization (thermal equilibrium) corresponds to having all A_i equal, so the $\alpha_{2,i}$ vanish in that case.

Now take the case of a non-isotropic, uniaxial solvent. Eq. (VI-85) still holds, but now Ω is not identical with Ξ. If Ψ stands for the Euler angles specifying the orientation of the director relative to the magnetic field, then:

$$D^L{}_{M,0}(\Xi) \equiv \sum_{M'} D^L{}_{M',0}(\Omega)\, D^L{}_{M',M}(\Psi^{-1}) = \sum_{M'} D^L{}_{M',0}(\Omega) \left(D^L{}_{M,M'}(\Psi)\right)^* =$$

$$= \sum_{M'} D^L{}_{M',0}(\Omega)\, d^L{}_{M,M'}(\psi) \tag{VI-86}$$

The last equality follows from the fact that for a uniaxial solvent: $\Psi = (0,\psi,0)$. Thus

$$\alpha(\Xi) = \alpha(\Omega,\psi) = \sum_{M,M'} \alpha^2{}_M D^2{}_{M',0}(\Omega)\, d^2{}_{M,M'}(\psi) = \sum_{M'} \alpha^{(M')} D^2{}_{M',0}(\Omega) \tag{VI-87}$$

using the definition

$$\alpha^{(n)} \equiv \sum_k \alpha^2{}_k d^2{}_{k,n}(\psi) = \alpha^2{}_0 d^2{}_{0,n}(\psi) + \alpha^2{}_2 \left(d^2{}_{2,n}(\psi) + d^2{}_{-2,n}(\psi)\right) \tag{VI-88}$$

For the case of $\psi = 0$ this reduces to the expression for isotropic solvents. The other practically important case is that of $\psi = \pi/2$, in which Eq. (VI-87) results in

$$\alpha(\Omega,\psi) = \left[-\frac{1}{2}\alpha^2_{\ 0} + \sqrt{3/2}\ \alpha^2_{\ 2} \right] D^2_{\ 0,0}(\Omega)$$

$$+ \left[\sqrt{3/8}\ \alpha^2_{\ 0} + \frac{1}{2}\alpha^2_{\ 2} \right] \left(D^2_{\ 2,0}(\Omega) + D^2_{\ -2,0}(\Omega) \right) \qquad \text{(VI-89)}$$

Eqs. (VI-87)-(VI-89) here apply also to the orientation dependence of the depopulation constants k_α , except that in k_α there is also a non-zero constant term (see Ref. 16). From these expressions it is clear that the orientation of the director relative to the magnetic field will have a very pronounced effect on the line shape, through the change in the population anisotropy factor. This can be demonstrated as follows. For a typical case, in which the ZFS parameters satisfy the relations: D > 3E > 0, the six resonance positions are: z_+, y_-, x_-, x_+, y_+ and z_- in the order of increasing resonance field ("+" denotes a 0-->1 transition, "-" a -1-->0 transition). The basic population parameters A_x, A_y, A_z can be expressed using the expansion coefficients of α (Eq. (VI-88)) as:

$$\frac{A_x}{A_T} = \frac{1}{3}\left(1 + \alpha^{(0)} - \sqrt{6}\ \alpha^{(2)}\right) \qquad \text{(VI-90a)}$$

$$\frac{A_y}{A_T} = \frac{1}{3}\left(1 + \alpha^{(0)} + \sqrt{6}\ \alpha^{(2)}\right) \qquad \text{(VI-90b)}$$

$$\frac{A_z}{A_T} = \frac{1}{3}\left(1 - 2\,\alpha^{(0)}\right) \qquad \text{(VI-90c)}$$

Then the expected intensity factors $(A_1\text{-}A_0)/A_T$ for these resonances are:

$$x_\pm \ : \quad \frac{\pm 1}{A_T}\left[\frac{A_y + A_z}{2} - A_x \right] = \mp \frac{1}{2}\left(\alpha^{(0)} - \sqrt{6}\ \alpha^{(2)} \right) \qquad \text{(VI-91a)}$$

$$y_\pm \ : \quad \frac{\pm 1}{A_T}\left[\frac{A_z + A_x}{2} - A_y \right] = \mp \frac{1}{2}\left(\alpha^{(0)} + \sqrt{6}\ \alpha^{(2)} \right) \qquad \text{(VI-91b)}$$

$$z_\pm \ : \quad \frac{\pm 1}{A_T}\left[\frac{A_x + A_y}{2} - A_z \right] = \pm\, \alpha^{(0)} \qquad \text{(VI-91c)}$$

respectively. For isotropic solvents or for $\psi=0$, $\alpha^{(n)}$ is equal to $\alpha^2_{\ n}$ (n=0 or 2), but for $\psi=\pi/2$, for example, $\alpha^{(n)}$ has a very different value, defined through Eq. (VI-88).

(b) Line shape theory for the general case

The general derivation of the line shape will now be given. It is convenient to define several operators in Liouville space (superoperators), through their operation on a general Liouville space vector (supervector - which is an operator on Hilbert space). If such a vector is denoted by **f**, the superoperators are defined by:

$$A\{f\} \equiv \frac{1}{\hbar}[H_{mol},f] = \frac{1}{\hbar}[H_0 + H_1(\Omega),f] \tag{VI-92a}$$

$$B(t)\{f\} \equiv 2\omega_1 \cos(\omega t) B_1\{f\} = \frac{1}{\hbar}[H_2(t),f] \tag{VI-92b}$$

$$C\{f\} \equiv (\Gamma_\Omega + \Gamma_k)\{f\} \tag{VI-92c}$$

A distinction is made here between the time independent molecular Hamiltonian $\mathbf{H}_{\mathbf{mol}}$ and the time dependent Hamiltonian $\mathbf{H}_2$(t) . One also defines the following vector in Liouville space:

$$d \equiv \Gamma_\Omega \rho_{eq} \tag{VI-93}$$

In all formulas of this kind a vector is indexed by a pair of indices, and a matrix element by two pairs of indices, as in

$$\left(A\{f\}\right)_{ij} = \sum_{kl} A_{ij,kl} f_{kl} \tag{VI-94}$$

Then Eq. (VI-75) may be written as follows:

$$\frac{\partial}{\partial t}\rho = -i\left(A + B(t)\right)\{\rho\} - C\{\rho\} + d \tag{VI-95}$$

In the solution of such an equation, one encounters so called "propagators" which are exponents of i**A**t and of a time integral over i**B**(t). These propagators are unitary, expressing evolution without decay. On the other hand, the propagator related to **C** is a relaxation operator, expressing decay to equilibrium through a dissipative process. In order to focus on the interesting component of the evolution, a pseudo-interaction picture is defined as follows:

$$\sigma = \exp\left((iA + C)t\right)\rho \tag{VI-96}$$

This is not a true interaction picture, because of the non-unitary evolution due to the decay operator **C**. Then, using Eq. (VI-95),

$$\frac{\partial}{\partial t}\sigma = \exp\big((iA+C)t\big)\big\{(iA+C)\{\rho\}\big\} + \exp\big((iA+C)t\big)\big\{-i\big(A+B(t)\big)\{\rho\} - C\{\rho\} + d\big\} =$$

$$= -i\exp\big((iA+C)t\big)\{B(t)\rho + id\} \qquad \text{(VI-97)}$$

This equation has to be solved with an initial condition for the density matrix at time t=0, the moment in which the experiment starts. For the case of photoexcited triplets, this initial time is the moment in which the triplet state is created, typically with populations that are very different from Boltzmann populations. The equation is formally solved by:

$$\sigma(t) = \sigma(0) - i\int_0^t \exp\big((iA+C)t'\big)\{B(t')\rho(t') + id\}\,dt'$$

$$\approx \sigma(0) - i\int_0^t \exp\big((iA+C)t'\big)\{B(t')\rho(0) + id\}\,dt' \qquad \text{(VI-98)}$$

where the approximate equality is based on a first order iteration. Transforming back to the original variable $\boldsymbol{\rho}$ and remembering that at time t=0: $\boldsymbol{\rho}(0) = \sigma(0)$, we obtain:

$$\rho(t) \approx \exp\big(-(iA+C)t\big)\rho(0) - i\int_0^t \exp\big(-(iA+C)(t-t')\big)\{B(t')\rho(0) + id\}\,dt' \qquad \text{(VI-99)}$$

Now express **B**(t') in terms of $\mathbf{B_1}$ as in Eq. (VI-92b), and define a new integration variable s ≡ t - t'. The term with the integral is then transformed to:

$$-2i\omega_1\int_0^t \exp\big(-(iA+C)s\big)\{\cos(\omega t)\cos(\omega s) + \sin(\omega t)\sin(\omega s)\}B_1\rho(0)\,ds$$

$$+\int_0^t \exp\big(-(iA+C)s\big)\{d\}\,ds \qquad \text{(VI-100)}$$

Here the upper integration limit (s = t) corresponds to t' = 0, which is the time of the initial condition, and the lower limit (s = 0) corresponds to t' = t, i.e., to the time for which $\boldsymbol{\rho}$ is calculated.

Using the standard expression, the energy absorption intensity due to the oscillating irradiation field is proportional to:

$$\chi''(\omega) = -2i\omega_1 \int_0^t \sin(\omega s)\, Tr\left\{ S_x \left\{ \exp\left(-(iA + C)s\right) B_1 \rho(0) \right\} \right\} ds$$

$$= -2i\omega_1 \int_0^t \sin(\omega s)\, Tr\left\{ S_x \left\{ \exp\left(-(iA + C)s\right) [\, S_x , \rho(0)\,] \right\} \right\} ds \qquad \text{(VI-101)}$$

A simplification occurs if all states (i.e., all energy levels) relax with the same relaxation time, and longitudinal relaxation has the same rate as transverse relaxation. Then a single relaxation constant: $T_1 = T_2 = \tau$ characterizes all elements of the density matrix. In such a case the relaxation superoperator is a trivial one, being proportional to the unit superoperator. In that special case:

$$\chi''(\omega) = -2i\omega_1 \times$$

$$\times \int_0^t e^{-s/\tau} \sin(\omega s)\, Tr\left\{ S_x \left\{ \exp(-i H_{mol} s)\, [\, S_x , \rho(0)\,]\, \exp(i H_{mol} s) \right\} \right\} ds \qquad \text{(VI-102)}$$

In the general case, however, relaxation must be taken into account inside the Trace expression. It is convenient to define:

$$U_0 = 2\omega_1 B_1 \rho(0) = 2\omega_1 [\, S_x , \rho(0)\,] \qquad \text{(VI-103)}$$

In most photoexcitation experiments it is assumed that the ISC process creates the triplet with no coherence between the different states, so that the initial density matrix is diagonal in the spin basis. With this assumption, $\mathbf{U_0}$ is given in that basis by a simple expression:

$$(U_0)_{i,j} = 2\omega_1 (S_x)_{i,j} \left(\rho(0)_{j,j} - \rho(0)_{i,i}\right) \qquad \text{(VI-104)}$$

In special cases there may be an initial coherence, and then Eq. (VI-103) must be used to get the appropriate form of $\mathbf{U_0}$.

The discussion will now continue with the general case, making no special assumptions about the form of $\mathbf{U_0}$. Defining the Liouville matrix $\mathbf{X} = i\mathbf{A}+\mathbf{C}$, we assume it is diagonalized by $\mathbf{\Theta}^{-1} \mathbf{X} \mathbf{\Theta} = \mathbf{\Lambda}$ so that: $\exp(-Xt) = \mathbf{\Theta} \exp(-\Lambda t)\, \mathbf{\Theta}^{-1}$ or, in terms of matrix elements:

$$\left(\exp(-Xt)\right)_{ij,kl} = \sum_{m,n} \Theta_{ij,mn} \exp(-\lambda_{mn} t) \left(\Theta^{-1}\right)_{mn,kl} \qquad \text{(VI-105)}$$

The complex numbers $\lambda_{mn} = \Lambda_{mn,mn}$ are the eigenvalues of $\mathbf{X}$, in which the imaginary part is the frequency (position) of a line, and the real part is the relaxation constant (width) of that line. This real part must be positive (when it is preceded by a minus sign in the exponent, as here !) if the solution corresponds to decay rather than to exponential growth of the magnetization. It is therefore possible to write:

$$\chi''(\omega) = -i \int_0^t \sin(\omega s) \sum_{i,j} \left\{ (S_x)_{j,i} \sum_{m,n,k,l} \boldsymbol{\Theta}_{ij,mn} \exp(-\lambda_{mn} s) (\boldsymbol{\Theta}^{-1})_{mn,kl} (U_0)_{k,l} \right\} ds \quad \text{(VI-106)}$$

In principle the initial time is t' = 0 , due to the finite time from the creation of the triplet to the measurement. Consequently, the upper limit of the integral in Eqs. (VI-100) and (VI-106) is equal to: t. However, in practice this time interval is sufficiently long for the integrand to decay, so the value of the integral will change very little if the upper limit is extended to infinity. Integration by parts yields, assuming $\lambda t \gg 1$ as explained here:

$$\int_0^t \sin(\omega s) \exp(-\lambda s) ds \approx \frac{\omega}{\lambda} \int_0^\infty \cos(\omega s) \exp(-\lambda s) ds \quad \text{(VI-107)}$$

as in the standard calculation of the magnetic resonance absorption line shape. Thus the line shape in the present problem may be written as:

$$\chi''(\omega) = -i\omega \int_0^\infty \cos(\omega s)\, Tr\{S_x U(s)\} ds$$

$$= -i\omega \int_0^\infty \cos(\omega s)\, Tr\left\{ S_x \left\{ \exp(-Xs) \left\{ X^{-1} [S_x, \rho(0)] \right\} \right\} \right\} ds \quad \text{(VI-108)}$$

where **U**(s) is defined as:

$$U(s) \equiv \exp(-Xs) \left\{ X^{-1} U_0 \right\} = \boldsymbol{\Theta} \exp(-\boldsymbol{\Lambda} s) \boldsymbol{\Lambda}^{-1} \boldsymbol{\Theta}^{-1} U_0 \quad \text{(VI-109)}$$

The matrix elements of **U**(s) are calculated as:

$$(U(s))_{i,j} = \sum_{m,n,k,l} \boldsymbol{\Theta}_{ij,mn} \frac{\exp(-\lambda_{mn} s)}{\lambda_{mn}} (\boldsymbol{\Theta}^{-1})_{mn,kl} (U_0)_{k,l} \quad \text{(VI-110)}$$

Eq. (VI-108) is similar to the standard line shape formula, except that the operator $\mathbf{S_x}(t) \equiv e^{-iHt}\, \mathbf{S_x}\, e^{iHt}$ is replaced by **U**(t). The difference results from the special non-equilibrium initial condition and from the stochastic process. If the medium is not isotropic, $\boldsymbol{\rho}_0$ has to be multiplied by the equilibrium distribution of molecular orientations $p_0(\Omega)$ as desribed in the previous Section, and therefore one has to multiply $\mathbf{U_0}$ by this function. It is possible to make the formula more symmetrical by multiplying both $\mathbf{S_x}$ and $\mathbf{U_0}$ by $p_0(\Omega)^{1/2}$ rather than by $p_0(\Omega)$ (see Eq. (VI-69)). If all operators are expanded in the normalized eigenfunctions of $\boldsymbol{\Gamma}_\Omega$, denoted as $|LMK\rangle$, then each row or column index for the superoperators is a set of the form (LMKij), as in Eq. (V-117). Finally the line shape may be written as:

$$\chi''(\omega) = -i\omega \int_0^\infty \cos(\omega t) \sum_k V_k \left\{ \sum_{l,m,n} \Theta_{k,l} \frac{\exp(-\lambda_l s)}{\lambda_l} (\Theta^{-1})_{l,m} T_{m,n} V_n \right\} dt \qquad \text{(VI-111)}$$

In this expression every index (k,l,m and n) stands for a set of indices of the form (ijLMK). The supervector **V** stands for a generalized "starting vector": $\mathbf{V} \equiv \mathbf{S}_x\, p_0(\Omega)^{1/2}$ and the superoperator **T** is defined here by $\mathbf{U_0} \equiv \mathbf{TV}$ (if the solvent is isotropic, $p_0(\Omega)$ is simply a constant). In the absence of initial quantum coherence, the matrix elements of **T** are equal to:

$$\langle LMKij|T|L'M'K'i'j'\rangle = \delta_{i,i'}\delta_{j,j'} \int d\Omega \left(D^L{}_{M,K}(\Omega)\right)^* \left(\rho(0)_{j,j} - \rho(0)_{i,i}\right) D^{L'}{}_{M',K'}(\Omega) \qquad (VI\text{-}112)$$

The initial value of the density matrix elements depends on the initial state of the triplet, created by populating the triplet levels via an ISC mechanism leading from an excited singlet state to the triplet. Therefore the coefficients A_α, expressing the rate at which the triplet levels are populated, determine the initial values of the density matrix elements.

If one works directly in Liouville space, as is natural here, it is convenient to write the equations in a basis which is labeled by specific transitions and not by Hilbert space states. This change in labeling can be carried out through the definitions:

$$p = i - j \qquad ; \qquad q = i + j \qquad \text{(VI-113)}$$

where p is the quantum order of the transition. In addition to this, a symmetrizing transformation which combines basis functions with K and -K leads to simpler expressions for the matrix elements of **X**. Using Eq. (VI-87) it can be shown that the matrix elements of **T** , symmetrized over K, are given by:

$$\langle LMKpq|T|L'M'K'p'q'\rangle = \delta_{p,p'}\delta_{q,q'}\delta_{L,L'}\delta_{M,M'}\delta_{K,K'} \frac{Im(\lambda_{p,q})}{3k_B T}$$

$$- \delta_{p,p'}\delta_{q,q'}(-1)^M \frac{\sqrt{(2L+1)(2L'+1)}}{\sqrt{(1+\delta_{K,0})(1+\delta_{K',0})}} \sum_{j,k} q\,p\,a^j{}_k \begin{pmatrix} L & j & L' \\ -M & k & M' \end{pmatrix} \times$$

$$\times \left\{ \begin{pmatrix} L & j & L' \\ -K & 0 & K' \end{pmatrix} + (-1)^{L'} \begin{pmatrix} L & j & L' \\ -K & 0 & -K' \end{pmatrix} \right\} \qquad \text{(VI-114)}$$

where

$$a^j_k = \delta_{j,2}\left(\alpha^{(0)}\delta_{k,0} + \alpha^{(2)}\delta_{|k|,2}\right) \tag{VI-115}$$

The first term is always present in **T**, even at thermal equilibrium, whereas the second term reflects the anisotropy of the special initial state of the photoexcited triplet. At ordinary temperatures the first term is negligible in comparison with the second term.

Doing the Fourier transform (the integration over t), neglecting as usual the term far from resonance and including iω**I** in **X** (and thus also in **Λ**) one obtains (keeping the operators in the original order of Eq. (VI-109)):

$$\int_0^\infty \cos(\omega t)\sum_{k,l,m} \Theta_{k,l}\exp(-\lambda_l t)\left(\Theta^{-1}\right)_{l,m} dt = -\sum_{k,l,m}\Theta_{k,l}(\lambda_l)^{-1}\left(\Theta^{-1}\right)_{l,m} = -\left(X^{-1}\right)_{k,m} \tag{VI-116}$$

It follows that, neglecting overall constants, the energy absorption (or the EPR signal) is equal to:

$$\chi''(\omega) = \sum_{k,m,n} V_k\left(X^{-2}\right)_{k,m} T_{m,n} V_n \tag{VI-117}$$

or:

$$\chi''(\omega) = \int_0^\infty \langle V|\exp(-Xt)X^{-1}T|V\rangle\cos(\omega t)\,dt = \langle V|X^{-2}T|V\rangle \tag{VI-118}$$

In this calculation it is convenient to transform |**V**⟩ to the level-shift basis in Liouville space, which is the exact counterpart of the Hilbert space basis of $\mathbf{H}_0+\mathbf{H}_1(\Omega)$ (see Chapter I). After operating on it with **T**, the resulting supervector is transformed back to the |LMKpq⟩ basis, where the rest of the calculation is done.

From Eqs. (VI-83),(VI-114) it is clear that in the absence of triplet spin polarization, **T** is proportional to the unit superoperator, and therefore trivially commutes with $\mathbf{X}^{-1}$. As will now be shown for more general cases, if **T** is symmetric the two operators may be interchanged in the matrix element in Eq. (VI-118). We first observe that **X** has a complex symmetric matrix in the |LMKij⟩ basis. It is therefore diagonalized by an orthogonal matrix, so $\Theta^{-1} = \Theta^{T}$ where Θ^{T} is the transpose matrix of **Θ**. Thus:

$$\left(X^{-1}\right)_{i,j} = \sum_k \Theta_{i,k}\left(\Lambda^{-1}\right)_{k,k}\left(\Theta^{-1}\right)_{k,j} = \sum_k \Theta_{i,k}\left(\Lambda^{-1}\right)_{k,k}\Theta_{j,k} =$$

$$= \sum_k \Theta_{j,k}\left(\Lambda^{-1}\right)_{k,k}\left(\Theta^{-1}\right)_{k,i} = \left(X^{-1}\right)_{j,i} \tag{VI-119}$$

So that $\mathbf{X}^{-1}$ (and $\mathbf{X}^{-2}$) is symmetric. Thus, if also **T** is symmetric, the line shape can be calculated by changing the order of operators in the following way:

$$\chi''(\omega) = \langle V | X^{-2} T | V \rangle = \sum_{i,j,k} V_i \left(X^{-2}\right)_{i,j} T_{j,k} V_k =$$

$$= \sum_{i,j,k} V_k T_{k,j} \left(X^{-2}\right)_{j,i} V_i = \langle V | T X^{-2} | V \rangle \qquad \text{(VI-120)}$$

In this equation we have relied on the fact that the "starting vector" **V** has only real (and not complex) elements.

The symmetry of the matrix of **T** depends on the form of the anisotropy coefficient α. Using the properties of the 3-j coefficients in Eq. (VI-114) it is easy to see that, in the presence of polarization, the terms in **T** with M-M' = 0, K-K' = 0 (k = 0) form a symmetric matrix, whereas the other terms will in general contribute non-symmetric additions to the matrix. The coefficient $\alpha^{(0)}$ (having k = 0 or M-M' = 0) is the coefficient of the symmetric part, and $\alpha^{(2)}$ (having |k| = 2 or |M-M'| = 2) is the coefficient of the non-symmetric part. Therefore, if

$$|\alpha^{(0)}| \gg |\alpha^{(2)}| \qquad \text{(VI-121)}$$

the main part of **T** is symmetric, so that Eq. (VI-120) would be approximately satisfied. If $\alpha^{(2)} = 0$ the equation holds exactly. For isotropic solvents the requirement is simply $|\alpha^2{}_0| \gg |\alpha^2{}_2|$, or

$$\left| \frac{A_x + A_y}{2} - A_z \right| \gg |A_x - A_y| \qquad \text{(VI-122)}$$

However, for a non-isotropic solvent with $\psi \neq 0$ the requirement $|\alpha^{(0)}| \gg |\alpha^{(2)}|$ is not necessarily satisfied, even when $|\alpha^2{}_0| \gg |\alpha^2{}_2|$. Therefore in calculations for non-isotropic solvents the general expression, Eq. (VI-118) above, must be used.

The specific Hamiltonian relevant for the triplet can be used to obtain the supermatrix **X** needed in the line shape expression. The calculation is done in the same basis mentioned above in connection with Eq. (VI-114). The matrix **X** is obtained as a sum of two parts, one of which is diagonal and the other one having elements both on the diagonal and off-diagonal:

$$X = X_d + X_{nd} \qquad \text{(VI-123)}$$

The following formulas are derived for the Fourier order n = 1:

$$\langle LMKpq | X_d | L'M'K'p'q' \rangle = \delta_{L,L'} \delta_{M,M'} \delta_{K,K'} \delta_{p,p'} \delta_{q,q'} \times$$

$$\times \left\{ i\left(\omega + \omega_p\right) + \gamma_{LMK} + \delta_{p,0}\left((-1)^{M-K} k^{ave} + 3W \right) \right\} \qquad \text{(VI-124)}$$

and

$$\langle LMKpq\,|\,X_{nd}\,|\,L'M'K'p'q'\rangle = (-1)^M\,\delta_{p,0}\,\delta_{p,p'}\,\delta_{q,q'}\,\delta_{K,K'}\,\sqrt{(2L+1)(2L'+1)}\;\times$$

$$\times\begin{pmatrix} L & 2 & L' \\ -K & 0 & K' \end{pmatrix}\left\{k^{(0)}{}_{q/2}\,\delta_{M,M'}\begin{pmatrix} L & 2 & L' \\ -M & 0 & M' \end{pmatrix} + k^{(2)}{}_{q/2}\,\delta_{|M-M'|,2}\begin{pmatrix} L & 2 & L' \\ -M & M-M' & M' \end{pmatrix}\right\}$$

$$-\;\delta_{L,L'}\,\delta_{M,M'}\,\delta_{K,K'}\,\delta_{p,0}\,\delta_{p,p'}\,\delta_{q,q'}\,W$$

$$+\;i\,(-1)^M\,\frac{\sqrt{(2L+1)(2L'+1)}}{\sqrt{(1+\delta_{K,0})(1+\delta_{K',0})}}\sum d^2{}_{m,M-M'}(\psi)\;\xi(m,p,q,p',q')\begin{pmatrix} L & 2 & L' \\ -M & M-M' & M' \end{pmatrix}\times$$

$$\times\left\{F^{(2,K-K')}\begin{pmatrix} L & 2 & L' \\ -K & K-K' & K' \end{pmatrix} + (-1)^{L'}F^{(2,K+K')}\begin{pmatrix} L & 2 & L' \\ -K & K+K' & -K' \end{pmatrix}\right\} \qquad \text{(VI-125)}$$

In this expression only non-negative even values of K and K' can appear. The constant k^{ave} is an average over the depopulation coefficients k_α of Eq. (VI-80), and the coefficient ξ is defined as:

$$\xi(m,p,q,p',q') = \delta_{p'p}\,\delta_{q'q}\,\delta_{m,0}\,\sqrt{(3/2)}\,p\,q$$

$$+\;\delta_{p',p-m}\,\delta_{q',q-m}\left\{\delta_{|m|,1}\,\frac{1}{\sqrt{2}}\left(\delta_{p,-q}-\delta_{p+q,2m}\right)+\delta_{|m|,2}\,\delta_{p+q,m}\right\}$$

$$+\;\delta_{p',p-m}\,\delta_{q',q+m}\left\{\delta_{|m|,1}\,\frac{1}{\sqrt{2}}\left(\delta_{p,q}-\delta_{p-q,2m}\right)-\delta_{|m|,2}\,\delta_{p-q,m}\right\} \qquad \text{(VI-126)}$$

For an isotropic solvent, $d^2{}_{m,M-M'}(\psi)$ is replaced in Eq. (VI-125) by $\delta_{m,M-M'}$. For the "starting vector" **V** the following result is found, assuming an isotropic solvent:

$$\langle V|LMKpq\rangle = \delta_{p,0}\frac{1}{3}\frac{(k_{q/2})^L{}_{MK}}{\sqrt{2L+1}} - \frac{i\,\omega_1}{2\,k_B T}\Big\{\omega_p\,\delta_{L,0}\,\delta_{M,0}\,\delta_{K,0}\sqrt{2}\,\delta_{|p|,1} - \delta_{L,2}\frac{1}{\sqrt{2L+1}}\times$$

$$\times F^{(2,K)}\Big(\delta_{p,0}(\delta_{M,1}+\delta_{M,-1})\frac{3q^2-8}{4} + \delta_{p,-1}(\sqrt{2}\,\delta_{M,-2}+\sqrt{3}\,\delta_{M,0})\,q$$

$$-\,\delta_{p,1}\big(\sqrt{2}\,\delta_{M,2}+\sqrt{3}\,\delta_{M,0}\big)\,q + 2\,\delta_{p,2}\,\delta_{M,1} + 2\,\delta_{p,-2}\,\delta_{M,-1}\Big)\Big\} \qquad \text{(VI-127)}$$

In the presence of an ordering potential U(Ω) as defined in Eqs. (VI-64) a general matrix element of the starting vector **V** is:

$$\langle V|LMKpq\rangle = \sqrt{\frac{2L+1}{8\pi^2}}\int d\Omega\,\big(D^L{}_{MK}(\Omega)\big)^*\exp(-U(\Omega)/k_BT)\Big\{\delta_{p,0}\frac{1}{3}\frac{(k_{q/2})^L{}_{MK}}{\sqrt{2L+1}} - \frac{i\,\omega_1}{2k_BT}$$

$$\times\Big\{\omega_0\,\delta_{L,0}\,\delta_{M,0}\,\delta_{K,0}\sqrt{2}\,(\delta_{p,1}-\delta_{p,-1}) - \sum_{m',m''} D^2{}_{m',m''}(\Omega)\,F^{(2,m'')}\Big(\delta_{p,0}\big(d^2{}_{1,m'}(\psi)+d^2{}_{-1,m'}(\psi)\big)$$

$$\times\frac{3q^2-8}{4} + \delta_{p,-1}\big(\sqrt{2}d^2{}_{-2,m'}(\psi)+\sqrt{3}d^2{}_{0,m'}(\psi)\big)\,q - \delta_{p,1}\big(\sqrt{2}d^2{}_{2,m'}(\psi)+\sqrt{3}d^2{}_{0,m'}(\psi)\big)\,q$$

$$+\,2\,\delta_{p,2}\,d^2{}_{1,m'}(\psi) + 2\,\delta_{p,-2}\,d^2{}_{-1,m'}(\psi)\Big)\Big\}\Big\} \qquad \text{(VI-128)}$$

The frequency ω_0 in this formula is just the Larmor frequency of the electron. One may then define:

$$G^L{}_{MK} = \sqrt{\frac{2L+1}{8\pi^2}}\int d\Omega\,\big(D^L{}_{MK}(\Omega)\big)^*\exp(-U(\Omega)/k_BT) \qquad \text{(VI-129)}$$

In practice, due to the structure of the potential, $G^L{}_{M,K}$ is non-zero only if M = 0 and L and K are even numbers. The summation in Eq. (VI-128) is over integrals of the form:

$$\sqrt{\frac{2L+1}{8\pi^2}}\int d\Omega\,\left(D^L{}_{MK}(\Omega)\right)^* D^2{}_{m',m''}(\Omega)\exp\left(-U(\Omega)/k_BT\right) = \sqrt{\frac{2L+1}{8\pi^2}}\,(-1)^{M-K}\times$$

$$\times\int d\Omega \sum_{j,k,k'}(2j+1)\begin{pmatrix} L & 2 & j\\ -M & m' & k\end{pmatrix}\begin{pmatrix} L & 2 & j\\ -K & m'' & k'\end{pmatrix}\left(D^j{}_{kk'}(\Omega)\right)^*\exp\left(-U(\Omega)/k_BT\right) =$$

$$= \sum_{j,k,k'}\{(2L+1)(2j+1)\}^{1/2}\begin{pmatrix} L & 2 & j\\ -M & m' & k\end{pmatrix}\begin{pmatrix} L & 2 & j\\ -K & m'' & k'\end{pmatrix}(-1)^{M-K}G^j{}_{kk'} =$$

$$= \sum_{j,k'}\{(2L+1)(2j+1)\}^{1/2}\begin{pmatrix} L & 2 & j\\ -M & M & 0\end{pmatrix}\begin{pmatrix} L & 2 & j\\ -K & m'' & k'\end{pmatrix}(-1)^{M}G^j{}_{0k'} \qquad \text{(VI-130)}$$

Symmetrizing over K multiplies each term by $(2/(1+\delta_{K,0}))^{1/2}$. Thus Eq. (VI-128) becomes:

$$\langle V|LMKpq\rangle = \delta_{p,0}\frac{1}{3}\frac{\sqrt{2}}{\sqrt{1+\delta_{K,0}}}\frac{\left(k_{q/2}\right)^L{}_{MK}}{\sqrt{2L+1}} - \frac{i\,\omega_1}{2k_BT}\frac{\sqrt{2}}{\sqrt{1+\delta_{K,0}}}\left\{\omega_0\sqrt{2}\left(\delta_{p,1}-\delta_{p,-1}\right)G^L{}_{0,k}\right.$$

$$-\sum_{m',m'',j,k'}F^{(2,m'')}\{(2L+1)(2j+1)\}^{1/2}\begin{pmatrix} L & 2 & j\\ -M & M & 0\end{pmatrix}\begin{pmatrix} L & 2 & j\\ -K & m'' & k'\end{pmatrix}G^j{}_{0,k'}\times$$

$$\times\left(\delta_{p,0}\left(d^2{}_{1,m}(\psi)+d^2{}_{-1,m}(\psi)\right)\frac{3q^2-8}{4}+\delta_{p,-1}\left(\sqrt{2}d^2{}_{-2,m}(\psi)+\sqrt{3}d^2{}_{0,m}(\psi)\right)q-\delta_{p,1}\times\right.$$

$$\left.\left.\times\left(\sqrt{2}d^2{}_{-2,m}(\psi)+\sqrt{3}d^2{}_{0,m}(\psi)\right)q+2\delta_{p,2}d^2{}_{1,m}(\psi)+2\delta_{p,-2}d^2{}_{-1,m}(\psi)\right)\right\}\Big\} \qquad (VI-131)$$

Example VI.6: *A microwave pulse applied to a triplet in equilibrium*

One of the major features in the discussion above, concerning photoexcited triplets, was the unusual initial state of such triplets. This was the main reason that a general derivation of the line shape was necessary. In connection with this treatment of triplets it is interesting to consider a particular non-equilibrium state of the triplet, which strictly speaking does not belong to the theory of CW EPR. Assume that a certain triplet state is at thermal equilibrium, and then a microwave pulse is applied to it, as in conventional NMR and EPR of stable species. Following the pulse the system evolves freely, which is the same as in the case of ISC after a photoexcitation, if triplet kinetics is ignored. The difference is in the initial values of the elements of the density matrix $\rho(0)$. For example, suppose that before the pulse the non-trivial part of the density matrix was proportional to

$$\sigma(0) = \frac{H_0 + H_1(\Omega)}{k_B T} = a\,S_z + \sum_m b_m A^{(2,m)} \tag{E-1}$$

which is the actual situation in thermal equilibrium at "high" temperatures. The coefficients a, b_m are ordinary numbers (not operators), which would be orientation dependent in the general case. The transformation of such a sum by a $\pi/2$ pulse is found in the following way. A pulse along **x** transforms it to:

$$\big(\sigma(0)\big)' = P\big(\sigma(0)\big) \equiv \exp\big(-i\,(\pi/2)\,S_x\big)\big(\sigma(0)\big)\exp\big(i\,(\pi/2)\,S_x\big) \tag{E-2}$$

which is equivalent to a rotation in spin space around the **x**-axis. Therefore spherical irreducible tensors would be transformed by the pulse as

$$(A')^k_q = \sum_p A^k_p D^k_{p,q}(\Omega_p) \tag{E-3}$$

where $\Omega_P = (\alpha = -\pi/2, \beta = \pi/2, \gamma = \pi/2)$ is the set of Euler angles describing the rotation caused by the pulse. Specifically, the angular momentum operators relevant to the problem would be transformed as:

$$P\big(S_z\big) = -S_y \tag{E-4a}$$

$$P\big(A^{(2,0)}\big) = \sqrt{3/2}\,\big(S_y^2 - S^2\big) = -\frac{1}{2}\Big\{A^{(2,0)} + \sqrt{3/2}\,\big(A^{(2,2)} + A^{(2,-2)}\big)\Big\} \tag{E-4b}$$

$$P\big(A^{(2,\pm1)}\big) = -\frac{1}{2}\Big\{\big(A^{(2,1)} + A^{(2,-1)}\big) \pm i\big(A^{(2,2)} - A^{(2,-2)}\big)\Big\} \tag{E-4c}$$

$$P\big(A^{(2,\pm2)}\big) = -\frac{1}{2}\Big\{\sqrt{3/2}\,A^{(2,0)} \pm 2i\big(A^{(2,1)} - A^{(2,-1)}\big) - \frac{1}{2}\big(A^{(2,2)} + A^{(2,-2)}\big)\Big\} \tag{E-4d}$$

Operating with the pulse will thus create some non-zero populations (diagonal elements) due to the $\mathbf{A}^{(2,0)}$ and $\mathbf{A}^{(2,\pm2)}$ terms, and these will be responsible for off-diagonal elements in $\mathbf{U}_0$, which will make a non-zero contribution to the signal.

In standard high field EPR the dominant term in the Hamiltonian is the Larmor term $\mathbf{H}_0 = \omega_0 \mathbf{S}_z$. Then after a $\pi/2$ pulse along the **x** direction the main non-trivial term in the density matrix is proportional to $\omega_0 S_y/(k_B T)$. This operator has zero elements on the diagonal, and non-zero "coherences", which is very different from the typical initial condition created by photoexcitation and ISC. Thus $\mathbf{U}_0$ in Eq. (VI-103) would be proportional to $\mathbf{S}_z$. If the ZFS is axially symmetric, i.e. E=0, **X** will commute with $\mathbf{S}_z$. Then the operator **X** would not affect $\mathbf{S}_z$, except by multiplying it with a time dependent scalar exponent. Thus the contribution of the main term to the final signal would be zero, because

$$\langle S_x \rangle \ \alpha \ Tr(S_x S_z) = 0 \tag{E-5}$$

Even if E is not zero, it is usually relatively small, and would thus lead to a relatively small signal. In any case it is clear that if any signal is observed, it is only due to the ZFS interaction.

****

(c) Rate and Anisotropy Dependence of Lineshapes

Experimentally, in the diode detection method the photoexcited triplet state is usually detected prior to equilibration, so observed spectra are almost always taken in

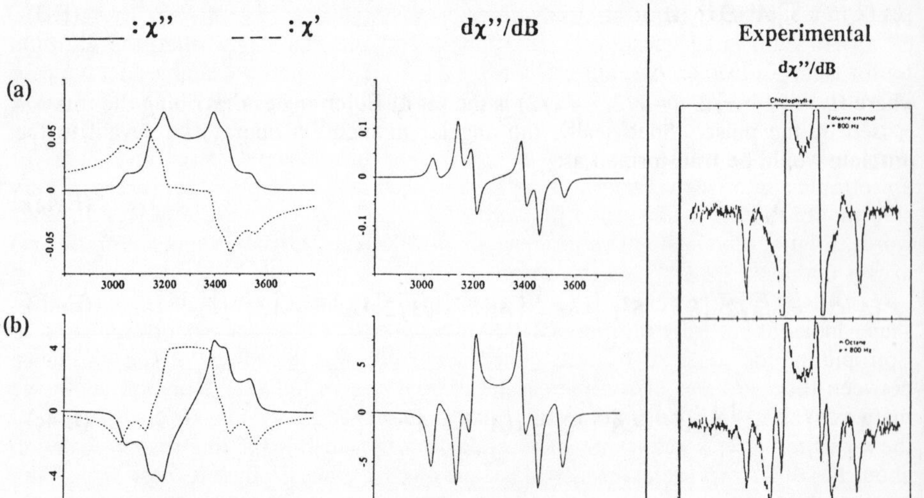

Fig. VI.7. Simulated absorption (left column) and first derivative (middle column) dynamic line shapes of the triplet state of Chl a (see text for parameters), with rotational diffusion rates of $R_\parallel = R_\perp = 10^7\ s^{-1}$. The top spectra belong to a thermally populated triplet and the bottom spectra to the same triplet when populated via spin-selective ISC, and thus exhibiting electron spin polarization. In the right hand column there are two experimental line shapes, taken on Chl a in different solvents. Our parameters are based on the bottom experimental spectrum.

a spin polarized state. In Fig. VI.7, a simulated thermal line shape is compared with the corresponding "polarized" line shape for Chl *a*, both calculated with ZFS interaction constants D = 262 and E = 22 (both in 10^{-4} cm^{-1}) and the same diffusion rates: $R_\parallel = R_\perp = 10^7\ s^{-1}$. The rate constants $R_\parallel$ and $R_\perp$ specify the rate of rotational diffusion parallel and perpendicular, respectively, to the **z** axis chosen in the molecular frame of reference. For the polarized line shape, the population constants were taken as: $A_x : A_y : A_z = 1.0 : 0.9 : 0.4$.

The six absorption lines of the thermal equilibrium state are known as a pattern of the form: **a,a,a,a,a,a** . This pattern is changed in the present case to **e,e,e,a,a,a** (three lines are turned to emission lines), and except for the line frequencies the overall line shape is completely different. The "polarized" absorption line shapes should be compared to the bottom experimental spectrum, in the right hand column of Fig. VI.7. This particular experimental result is the source of our parameters for Chl *a*.

In Fig. VI.8 line shapes are presented for polarized Chl *a* with isotropic rotational diffusion at several different rates. It is seen that for $R_\parallel = R_\perp = 10^6\ s^{-1}$ the line shape is only slightly different from that for $R_\parallel = R_\perp = 10^5\ s^{-1}$. This means that at such rates the diffusion is so slow that it is effectively at the "rigid limit". Slower molecular motions will not be detectable with the present technique. In fact, for $R_\parallel = R_\perp = R = 10^5\ s^{-1}$: $F^{(2,0)}/R \approx 5\times10^4$ in this case, which corresponds indeed to extremely slow motion. For this case one also notes slight wiggles in the line shape, which simply means that the computational parameters employed in this case ($L_{max} = 64$, $K_{max} = 32$, $M_{max} = 2$) do not result in a sufficiently accurate truncation.

An interesting question is what happens to the line shape when the diffusion tensor is not isotropic. Assuming axially symmetric diffusion, we examine first the case in which rotation is slower about the main axis, namely $R_\perp > R_\parallel$. The absorption spectra are presented in the left hand column in Fig. VI.9, and the corresponding derivative spectra in the right hand column in the same Figure. Each line shape for anisotropic motion with given values of $R_\parallel$ and $R_\perp$ is practically identical (except at very slow motion) with the line shape for isotropic motion at the given rate of $R_\perp$. In other words, rotation about the main axis has no effect. Thus, rotation about the transverse axes is dominant for Chl *a* in this case.

The opposite case, i.e., $R_\perp < R_\parallel$, is examined in Fig. VI.10. In this case each line shape looks like a truly intermediate form between the line shapes corresponding to isotropic motion at the two given rates, that of $R_\parallel$ and that of $R_\perp$. The difference between the two types of results may be rationalized as follows. Rotation about the main axis (z axis) at a rate of $R_\parallel$ tends to average out features in the x-y plane, reducing the x-y asymmetry, which is only of secondary importance for the line shape. Rotation about the other two axes averages in the y-z and x-z planes, which is more significant, because the main term in the ZFS is related to the z-axis. Therefore, when $R_\perp > R_\parallel$ the main effect is achieved by rotation about the transverse axes. In the opposite case, however, there is a competition between averaging by the "parallel" rotation, which is fast in this case, and averaging by the "perpendicular" rotation, which is slower but is inherently more effective.

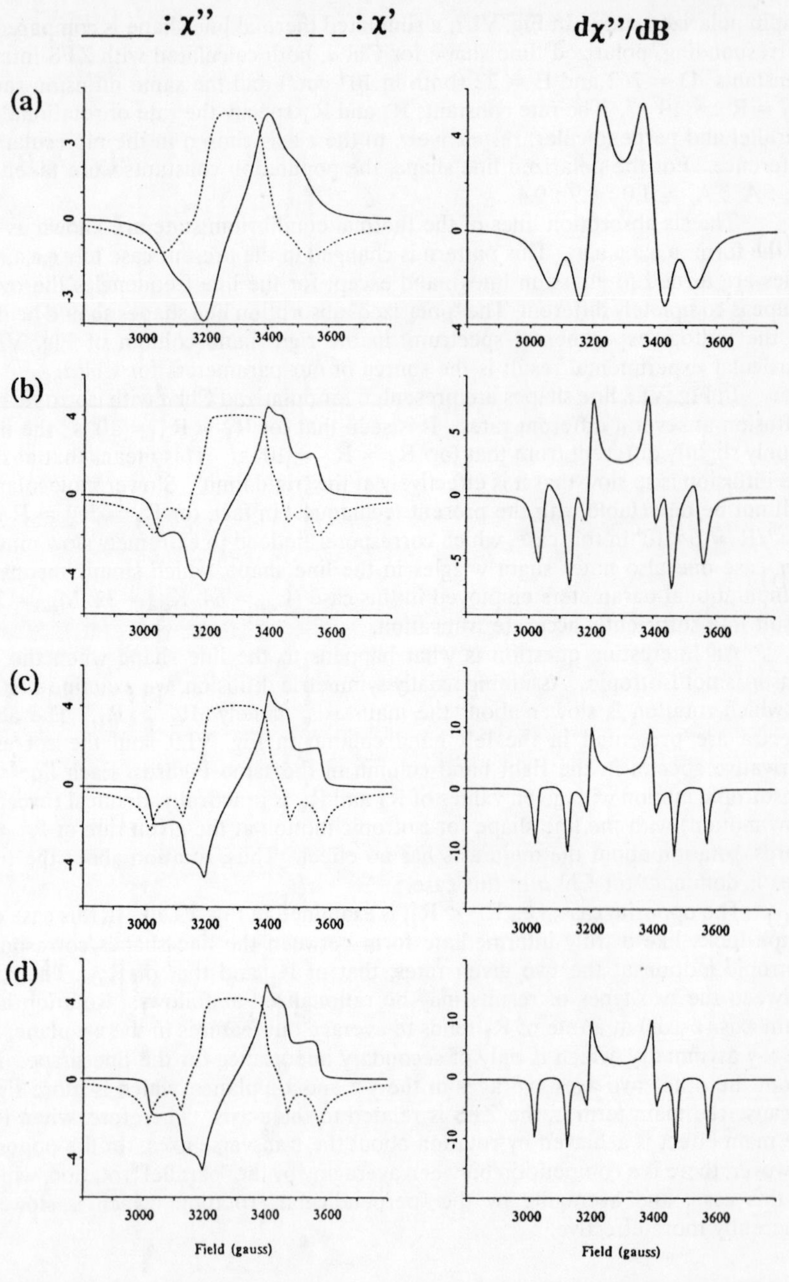

Fig. VI.8. Simulated absorption (left column) and derivative (right column) dynamic line shapes for electron spin polarized Chl a with isotropic rotational diffusion, at the rates of: (a) $R_\parallel = R_\perp = 10^8$, (b) $R_\parallel = R_\perp = 10^7$, (c) $R_\parallel = R_\perp = 10^6$, (d) $R_\parallel = R_\perp = 10^5$ (s^{-1}).

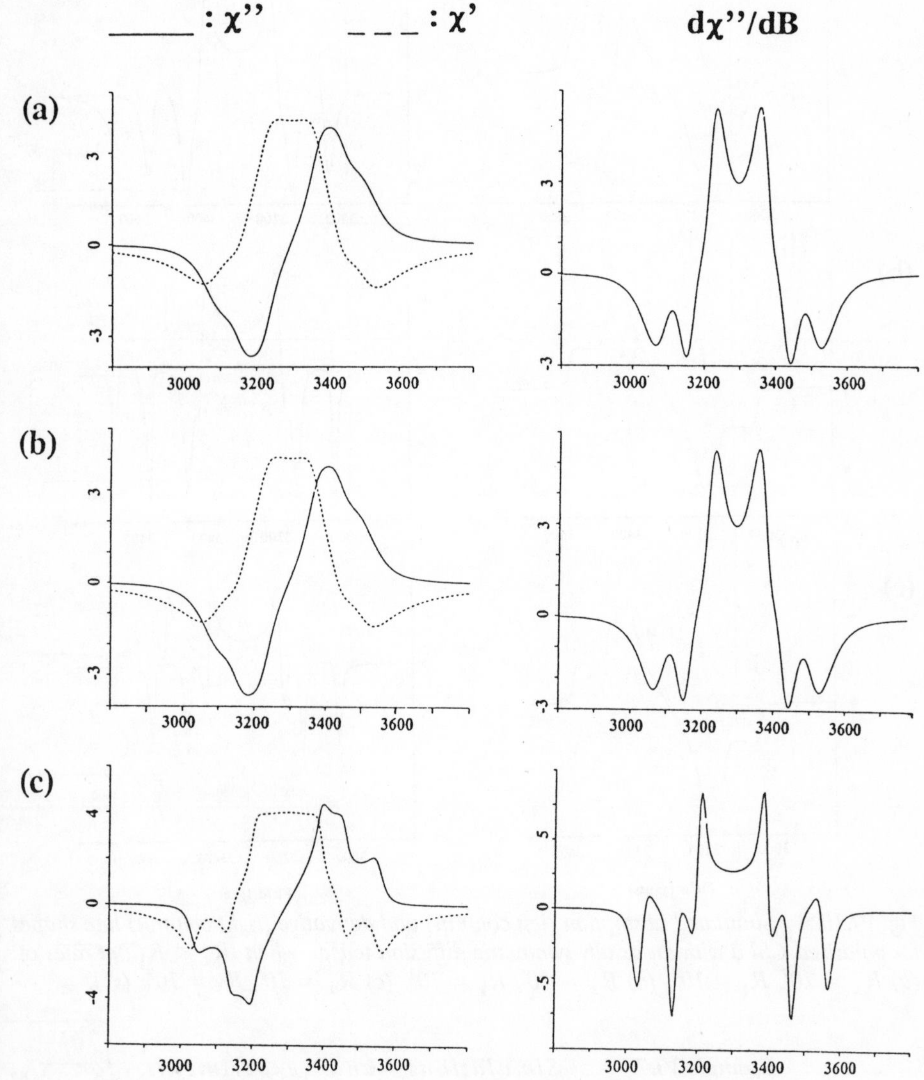

Fig. VI.9. Simulated absorption (left column) and derivative (right column) line shapes for polarized Chl a with an axially symmetric diffusion tensor, when $R_\perp > R_\parallel$, at rates of: (a) $R_\perp = 10^8$, $R_\parallel = 10^7$, (b) $R_\perp = 10^8$, $R_\parallel = 10^6$ and (c) $R_\perp = 10^7$, $R_\parallel = 10^6$ (s^{-1}).

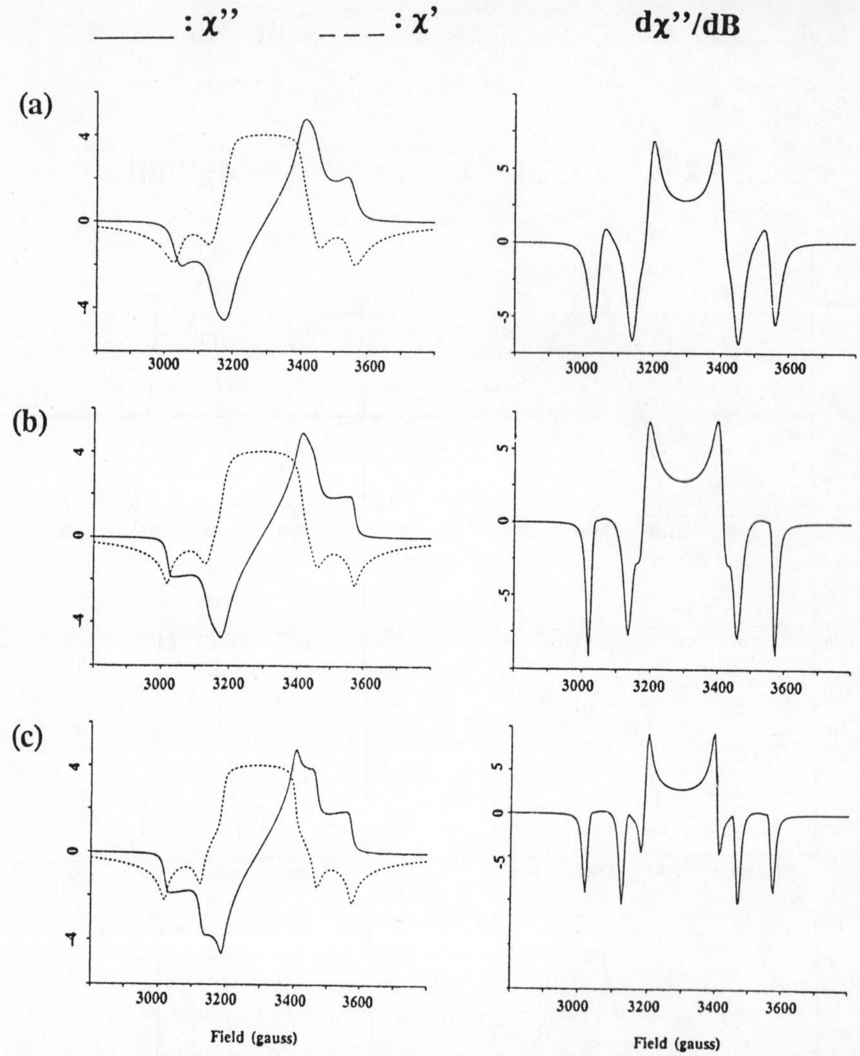

Fig. VI.10. Simulated absorption (left column) and derivative (right column) line shapes for polarized Chl a with an axially symmetric diffusion tensor, when $R_\perp < R_\parallel$, at rates of: (a) $R_\perp = 10^7$, $R_\parallel = 10^8$ (b) $R_\perp = 10^6$, $R_\parallel = 10^8$ (c) $R_\perp = 10^6$, $R_\parallel = 10^7$ (s^{-1}).

Example VI.7: Simulations and experiments for C_{60} (buckminsterfullerene)

Several simulations were performed for the photoexcited triplet state of a novel molecule of great interest, i.e., C_{60} (buckminsterfullerene) (see Ref. 16). The parameters used for this molecule were: D = 114, E = 6.9 (both in 10^{-4} cm^{-1}) and $A_x : A_y : A_z$ = 0.0 : 0.0 : 1.0.

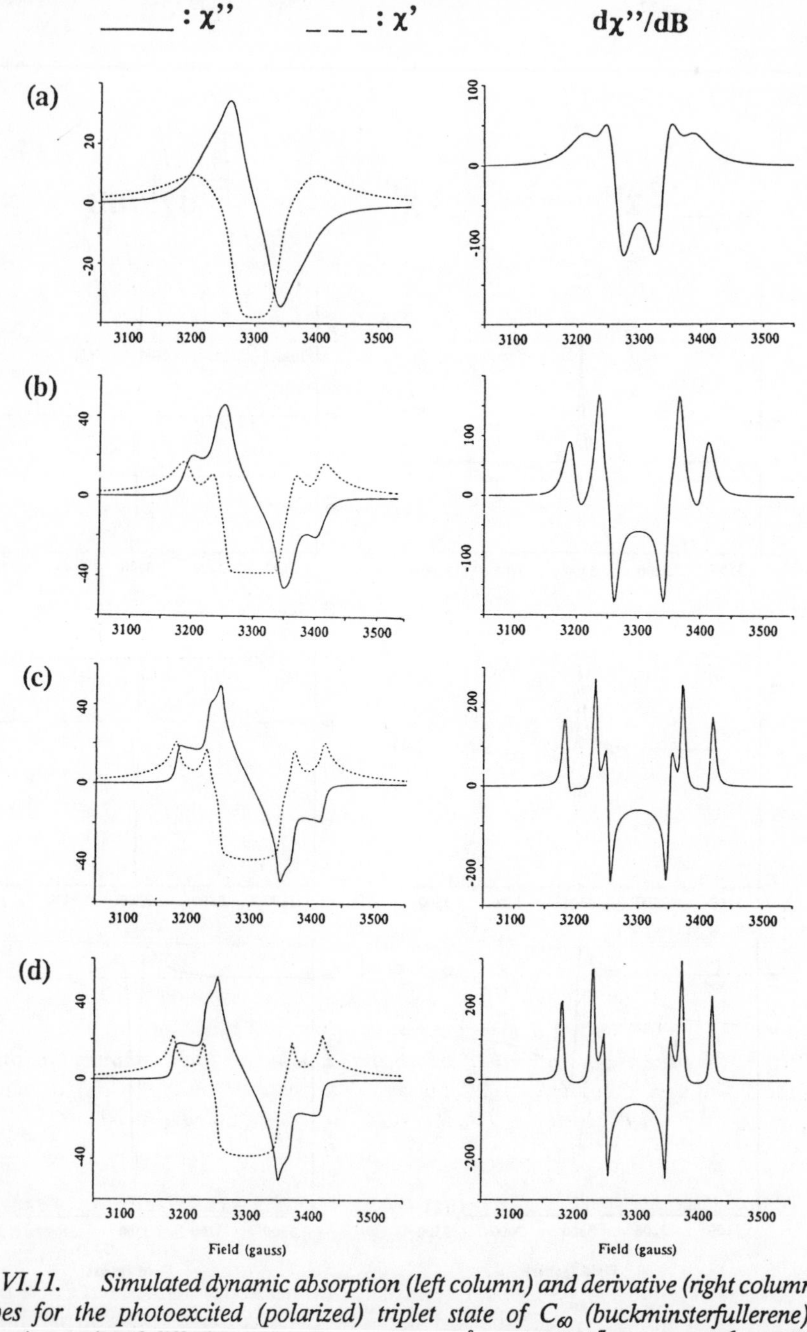

Fig. VI.11. Simulated dynamic absorption (left column) and derivative (right column) line shapes for the photoexcited (polarized) triplet state of C_{60} (buckminsterfullerene) with isotropic rotational diffusion, at rates of: (a) $R = 10^8$, (b) $R = 10^7$, (c) $R = 10^6$, (d) $R = 10^5$ (s^{-1}). The linewidth parameter is $1/T_2$ = 2 gauss.

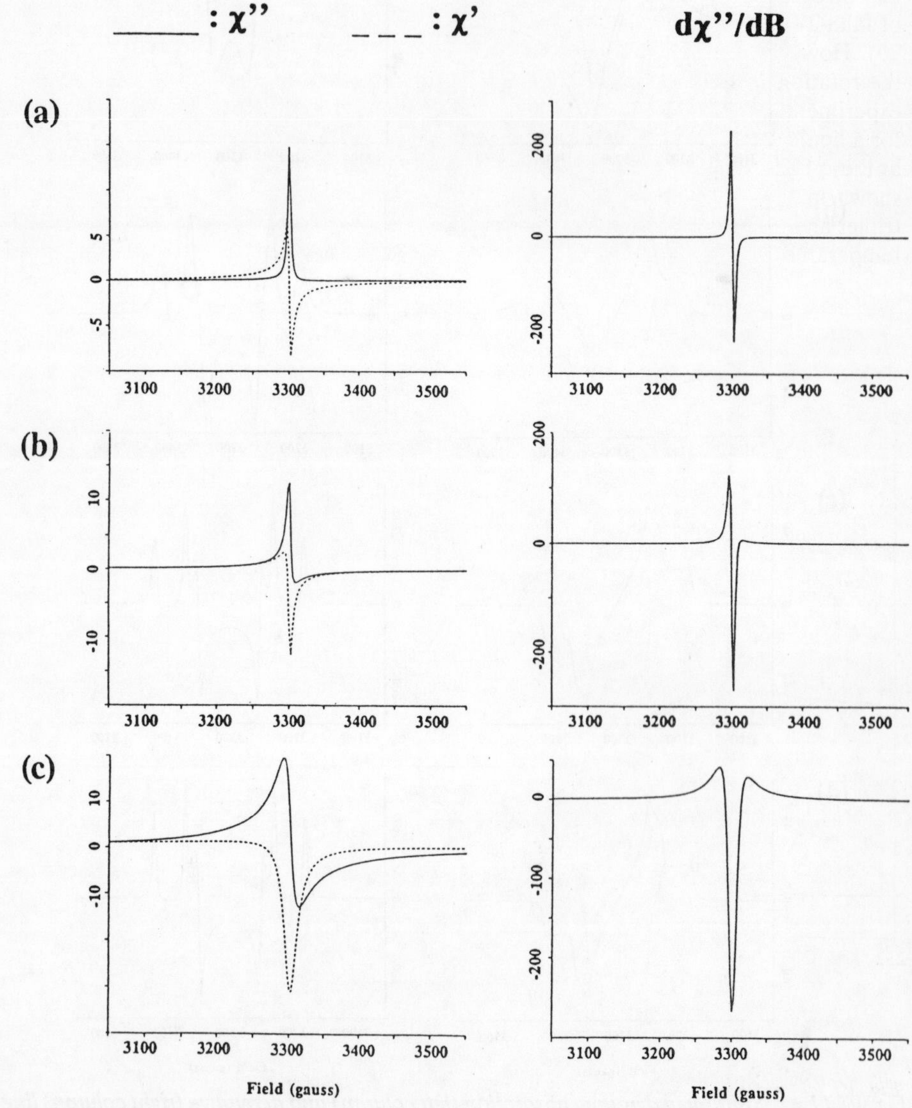

Fig. VI.12. Same as Fig. VI.11, with rates of: (a) $R = 10^{11}$, (b) $R = 10^{10}$, (c) $R = 10^{9}$ (s^{-1}).

The results, presented in Figs. VI.11 (slow motional regime) and VI.12 (fast motional regime) show the predicted behavior of the line shape over the complete dynamic range. The slowest motional rate gives a line shape which corresponds to experimental low temperature (T = 5-7 K, which is effectively the rigid limit) results in isotropic solvents. For higher temperatures the experimental behavior of the system in an isotropic solvent is more complex, and a better fit to the spectra was given by a model of Jahn-Teller jumps between nearly degenerate configurations of the molecule (see Ref. 20). However, at still higher temperatures, when the solvent becomes an isotropic liquid, the rotational diffusion model is found to be the appropriate model for describing the experimental results (Ref. 20). This is true both for an isotropic solvent (toluene) and for a liquid crystalline solvent (E-7). It is also found here that the triplet is thermalized at these high temperatures. The high temperature spectra for the isotropic solvent are shown in Fig. VI.13, compared with simulated line shapes both for a spin polarized triplet and for a thermalized triplet. It is clear that the triplet is thermalized at this high temperature.

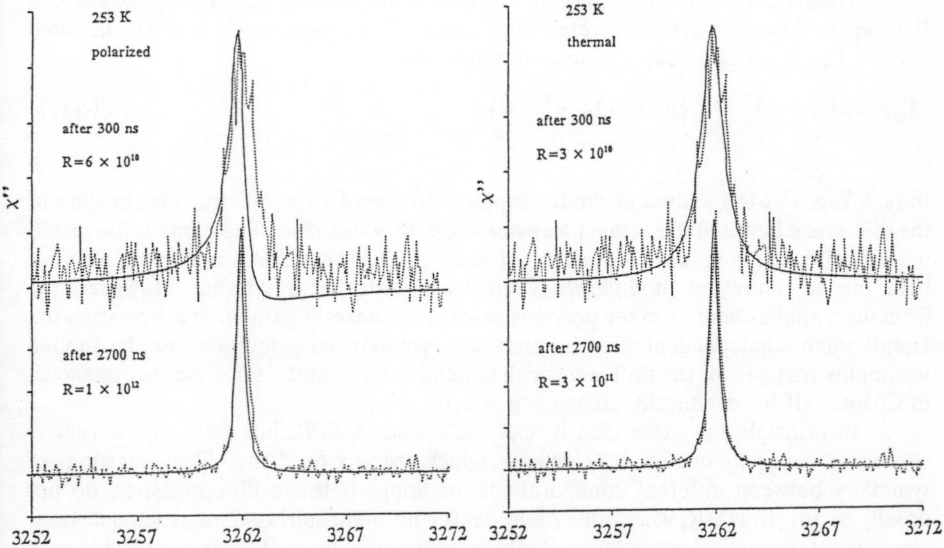

Fig. VI.13. Time resolved EPR spectra (dashed lines) of $^3C_{60}$ in liquid toluene at T = 253 K, 300 ns and 2700 ns after the laser excitation pulse. The solid lines were simulated with the Brownian diffusion model for spin polarized triplets (left), and for thermalized triplets (right). The line width parmeter used is $1/T_2$ = 0.1 G and diffusion rate constants R are indicated on the Figure.

VI.4 NMR line shapes for rotational diffusion in symmetrical systems

(a) *Rotational diffusion combined with discrete jumps*

Equations (V-4) or (V-78) apply equally to EPR and to NMR. If one is interested in relaxation due to a rotational diffusion process, an equation of the same form as (V-101) or (VI-75) will be appropriate also for NMR. Moreover, interaction Hamiltonians in NMR, as in EPR, normally involve either scalar interactions or second rank tensor interactions, so that Eq. (V-97) is applicable also to NMR. Eqs. (V-116) - (V-118) can therefore be used in the context of NMR. They are useful when the process is sufficiently slow on the NMR time scale. Equations of this type have been used for particular applications, studying the NMR line shapes of symmetrical systems with rotational diffusion. In one application, a study was made on systems undergoing simultaneously two different processes (see Ref. 23). One is a rotational diffusion process as discussed above, and the other is a process of jumps between discrete orientations, as in chemical exchange. The first process involves a change in a continuous parameter Ω, and the second process involves a change in a discrete parmeter ω_j. If the system is symmetrical, i.e., the Hamiltonian does not depend on the index j, information on symmetry can facilitate the treatment of the problem, as in cases which involve only chemical exchange.

The basic equation of motion in this case is the same as Eq. (VI-75), except that $\mathbf{\Gamma_k}$ is replaced by an operator representing a discrete jump process, which will be denoted here by $\mathbf{\Gamma_d}$. It is defined by a transition matrix W:

$$\Gamma_d \rho(\omega_j,\Omega) = \sum_k W_{j,k}\{\rho(\omega_k,\Omega) - \rho(\omega_j,\Omega)\} \qquad \text{(VI-132)}$$

In fact, Eqs. (VI-80) written above for triplet EPR are of a similar structure, in spite of the difference between the actual processes which they describe. There one is interested in three different quantum states of a system, unrelated to a spatial orientation, whereas here one is interested in different geometrical states of a system. However, the formalism applies here as in the previous cases. An interesting situation arises when the Hamiltonian is independent of ω_j, because then symmetry considerations may be applied to simplify that part of the problem which depends on $\mathbf{\Gamma_d}$, in the same manner as shown in Chapter III for chemically exchanging systems.

In principle the same idea is applicable also to EPR, but there one is usually concerned with only one or two electrons, which are at a fixed site. Thus questions of symmetry between different configurations, or jumps between different sites, do not usually occur. In NMR, where one often deals with multi-spin systems, such questions are of greater practical importance. Only in exceptional cases is there a molecule with a highly symmetrical electronic structure, in which discrete jumps between symmetry-related configurations are relevant for EPR line shapes. Such a situation was in fact encountered in EPR studies on the highly symmetrical C_{60} and C_{70} molecules at low temperatures (Refs. 20,21), but rotational diffusion was negligible in that low temperature regime.

(b) NMR line shapes for rotational diffusion in a symmetric potential

Another application of the SLE to NMR was made for a system in which rotational diffusion is the only dynamic process, but a geometrical symmetry in the system modifies the dynamics (see Ref. 24). The system is assumed to evolve in an an orientation dependent potential, as discussed above in subsection VI.2.(b), but with a discrete symmetry. In principle one may use the expressions given there for the potential energy, and then impose the appropriate symmetry conditions on the problem. In general one is concerned with three rotation angles (Euler angles) and with possibly complicated symmetries. In a simple example of this situation, the molecule may be dissolved in a solvent which forces molecules to be strictly parallel to one axis - the director, so that the molecule cannot rotate about the two transverse directions. During the rotation with azimuthal angle ϕ about the longitudinal direction, the molecule encounters an orientation dependent potential $V(\phi)$, which has the following symmetry:

$$V(\phi) = V(\phi + 2\pi/n) \qquad \text{(VI-133)}$$

The trivial case in which $V(\phi)$ is a constant is obviously uninteresting, since such a potential will not affect the Boltzmann distribution and will consequently have no effect on the dynamics. For a non-constant potential, the equation implies there are n equivalent potential minima, which are the preferred orientations of the system.

The full Hamiltonian thus consists of a part which is independent of orientation, and another part which is just the potential $V(\phi)$. In a generalized form of Eq. (VI-76b), the angle-dependent Hamiltonian can be written as:

$$H_1(\Omega) = \sum_{m,m',m'',\mu} D^2_{m,m'}(\Psi)\, D^2_{m',m''}(\Omega)\, D^2_{m'',\mu}(\Phi)\, F^{(2,\mu)} A_{lab}^{(2,m)} \qquad \text{(VI-134)}$$

Here Ω is the orientation of the diffusion tensor of the molecule relative to the director, and $\Psi = (\alpha,\phi_0)$ is the orientation of the director relative to the magnetic field. $\Phi = (\beta,\phi)$ is the orientation of the magnetic interaction tensor of the molecule relative to the PAS of its diffusion tensor. In this case the **z** axis of the diffusion tensor is assumed to coincide with the director, so we remain with

$$H_1 = \sum_{m,m',m''} D^2_{m,m'}(\Psi)\, D^2_{m',m''}(\Phi)\, F^{(2,m'')} A_{lab}^{(2,m)} \qquad \text{(VI-135)}$$

In this formula the dependence on the rotation angle ϕ is through the exponents $\exp(\pm i\phi)$, $\exp(\pm 2i\phi)$ only.

The symmetrized form of the rotational diffusion operator for this case is derived from the equations (V-88) and (VI-67),(VI-73) above, with $R_\parallel \equiv R$, $R_\perp = 0$, and $\mathbf{J_z} = -i\hbar\, \partial/\partial\phi$:

$$\boldsymbol{\Gamma}_V = -(\hbar^2 R)\left\{\left[\frac{\partial^2}{\partial\phi^2}\right] + \frac{1}{2k_BT}\left[\frac{\partial^2}{\partial\phi^2}V(\phi)\right] - \frac{1}{(2k_BT)^2}\left[\frac{\partial V(\phi)}{\partial\phi}\right]^2\right\} \qquad (VI\text{-}136)$$

Choosing, for example, the form

$$V(\phi) = -V_0 \cos(n\phi) \qquad \text{(VI-137)}$$

one can apply to it Γ_V, and use the formalism discussed in this Chapter (Sec. VI.2) to solve the diffusion equation. Expanding the density matrix in terms of its ϕ-dependent Fourier components, one may solve the problem along the same lines as in EPR examples discussed previously.

Suggested References

* On application of the SLE to doublet EPR:

1. J.H. Freed, G.V. Bruno and C.F. Polnaszek, *J. Phys. Chem.* **75**, 3385 (1971).
2. S.A. Goldman, G.V. Bruno, C.F. Polnaszek and J.H. Freed, *J. Chem. Phys.* **56**, 716 (1972).
3. C.F. Polnaszek, G.V. Bruno and J.H. Freed, *J. Chem. Phys.* **58**, 3185 (1973).
4. C.F. Polnaszek and J.H. Freed, *J. Phys. Chem.* **79**, 2283 (1975).
5. J.H. Freed in *"Spin Labeling - Theory and Applications"*, Vol. I of *"Magnetic Resonance in Biology"*, edited by L.J. Berliner (Academic Press, New York, 1976), Ch. 3.
6. K.V.S. Rao, C.F. Polnaszek and J.H. Freed, *J. Phys. Chem.* **81**, 449 (1977).
7. E. Meirovitch, D. Igner, E. Igner, G. Moro and J.H. Freed, *J. Chem. Phys.* **77**, 3915 (1982).
8. K. Hensen, W.-O. Riede, H. Sillescu and A.v. Wittgenstein, *J. Chem. Phys.* **61**, 4365 (1974).
9. N. Sistovaris, W.-O. Riede and H. Sillescu, *Ber. Bunsen-Ge.* **79**, 882 (1975).
10. K.-H. Wassmer, E. Ohmes, M. Portugall, H. Ringsdorf and G. Kothe, *J. Am. Chem. Soc.* **107**, 1511 (1985).
11. A. Kumar, *J. Chem. Phys.* **91**, 1232 (1989).
12. (a) D.J. Schneider and J.H. Freed in *Adv. Chem. Phys.*, vol. 73 (1989), p. 387.
    (b) D.J. Schneider and J.H. Freed in *Biological Magnetic Resonance,* Vol. 8, edited by L.J. Berliner and J. Reuben (Plenum Publishing Corporation, 1989), p. 1.
13. L.D. Favro in *"Fluctuation Phenomena in Solids"*, edited by R.E. Burgess (Academic Press, New York, 1965), p.79.

* On application of the SLE to triplet EPR:

14. S.P. McGlynn, T. Azumi, and M. Kinoshita, *"The Triplet State"* (Prentice Hall, Inc., Englewood Cliffs, NJ, 1969).
15. J.H. Freed, G.V. Bruno and C. Polnaszek, *J. Chem. Phys.* **55**, 5270 (1971).
16. D. Gamliel and H. Levanon, *J. Chem. Phys.* **97**, 7140 (1992).
17. H. Levanon and S. Vega, *J. Chem. Phys.* **61**, 2265 (1974).

18. (a) J.H. van der Waals and M.S. de Groot in *"The Triplet State"*, edited by A.B. Zahlan (Cambridge University Press, 1967), p. 101.
(b) C.C. Felix and S.I. Weissman, *Proc. Nat. Acad. Sci. USA* **72**, 4203 (1975).
19. (a) R.A. Schadee, J. Schmidt and J.H. van der Waals, *Chem. Phys. Lett.* **41**, 435 (1976).
(b) C.J. Nonhof, F.L. Plantenga, J. Schmidt, C.A.G.O. Varma and J.H. van der Waals, *Chem. Phys. Lett.* **60**, 353 (1979).
20. A. Regev, D. Gamliel, V. Meiklyar, S. Michaeli and H. Levanon, *J. Phys. Chem.* **97**, 3671 (1993).
21. H. Levanon, V. Meiklyar, S. Michaeli and D. Gamliel, *J. Am. Chem. Soc.* **115**, 8722 (1993).
22. G. Kothe, S. Weber, E. Ohmes, M.C. Thurnauer and J.R. Norris, *J. Phys. Chem.* **98**, 2706 (1994).

* On application of the SLE to NMR:

23. S. Alexander, A. Baram and Z. Luz, *Mol. Phys.* **27**, 441 (1974).
24. S. Zamir, R. Poupko, Z. Luz and S. Alexander, *J. Chem. Phys.* **94**, 5939 (1991).

CHAPTER VII

APPLICATIONS TO MULTIPLE EXCITATION METHODS

The SLE formalism has been applied to a variety of experimental techniques in magnetic resonance, in addition to the basic CW or FID modes. All these other modes can be characterized as involving more than a single source of excitation. The two main classes of such experimental methods are multiple resonance and multiple pulse methods. The first class includes all experiments in which more than one irradiation source (microwave or rf) is used for perturbing the spins. This includes methods of double resonance, where one excites simultaneously two electronic resonances or an electronic resonance and a nuclear resonance. More elaborate methods, such as various triple resonance techniques, have also been developed. The second class encompasses all those experiments in which a single irradiation source is applied in two or more pulses, and the time intervals between the pulses are manipulated to control the final signal. These methods are generally known as time domain magnetic resonance, and since they involve at least two independent time intervals (including the time of measurement) the signals are presented as two dimensional (or higher dimensional) line shapes. In a wider sense, multiple excitation methods include also experiments in which there is only one kind of microwave or rf excitation, but the system is also excited in some other manner. This includes CIDEP and CIDNP, where a molecule is excited optically or chemically and then also excited with some microwave or rf irradiation (either CW or pulsed), and a magnetic resonance response is detected. This also includes, e.g., ODMR, where the system is excited magnetically, but an optical measurement is made. However, the theory of such techniques is quite different from that of the two main categories of experiments mentioned above. This Chapter treats the main types of experiments to which the SLE has been applied, and which belong to one of those two types.

VII.1 Multiple Resonance Methods

In multiple resonance methods one excites some quantum transition (or transitions) appreciably, so as to influence the response of the system to another excitation mode. This means that some degree of saturation is always involved in the experiment. If one is interested in relaxation effects in the presence of multiple resonance irradiation sources, these effects must be considered in the presence of saturation. It is therefore necessary to return to the subject of relaxation in the presence of saturation, which was discussed for a specific case in the previous Chapter (subsection VI.1.(d)), and present it in a more general context, before entering into the detailed treatment of multiple resonance experiments. This will be done here in two stages. In the first stage, relaxation will be taken in its simplest form, i.e., in the motional narrowing regime, in order to focus on the typical characteristics of multiple resonance. After an introductory discussion, some important double resonance techniques will be considered in this limit. After the basic features of multiple resonance are clear, the more general situation of the slow motional regime will be treated in the second stage (in subsection VII.1.(d)).

(a) Saturation and relaxation in the motional narrowing regime

In Chapter II a relaxation equation, Eq. (II-102) was derived for the motional narrowing limit, where coherent (and possibly strong) irradiation is applied to the system. The same equation can also be derived directly from the Stochastic Liouville formalism (see Ref. 5). If the eigenfunctions of the constant Hamiltonian $\mathbf{H_0}$ appearing in that equation are used as a basis for Hilbert space, the equation for individual elements of the density matrix is:

$$\frac{d}{dt}\rho_{i,j} = -i\omega_{ij}\rho_{i,j} - i\left[\frac{1}{\hbar}H_2(t)\,,\rho\right]_{i,j} + \sum_{m,n} R_{ij,mn}\left(\rho_{mn} - \rho_{(eq.)mn}\right) \tag{VII-1}$$

where ω_{ij} is the transition frequency between eigenstates $|i\rangle$, $|j\rangle$ of $\mathbf{H_0}$. Here the coherent irradiation Hamiltonian $\mathbf{H_2}(t)$ represents the interaction of the spins with one or more sources of microwave or rf radiation. No restriction is assumed for the strength of these sources. The equilibrium density matrix $\rho_{(eq.)}$ is included in the equation so that the system will relax to the correct, finite temperature equilibrium state (see Eq. (II-100)). The motional narrowing assumption means that $\|\mathbf{H_1}\|\tau_C \ll 1$ where $\mathbf{H_1}$ is the perturbation (e.g., rotational diffusion) leading to relaxation, and τ_C is the correlation time of this motion. In the presence of radiation, also $\|\mathbf{H_2}\|\tau_C \ll 1$ must hold, in order that the radiation field will not affect the relaxation matrix R.

The simplest case for which the equation is applicable is that of a two level system, where the spectrum consists of a single line. Such a system is mathematically equivalent to a single S = ½ spin, and therefore in the absence of relaxation and radiation the two non-diagonal elements of ρ simply oscillate in time as $\exp\{(\pm i\,\omega_{12})\,t\}$ (see Eq. (I-89)). Now assume that relaxation is also present, but only in the simplest form, so that each transition is uncoupled from any other transition (including its own inverse) and from any spin population. Then **R** does not have off-diagonal elements which connect ρ_{12} with ρ_{21}. Therefore the time dependence of one off-diagonal element is modified to:

$$\rho_{12}(t) = \exp\left(\left(-i\omega_{12} + R_{12,12}\right)t\right)\rho_{12}(0) \tag{VII-2}$$

This phenomenological form of relaxation can be written in more familiar notation with the definition:

$$R_{12,12} = R_{21,21} = -\frac{1}{T_2} \tag{VII-3}$$

If the system is actually a single S = ½ spin, where $|1\rangle = |+\rangle$ and $|2\rangle = |-\rangle$ then the observed magnetization is proportional to:

$$\langle S_+ \rangle \;=\; Tr\{\rho(t)\, S_+\} \;=\; \rho_{21}(t)\left(S_+\right)_{12} \;=\; \exp\!\left(\left(-i\omega_{21} + R_{21,21}\right)t\right)\rho_{21}(0) \qquad \text{(VII-4)}$$

The positive value of T_2 guarantees decay of ρ_{21}, and the observed magnetization, to zero.

Steady state magnetization can be established by steady absorption of energy from a radiation source, and in the simplest case only one such source is operating, with:

$$H_2(t) \;=\; 2\gamma_e B_1 S_x \cos(\omega t) \;\equiv\; 2\,\omega_1 S_x \cos(\omega t) \qquad \text{(VII-5)}$$

Then:

$$\frac{d}{dt}\rho_{21} \;=\; -i\omega_{21}\rho_{21} - \frac{2i\omega_1}{\hbar}\cos(\omega t)\left\{(S_x)_{21}\rho_{11} - \rho_{22}(S_x)_{21}\right\} + R_{21,21}\rho_{21} \qquad \text{(VII-6a)}$$

$$\frac{d}{dt}\rho_{12} \;=\; -i\omega_{12}\rho_{12} - \frac{2i\omega_1}{\hbar}\cos(\omega t)\left\{(S_x)_{12}\rho_{22} - \rho_{11}(S_x)_{12}\right\} + R_{12,12}\rho_{12} \qquad \text{(VII-6b)}$$

Since $(S_x)_{12} = (S_x)_{12} = \tfrac{1}{2}\hbar$, the equation for ρ_{21} can be written as

$$\frac{d}{dt}\rho_{21} \;=\; \left\{i\omega_{12} + R_{21,21}\right\}\rho_{21} - 2id\cos(\omega t)\left\{\rho_{11} - \rho_{22}\right\} \qquad \text{(VII-7)}$$

with $d \equiv (\omega_1/\hbar)\,(S_x)_{12} = \tfrac{1}{2}\omega_1$, and a similar equation applies to ρ_{12}. This equation can be solved by expanding the density matrix in Fourier components, as in Eq. (V-102):

$$\rho \;=\; \sum_{n=-\infty}^{\infty} Z_n\, e^{in\omega t} \qquad \text{(VII-8)}$$

Substituting the expansion in Eq. (VII-7) and taking the n'th Fourier component:

$$in\omega\,(Z_n)_{21} \;=\; \left\{i\omega_{12} + R_{21,21}\right\}(Z_n)_{21}$$

$$-\,id\left\{\left((Z_{n-1})_{11} + (Z_{n+1})_{11}\right) - \left((Z_{n-1})_{22} + (Z_{n+1})_{22}\right)\right\} \qquad \text{(VII-9)}$$

The matrix element $(Z_n)_{21}$ is not coupled to any $(Z_k)_{21}$ for $k \neq n$. It is therefore sufficient to solve for those n values which contribute to the observed signal. If $\mathbf{M} = \langle \mathbf{S} \rangle$ is the magnetization in the laboratory frame, and $\mathbf{M}_{rot.}$ is the magnetization in the rotating frame, it is known that the power absorbed by the system is:

$$P = \omega\,\omega_1 \left(M_{rot.}\right)_y = \omega\,\omega_1 \{-\sin(\omega t)\,M_x + \cos(\omega t)\,M_y\} =$$

$$= -\frac{i}{2}\,\omega\,\omega_1 \left\{\exp(-i\omega t)\,M_+ - \exp(i\omega t)\,M_-\right\} \qquad \text{(VII-10)}$$

As seen in Eq. (VII-4) above, $\mathbf{M}_+$ is proportional to ρ_{21} , so in the time average over exp(-iωt) $\mathbf{M}_+$ the only non-zero contribution will come from the n=1 term. This is equivalent to stating that significant absorption occurs only close to resonance, when $\omega \approx \omega_0 \equiv \omega_{12}$. Similarly, $\mathbf{M}_-$ is proportional to ρ_{12} , and the only non-zero contribution to exp(iωt) $\mathbf{M}_-$ will come from n = -1. In other words, in the calculation of ρ_{21} one only needs the n = 1 term, and in its complex conjugate ρ_{12} one only needs the n= - 1 term.

In fact, this can be seen directly from the equations for these elements. For example, for n = 1 one obtains from Eq. (VII-9) and from the analogous equation for ρ_{12}:

$$\{i(\omega - \omega_{12}) - R_{21,21}\}\,(Z_1)_{21} = -i\,d\,\{(Z_0)_{11} - (Z_0)_{22}\} \qquad \text{(VII-11a)}$$

$$\{i(\omega + \omega_{12}) - R_{12,12}\}\,(Z_1)_{12} = i\,d\,\{(Z_0)_{11} - (Z_0)_{22}\} \qquad \text{(VII-11b)}$$

In this equation the elements $(\mathbf{Z}_2)_{ii}$ do not appear, because assuming a steady state has been reached the populations are constant, so that in fact $\rho_{ii} = (\mathbf{Z}_0)_{ii}$. Thus:

$$\frac{(Z_1)_{12}}{(Z_1)_{21}} = -\frac{i(\omega - \omega_{12}) - R_{21,21}}{i(\omega + \omega_{12}) - R_{12,12}} = -\frac{i(\omega - \omega_0) - \dfrac{1}{T_2}}{i(\omega + \omega_0) - \dfrac{1}{T_2}} \qquad \text{(VII-12)}$$

and this expression tends to zero when $\omega \to \omega_{12}$, because $\omega_0 \gg 1/T_2$ in typical experiments. Therefore $(\mathbf{Z}_1)_{12} \approx 0$, and similarly $(\mathbf{Z}_{-1})_{21} \approx 0$.

To calculate the populations one takes the (i,i) elements of Eq. (VII-1), which are

$$0 = -2i\omega_1 \cos(\omega t)\,\{(S_x)_{12}\,\rho_{21} - \rho_{12}\,(S_x)_{21}\} + R_{11,11}\,\chi_{11} + R_{11,22}\,\chi_{22} \qquad \text{(VII-13a)}$$

$$0 = -2i\omega_1 \cos(\omega t)\,\{(S_x)_{21}\,\rho_{12} - \rho_{21}\,(S_x)_{12}\} + R_{22,22}\,\chi_{22} + R_{22,11}\,\chi_{11} \qquad \text{(VII-13b)}$$

with $\chi \equiv \rho - \rho_{(eq.)}$. In these equations, the constant value of the diagonal elements of ρ implies that, in the off-diagonal elements of the density matrix, only the Fourier terms with n =1 and with n = -1 appear. The hermiticity of ρ requires

$$(Z_1)_{12} = (Z_{-1})_{21}^{*} \approx 0 \qquad \textit{and} \qquad (Z_1)_{21} = (Z_{-1})_{12}^{*} \qquad \text{(VII-14)}$$

so that the equations can be rewritten as:

$$R_{11,11}\,\chi_{11} + R_{11,22}\,\chi_{22} = id\left\{(Z_1)_{21} + (Z_1)_{12}{}^* - \left((Z_1)_{12} + (Z_1)_{21}{}^*\right)\right\} \qquad \text{(VII-15a)}$$

$$R_{22,11}\,\chi_{11} + R_{22,22}\,\chi_{22} = -id\left\{(Z_1)_{21} + (Z_1)_{12}{}^* - \left((Z_1)_{12} + (Z_1)_{21}{}^*\right)\right\} \qquad \text{(VII-15b)}$$

The relevant component $(\mathbf{Z}_1)_{21}$ will simply be written as Z , for which the following equation is obtained:

$$R_{11,11}\,\chi_{11} + R_{11,22}\,\chi_{22} = -2d\,Im(Z) \qquad \text{(VII-16a)}$$

$$R_{22,11}\,\chi_{11} + R_{22,22}\,\chi_{22} = 2d\,Im(Z) \qquad \text{(VII-16b)}$$

The physical significance of the elements $\mathbf{R}_{ii,mm}$ can be seen from their role in Eq. (VII-1), taken for the case i = j, m = n. If there is no irradiation, the diagonal elements of $\boldsymbol{\rho}$ still evolve in time due to those elements of **R**. In that case, the set of equations for the diagonal elements of $\boldsymbol{\rho}$ is a typical master equation (see Eq. (II-79)). Thus $\mathbf{R}_{ii,mm}$ (for i ≠ m) is the transition probability from state $|m\rangle$ to state $|i\rangle$, while - $\mathbf{R}_{ii,ii}$ is the probability of leaving state $|i\rangle$. In a steady state, the probability of leaving state $|i\rangle$ is equal to the total probability of reaching it, so that

$$-W_{i,i} = \sum_{m \neq i} W_{i,m} \qquad \text{(VII-17)}$$

using the conventional notation $W_{i,m} \equiv \mathbf{R}_{ii,mm}$. Assuming equal probability for transitions in opposite directions: $W_{i,m} = W_{m,i}$. In this case:

$$R_{11,22} = R_{22,11} = W \qquad \text{(VII-18a)}$$

$$R_{11,11} = -R_{11,22} = -W \qquad R_{22,22} = -R_{22,11} = -W \qquad \text{(VII-18b)}$$

Substituting these results into Eq. (VII-16) one obtains

$$W\left(\chi_{22} - \chi_{11}\right) = -2d\,Im(Z) \qquad \text{(VII-19a)}$$

$$W\left(\chi_{11} - \chi_{22}\right) = 2d\,Im(Z) \qquad \text{(VII-19b)}$$

Eq. (VII-19) can be solved together with Eq. (VII-11) to yield the time evolution of the system. With the notation: Z' ≡ Re(Z) and Z" ≡ Im(Z), Eq. (VII-11) can be written as:

$$\left\{ i(\omega - \omega_0) + \frac{1}{T_2} \right\} (Z' + iZ'') = -id \left\{ (\chi_{11} - \chi_{22}) - q\omega_0 \right\} \tag{VII-20}$$

The high temperature approximation (Eq. (V-99)) was used to get: $\rho_{(eq.)11} - \rho_{(eq.)22} = -\hbar\omega_0/(A\, k_B\, T) \equiv -q\omega_0$ (A is the normalization constant, equal to the trace of the Boltzmann exponential). The real and imaginary parts of this equation, together with Eq. (VII-19), form a set of three linear equations. These can be written in matrix form as:

$$\begin{pmatrix} \frac{1}{T_2} & -\Delta\omega & 0 \\ \Delta\omega & \frac{1}{T_2} & d \\ 0 & -2d & W \end{pmatrix} \begin{pmatrix} Z' \\ Z'' \\ \Delta\chi \end{pmatrix} = \begin{pmatrix} 0 \\ q\,\omega_0\, d \\ 0 \end{pmatrix} \tag{VII-21}$$

where $\Delta\omega \equiv \omega - \omega_0$ and $\Delta\chi \equiv \chi_{11} - \chi_{22}$. The solution of these equations is:

$$Z' = \Delta\omega\, T_2\, Z'' \tag{VII-22a}$$

$$Z'' = \frac{1}{F} q\omega_0\, d\, T_2 \tag{VII-22b}$$

$$\Delta\chi = \frac{1}{F} q\omega_0\, 4\, d^2\, T_1\, T_2 \tag{VII-22c}$$

with the definitions: $T_1 \equiv 1/(2W)$ and

$$F = 1 + (\Delta\omega)^2 (T_2)^2 + 4\, d^2\, T_1\, T_2 \tag{VII-23}$$

These solutions are the same as the steady state solutions of the Bloch equations, provided one makes the identification: $Z = M_+$ and $\Delta\chi = 2(M_0 - M_z)$. As in those solutions, saturation occurs when d is relatively strong, so that in F the dominant term is the one proportional to d^2 .

In a more general case, there is still only one relevant transition, but relaxation processes couples more states. The other transitions are far from resonance, so the corresponding off-diagonal elements in the density matrix are negligible. However, the two states involved in the transition are coupled to other states through relaxation processes. Then Eq. (VII-20) still holds, but Eq. (VII-19) is replaced by:

$$\sum_{j\neq 1} W_{1j}(\chi_{11} - \chi_{jj}) = 2d\, Im(Z) \tag{VII-24a}$$

$$\sum_{j\neq 2} W_{2j}(\chi_{22} - \chi_{jj}) = -2d\, Im(Z) \tag{VII-24b}$$

$$\sum_{j\neq k} W_{kj}(\chi_{kk} - \chi_{jj}) = 0 \qquad (k \neq 1,2) \tag{VII-24c}$$

General solutions have been obtained for this set of equations. The general solution will not be developed here, since it requires fairly complicated algebra, but the solutions for specific cases of two transitions with various relaxation networks will be derived below.

(b) *Electron-electron double resonance (ELDOR)*

Having reviewed some basic characteristics of relaxation in a system with a single transition, let us now consider the most elementary case in which two irradiation fields are present, exciting two different resonances in the system. It is simpler to start with two EPR transitions (resonances). Such a situation occurs in practice in electron-electron double resonance (ELDOR) experiments. In these experiments one usually operates with strong irradiation to achieve significant saturation in one resonance frequency, observing the influence of this "pumping" on a second resonance, to which only weak irradiation is applied. The general formalism will now be developed for such double resonance experiments. Special assumptions about the relative strength of the two microwave fields will be introduced at a later stage.

The two sets of states for transitions a, b are: $|1\rangle$, $|2\rangle$ and $|1'\rangle$, $|2'\rangle$, respectively (see Fig. VII.1). The radiation term in the Hamiltonian is now:

$$H_2(t) = \gamma_e B_a S_x \cos(\omega_a t) + \gamma_e B_b S_x \cos(\omega_b t) \tag{VII-25}$$

where ω_a , ω_b are the two irradiation frequencies and B_a , B_b are the amplitudes of the two time dependent magnetic fields. The corresponding resonance frequencies are ω_{12} and $\omega_{1'2'}$.

In ELDOR one deals with two transitions correponding to two different hyperfine lines, so $|\omega_{12} - \omega_{1'2'}|$ is of the order of the hyperfine constant. Its magnitude is assumed to be large relative to the strength of the two irradiation fields:

$$\gamma_e B_a \, , \, \gamma_e B_b \ll |\omega_{12} - \omega_{1'2'}| \tag{VII-26}$$

so that, in the absence of motion, each irradiation field affects only its corresponding resonance, and has no influence on the other resonance. Motional narrowing is assumed, as in the single line case. At the same time, the rate constant R (which determines the magnitude of the elements of the matrix **R**) may be assumed small compared with the separation in frequency between the two lines:

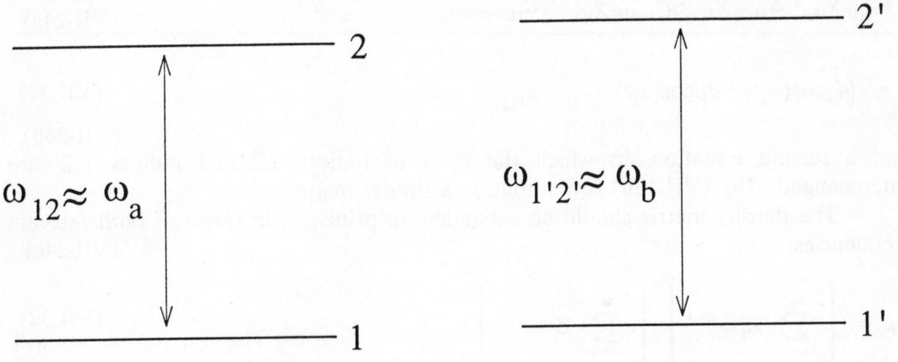

Fig. VII.1. A level scheme for ELDOR. Two electronic transitions are coupled via relaxation mechanisms. Here $E_2 - E_1 \neq E_{2'} - E_{1'}$, but the energies of states $|1\rangle$ and $|1'\rangle$ need not be equal.

$$R \ll |\omega_{12} - \omega_{1'2'}| \tag{VII-27}$$

so that the two lines keep their identity, without being mixed by the motion. In order to have any effect in such a case, the two irradiation frequencies must be close to their respective resonance frequencies:

$$|\omega_a - \omega_{12}|\,,\ |\omega_b - \omega_{1'2'}| \ll |\omega_{12} - \omega_{1'2'}| \tag{VII-28}$$

and therefore:

$$|\omega_a - \omega_{1'2'}|\,,\ |\omega_b - \omega_{12}| \approx |\omega_{12} - \omega_{1'2'}| \tag{VII-29}$$

Eq. (VII-7) is now replaced by:

$$\frac{d}{dt}\rho_{21} = \{i\omega_{12} + R_{21,21}\}\rho_{21} - 2i\left(d_a\cos(\omega_a t) + d_b\cos(\omega_b t)\right)\{\rho_{11} - \rho_{22}\} \tag{VII-30a}$$

$$\frac{d}{dt}\rho_{2'1'} = \{i\omega_{1'2'} + R_{2'1',2'1'}\}\rho_{2'1'} - 2i\left(d_a\cos(\omega_a t) + d_b\cos(\omega_b t)\right)\{\rho_{1'1'} - \rho_{2'2'}\} \tag{VII-30b}$$

where $d_\alpha \equiv \frac{1}{2}\gamma_e B_\alpha$. Eq. (VII-13a) is replaced by:

$$R_{11,11}\chi_{11} + R_{11,22}\chi_{22} + R_{11,1'1'}\chi_{1'1'} + R_{11,2'2'}\chi_{2'2'} =$$

$$= i\left(d_a\cos(\omega_a t) + d_b\cos(\omega_b t)\right)\left\{\rho_{21} - \rho_{12}\right\} \tag{VII-31}$$

and a second equation, in which the roles of indices 1,2 and indices 1',2' are interchanged. Eq. (VII-13b) is modified in a similar manner.

The density matrix should be expanded, in principle, in terms of both relevant frequencies:

$$\rho = \left[\sum_{k=-\infty}^{\infty} Z_k e^{ik\omega_a t}\right]\left[\sum_{n=-\infty}^{\infty} Z_n e^{in\omega_b t}\right] \tag{VII-32}$$

In practice, however, the calculation is much simpler. The only non-zero elements of $\mathbf{S}_x$ are $(\mathbf{S}_x)_{12} = (\mathbf{S}_x)_{21}^*$ and $(\mathbf{S}_x)_{1'2'} = (\mathbf{S}_x)_{2'1'}^*$. Then the magnetization $\mathbf{M}_+$ is:

$$M_+ = \langle S_+\rangle = \rho_{21}(S_+)_{12} + \rho_{2'1'}(S_+)_{1'2'} = \rho_{21} + \rho_{2'1'} \tag{VII-33}$$

If the signal is measured relative to ω_a, i.e., close to the resonance frequency ω_{12}, it is natural to calculate it in the rotating frame "a". Then Eq. (VII-10) with $\omega_0 = \omega_{12}$ is relevant. The time-independent contribution of $\mathbf{M}_+$ to the absorbed power will come from the term in $\boldsymbol{\rho}$ with $\exp(i\omega_a t)$, which is the k = 1, n = 0 term in Eq. (VII-32). If, on the other hand, the signal is measured relative to ω_b, then Eq. (VII-10) with $\omega_0 = \omega_{1'2'}$ is relevant. The important term in $\boldsymbol{\rho}$ is that with $\exp(i\omega_b t)$, which is the k = 0, n = 1 term in Eq. (VII-32).

As a consequence of this situation and on the basis of Eq. (VII-30), the only relevant terms in the Fourier expansion of $\boldsymbol{\rho}$ are $(\mathbf{Z}_1)_{21}\exp(i\omega_a t)$, $(\mathbf{Z}_1)_{2'1'}\exp(i\omega_b t)$ and $(\mathbf{Z}_0)_{jj}$ (j = 1, 2, 1' or 2'). Thus the analog of Eq. (VII-20) is:

$$\left\{i\Delta\omega_a + \frac{1}{(T_2)_a}\right\}\left(Z_a' + iZ_a''\right) + id_a\left(\chi_{11} - \chi_{22}\right) = iq\,\omega_{12}\,d_a \tag{VII-34a}$$

$$\left\{i\Delta\omega_b + \frac{1}{(T_2)_b}\right\}\left(Z_b' + iZ_b''\right) + id_b\left(\chi_{1'1'} - \chi_{2'2'}\right) = iq\,\omega_{1'2'}\,d_b \tag{VII-34b}$$

with the definitions $\Delta\omega_a \equiv \omega_a - \omega_{12}$ and $\Delta\omega_b \equiv \omega_b - \omega_{1'2'}$. For high magnetic fields (X-band or higher) the Zeeman term is much bigger than the hyperfine splitting:

$$\omega_{12} \gg \left|\omega_{12} - \omega_{1'2'}\right| \tag{VII-35}$$

It is then a good approximation to replace ω_{12} and $\omega_{1'2'}$ on the right hand side of Eq. (VII-34) by their average $\omega_e \equiv \frac{1}{2}(\omega_{12} + \omega_{1'2'})$. The generalized form of Eq. (VII-19) is:

$$\sum_{j\neq 1} W_{1,j}\left(\chi_{11} - \chi_{jj}\right) = 2 d_a Z_a'' \tag{VII-36a}$$

$$\sum_{j\neq 2} W_{2,j}\left(\chi_{22} - \chi_{jj}\right) = -2 d_a Z_a'' \tag{VII-36b}$$

$$\sum_{j\neq 1'} W_{1',j}\left(\chi_{1'1'} - \chi_{jj}\right) = 2 d_b Z_b'' \tag{VII-36c}$$

$$\sum_{j\neq 2'} W_{2',j}\left(\chi_{2'2'} - \chi_{jj}\right) = -2 d_b Z_b'' \tag{VII-36d}$$

The analog of Eq. (VII-21) is therefore:

$$\begin{pmatrix} -\boldsymbol{R}\boldsymbol{Z}' & -\boldsymbol{K}\boldsymbol{Z}'' & \\ \boldsymbol{K}\boldsymbol{Z}' & -\boldsymbol{R}\boldsymbol{Z}'' & +\boldsymbol{d}\boldsymbol{\chi} \\ & 2\,\boldsymbol{d}^{tr}\boldsymbol{Z}'' & +\boldsymbol{W}\boldsymbol{\chi} \end{pmatrix} = \begin{pmatrix} \boldsymbol{0}_2 \\ \boldsymbol{Q} \\ \boldsymbol{0}_4 \end{pmatrix} \tag{VII-37}$$

using the following definitions of vectors:

$$\boldsymbol{Z}' = \begin{pmatrix} Z_a' \\ Z_b' \end{pmatrix} \qquad \boldsymbol{Z}'' = \begin{pmatrix} Z_a'' \\ Z_b'' \end{pmatrix} \qquad \boldsymbol{Q} = q\,\omega_e \begin{pmatrix} d_a \\ d_b \end{pmatrix} \qquad \boldsymbol{0}_2 = \begin{pmatrix} 0 \\ 0 \end{pmatrix} \tag{VII-38}$$

$$\boldsymbol{\chi} = \begin{pmatrix} \chi_{11} \\ \chi_{22} \\ \chi_{1'1'} \\ \chi_{2'2'} \end{pmatrix} \qquad \boldsymbol{0}_4 = \begin{pmatrix} 0 \\ 0 \\ 0 \\ 0 \end{pmatrix} \tag{VII-39}$$

and matrices:

$$\boldsymbol{K} = \begin{pmatrix} \Delta\,\omega_a & 0 \\ 0 & \Delta\,\omega_b \end{pmatrix} \qquad \boldsymbol{R} = -\begin{pmatrix} \dfrac{1}{(T_2)_a} & 0 \\ 0 & \dfrac{1}{(T_2)_b} \end{pmatrix} \tag{VII-40}$$

$$\boldsymbol{d} = \begin{pmatrix} d_a & -d_a & 0 & 0 \\ 0 & 0 & d_b & -d_b \end{pmatrix} \qquad \boldsymbol{d}^{tr} = \begin{pmatrix} d_a & 0 \\ -d_a & 0 \\ 0 & d_b \\ 0 & -d_b \end{pmatrix} \tag{VII-41}$$

$$\mathbf{W} = -\begin{pmatrix} -W_{1,1} & W_{1,2} & W_{1,1'} & W_{1,2'} \\ W_{2,1} & -W_{2,2} & W_{2,1'} & W_{2,2'} \\ W_{1',1} & W_{1',2} & -W_{1',1'} & W_{1',2'} \\ W_{2',1} & W_{2',2} & W_{2',1'} & -W_{2',2'} \end{pmatrix} \tag{VII-42}$$

The transition probability matrix **W** is symmetric. Also the following relations hold between its elements:

$$W_{1,1} = W_{1,2} + W_{1,1'} + W_{1,2'} \qquad W_{2,2} = W_{2,1} + W_{2,1'} + W_{2,2'}$$

$$W_{1',1'} = W_{1',1} + W_{1',2} + W_{1',2'} \qquad W_{2',2'} = W_{2',1} + W_{2',2} + W_{2',1'} \tag{VII-43}$$

Due to these relations, the columns (and rows) of **W** are not independent, so it is not invertible. Thus the set of equations written in matrix form in the third row of Eq. (VII-37) cannot be solved as it stands. The difficulty is eliminated if one replaces any row of the matrix equation by another equation, independent of all others. Such an addditional equation is:

$$Tr(\chi) = \sum_j \chi_{jj} = 0 \tag{VII-44}$$

which is a consequence of the conservation of probability:

$$Tr(\rho) = Tr(\rho_{eq}) = 1 \tag{VII-45}$$

When Eq. (VII-44) replaces a row in the matrix equation, the corresponding row in **W** is replaced by: (1 1 1 1) and the corresponding row in $\mathbf{d}^{tr}$ is replaced by (0 0). This leads to a modified equation in the third row of Eq. (VII-37):

$$2\,\boldsymbol{d}^{tr} \boldsymbol{Z}'' + \boldsymbol{W}' \boldsymbol{\chi} = \boldsymbol{0}_4 \tag{VII-37c'}$$

which can be solved by inverting **W'**. In order to express the solutions of the modified Eq. (VII-37) in a simple form, it is convenient to define the 2 × 2 matrices:

$$\boldsymbol{S} = 2\,\boldsymbol{d}\,\boldsymbol{W}'^{-1}\,\boldsymbol{d}^{tr} \tag{VII-46}$$

$$\boldsymbol{F} = \boldsymbol{I} + \left(\boldsymbol{R}^{-1}\boldsymbol{K}\right)^2 - \boldsymbol{R}^{-1}\boldsymbol{S} \tag{VII-47}$$

where **I** is the 2 × 2 unit matrix. The formal solution of the modified Eq. (VII-37) is given by:

$$\boldsymbol{Z}' = -\boldsymbol{R}^{-1}\boldsymbol{K}\boldsymbol{Z}'' \tag{VII-48a}$$

$$\boldsymbol{Z}'' = -\boldsymbol{F}^{-1}\boldsymbol{R}^{-1}\boldsymbol{Q} \tag{VII-48b}$$

$$\boldsymbol{d}\boldsymbol{\chi} = -\boldsymbol{S}\boldsymbol{Z}'' \tag{VII-48c}$$

Since **d** cannot be inverted, one cannot re-write the last equation directly as an equation for χ itself. In order to get the solutions for a specific case, one has to start by inverting **W'**, then construct **S** and **F**, and then the measured magnetization can be calculated.

Example VII.1: A single time dependent field

Suppose a system consists of two lines (transitions) as described above, but only a single irradiating field is operating, i.e., $d_b = 0$. Then **S** depends only on the upper left 2 × 2 block in $\mathbf{W}'^{-1}$, and **d** χ involves only $d_a(\chi_{1,1} - \chi_{2,2})$. It is straightforward to verify that (omitting the prime)

$$S = 2d_a^2 \begin{pmatrix} (W^{-1})_{1,1} - (W^{-1})_{1,2} - (W^{-1})_{2,1} + (W^{-1})_{2,2} & 0 \\ 0 & 0 \end{pmatrix} \tag{E-1}$$

The expressions for **F** and for **Z'**, **Z"** and χ are then only slightly more general than the expressions in Eq. (VII-22), with $S_{1,1}$ replacing T_1 . This change results from the introduction of additional relaxation paths to the original single line problem.

****

Continuing with the general case, assume that row j in **W** and in $\mathbf{d}^{tr}$ was replaced by 1's and 0's, respectively, as explained above. From the definition of **S** it follows that $S_{11} \propto d_a^2$, $S_{12}, S_{21} \propto d_a d_b$ and $S_{22} \propto d_b^2$. It is therefore convenient to write:

$$S = \begin{pmatrix} d_a^2 \Omega_{a,a} & d_a d_b \Omega_{a,b} \\ d_a d_b \Omega_{b,a} & d_b^2 \Omega_{b,b} \end{pmatrix} \tag{VII-49}$$

with the following definitions (omitting the prime from here on):

$$\Omega_{a,a} = 2\{ W^{-1})_{1,1} - (W^{-1})_{1,2} - (W^{-1})_{2,1} + (W^{-1})_{2,2} \}$$

$$\Omega_{a,b} = 2\{ W^{-1})_{1,1'} - (W^{-1})_{1,2'} - (W^{-1})_{2,1'} + (W^{-1})_{2,2'} \}$$

$$\Omega_{b,a} = 2\{ W^{-1})_{1',1} - (W^{-1})_{1',2} - (W^{-1})_{2',1} + (W^{-1})_{2',2} \}$$

$$\Omega_{b,b} = 2\{ W^{-1})_{1',1'} - (W^{-1})_{1',2'} - (W^{-1})_{2',1'} + (W^{-1})_{2',2'} \} \tag{VII-50}$$

Each of the $\Omega_{\alpha,\beta}$ is essentially a T_1-like relaxation constant. The general form of **F** is:

$$F = \begin{pmatrix} 1 + ((T_2)_a \Delta\omega_a)^2 + (T_2)_a d_a^2 \Omega_{a,a} & (T_2)_a d_a d_b \Omega_{a,b} \\ (T_2)_b d_a d_b \Omega_{b,a} & 1 + ((T_2)_b \Delta\omega_b)^2 + (T_2)_b d_b^2 \Omega_{b,b} \end{pmatrix} \tag{VII-51}$$

Standard methods of matrix inversion yield:

$$F^{-1} = \frac{1}{\det(F)} \begin{pmatrix} F_{2,2} & -F_{1,2} \\ -F_{2,1} & F_{1,1} \end{pmatrix} \tag{VII-52}$$

where the determinant of **F** is equal to

$$\det(F) = F_{1,1} F_{2,2} - F_{1,2} F_{2,1} \tag{VII-53}$$

Then the magnetization is proportional to

$$Z_a'' = \frac{1}{\det(F)} q \omega_e \left(d_a (T_2)_a F_{2,2} - d_b (T_2)_b F_{1,2} \right) \tag{VII-54}$$

Now consider the typical ELDOR experiment, in which one observes line "a" while line "b" is strongly saturated. The saturation condition is:

$$(d_b)^2 (T_2)_b \Omega_{b,b} \gg 1 \tag{VII-55}$$

in analogy to the saturation in standard steady state magnetic resonance. One may also assume the "pumping" field is on resonance:

$$\Delta\omega_b = 0 \tag{VII-56}$$

The "observing" field is weak:

$$(d_a)^2 (T_2)_a \Omega_{a,a} \ll 1 \tag{VII-57}$$

Under these conditions:

$$F_{2,2} \approx (d_b)^2 (T_2)_b \Omega_{b,b} \tag{VII-58}$$

and

$$\det(F) \approx F_{2,2} \left\{ 1 + ((T_2)_a \Delta\omega_a)^2 - (T_2)_a d_a^{\,2} \frac{\Omega_{a,b} \Omega_{b,a}}{\Omega_{b,b}} \right\} \tag{VII-59}$$

The last term in the curly brackets is small if d_a is sufficiently small. The solution then becomes:

$$Z_a'' = \frac{q \omega_e d_a (T_2)_a}{1 + ((T_2)_a \Delta\omega_a)^2} \left(1 - \frac{\Omega_{a,b}}{\Omega_{b,b}} \right) \tag{VII-60}$$

The ratio $\Omega_{a,b} / \Omega_{b,b}$ thus determines the influence of the saturated line on the non-saturated line. In particular, the sign of this ratio determines whether the saturation of line "b" leads to an increase or to a decrease in the signal observed for line "a". These effects are related to the results of saturation transfer experiments in NMR. Those are double resonance NMR experiments which also involve saturation. However, there one observes the recovery of an NMR signal after a period of saturation of another line, whereas in ELDOR a signal is observed during the saturation period of the other line.

Example VII.2: Modification of a signal in ELDOR

In some cases the structure of **W** can be simplified as follows. The main relaxation mechanisms usually involve either an electron spin flip (1 ⇔ 2, 1' ⇔ 2') or a nuclear spin flip (1 ⇔ 1', 2 ⇔ 2') but not both together (1 ⇔ 2', 1' ⇔ 2). If all electronic transition rates have the same value, denoted by W_e, and all nuclear transition rates are equal to one value, denoted by W_n, the transition probability matrix becomes:

$$\boldsymbol{W} = \begin{pmatrix} W_e + W_n & -W_e & -W_n & 0 \\ -W_e & W_e + W_n & 0 & -W_n \\ -W_n & 0 & W_e + W_n & -W_e \\ 0 & -W_n & -W_e & W_e + W_n \end{pmatrix} \tag{E-1}$$

The last row can be replaced by 1's in order to make the matrix invertible. The modified matrix, which will be denoted here simply as **W**, can be inverted using standard methods, yielding

$$\mathbf{W}^{-1} = \frac{1}{4\,U_P}\begin{pmatrix} U_S^2 & W_e^2 & W_n^2 & U_P \\ U_D\,U_S & W_e(U_S + W_n) & -W_n^2 & U_P \\ -U_D\,U_S & -W_e^2 & W_n(U_S + W_e) & U_P \\ -U_S^2 & -W_e(U_S + W_n) & -W_n(U_S + W_e) & U_P \end{pmatrix} \tag{E-2}$$

with $U_S \equiv W_e + W_n$, $U_D \equiv W_e - W_n$ and $U_P \equiv W_e\,W_n\,U_S$. From this formula for $\mathbf{W}^{-1}$ one may calculate:

$$\Omega_{a,a} = \Omega_{b,b} = \frac{W_n + 2\,W_e}{W_e\,(W_e + W_n)} \tag{E-3}$$

$$\Omega_{a,b} = \Omega_{b,a} = \frac{W_n}{W_e\,(W_e + W_n)} \tag{E-4}$$

Here both $\Omega_{b,b} > 0$ and $\Omega_{a,b} > 0$, so that

$$1 - \frac{\Omega_{a,b}}{\Omega_{b,b}} = 1 - \frac{W_n}{W_n + 2W_e} < 1 \qquad \text{(E-5)}$$

This implies that saturating the "b" line will lead to a reduction in the signal of line "a". In particular, if $W_e \ll W_n$ the signal will be reduced almost to zero, while if $W_e \gg W_n$ the reduction of the signal will be insignificant. On the other hand, in cases in which $(\Omega_{a,b}/\Omega_{b,b})$ is negative, saturation of line "b" will enhance the signal measured for line "a".

****

(c) *Electron-nuclear double resonance (ENDOR)*

In the most basic case in ENDOR there is a single S = ½ electron interacting via a hyperfine interaction with a single I = ½ nucleus. The main Hamiltonian is the Zeeman Hamiltonian

$$\boldsymbol{H_0} = \gamma_e B_0 \boldsymbol{S_z} - \gamma_n B_0 \boldsymbol{I_z} \qquad \text{(VII-61)}$$

Its eigenfunctions are

$$|1\rangle_0 = |m_S = -\tfrac{1}{2}, m_I = \tfrac{1}{2}\rangle \qquad |2\rangle_0 = |m_S = \tfrac{1}{2}, m_I = \tfrac{1}{2}\rangle$$

$$|1'\rangle_0 = |m_S = -\tfrac{1}{2}, m_I = -\tfrac{1}{2}\rangle \qquad |2'\rangle_0 = |m_S = \tfrac{1}{2}, m_I = -\tfrac{1}{2}\rangle \qquad \text{(VII-62)}$$

with energies:

$$E_1^{(0)} = -\frac{1}{2}\left(\omega_e^{(0)} + \omega_n^{(0)}\right) \qquad E_2^{(0)} = \frac{1}{2}\left(\omega_e^{(0)} - \omega_n^{(0)}\right)$$

$$E_{1'}^{(0)} = -\frac{1}{2}\left(\omega_e^{(0)} - \omega_n^{(0)}\right) \qquad E_{2'}^{(0)} = \frac{1}{2}\left(\omega_e^{(0)} + \omega_n^{(0)}\right) \qquad \text{(VII-63)}$$

where $\omega_e^{(0)} \equiv \gamma_e B_0$ and $\omega_n^{(0)} \equiv \gamma_e B_0$. The hyperfine interaction, which is much smaller (for ordinary B_0 fields) than the Zeeman interaction, includes terms of the form: $\gamma_e a_0 \boldsymbol{S_z I_z}$, $\gamma_e a \boldsymbol{S_z I_\pm}$ and other terms. The first term is isotropic, and the others are multiplied by orientation dependent factors (which will not be given here). The first term is diagonal in the basis of $\boldsymbol{H_0}$, so it does not change the eigenfunctions, its only effect being a small modification of the energies. The modified energies are:

$$E_1 = -\frac{1}{2}\left(\omega_e^{(0)} + \omega_n^{(0)}\right) - \frac{\gamma_e a_0}{4} \qquad E_2 = \frac{1}{2}\left(\omega_e^{(0)} - \omega_n^{(0)}\right) + \frac{\gamma_e a_0}{4}$$

$$E_{1'} = -\frac{1}{2}\left(\omega_e^{(0)} - \omega_n^{(0)}\right) + \frac{\gamma_e a_0}{4} \qquad E_{2'} = \frac{1}{2}\left(\omega_e^{(0)} + \omega_n^{(0)}\right) - \frac{\gamma_e a_0}{4} \tag{VII-64}$$

The other terms are perturbations having only off-diagonal elements in the basis of $\mathbf{H}_0$, so to first order in perturbation theory they modify the wave functions without changing the energies. In principle, the modified wave functions are of the form:

$$|1\rangle = |1\rangle_0 + c_1|1'\rangle_0 + c_2|2\rangle_0 \qquad |2\rangle = |2\rangle_0 + c_1|2'\rangle - c_2|1\rangle_0$$

$$|1'\rangle = |1'\rangle_0 - c_1|1\rangle_0 + c_2|2'\rangle_0 \qquad |2'\rangle = |2'\rangle_0 - c_1|2\rangle - c_2|1'\rangle_0 \tag{VII-65}$$

where $c_1 \propto \gamma_e a/\gamma_n B_0$ and $c_2 \propto \gamma_e a/\gamma_e B_0$. Normally $c_1 \ll 1$, and since $\gamma_e/\gamma_n \gg 1$ it follows that $c_2 \ll c_1$ so that c_2 can be neglected.

In the ENDOR experiment one irradiates with one field close to the electronic resonance frequency and another field close to the nuclear resonance frequency. The interaction of the magnetic fields with the electron-nucleus system is:

$$H_2(t) = \gamma_e B_e(S_x + I_x)\cos(\omega_e t) + \gamma_n B_n(S_x + I_x)\cos(\omega_n t) \tag{VII-66}$$

This leads to four types of transitions, with relative transition moments:

$$|\langle 1|S_x|2\rangle|^2, \; |\langle 1'|S_x|2'\rangle|^2 = 1 \qquad (\omega_0 = \omega_e) \tag{VII-67a}$$

$$|\langle 1|I_x|1'\rangle|^2, \; |\langle 2|I_x|2'\rangle|^2 = 1 \qquad (\omega_0 = \omega_n) \tag{VII-67b}$$

$$|\langle 1|S_x|2'\rangle|^2, \; |\langle 1'|S_x|2\rangle|^2 = |c_1|^2 \qquad (\omega_0 = \omega_e \pm \omega_n) \tag{VII-67c}$$

$$|\langle 1|I_x|2'\rangle|^2, \; |\langle 1'|I_x|2\rangle|^2 = |c_2|^2 \qquad (\omega_0 = \omega_e \pm \omega_n) \tag{VII-67d}$$

The level scheme and the relevant transitions and relaxation constants are shown in Fig. VII.2. The main transitions are those of Eqs. (VII-67a), (VII-67b).

The analog of Eq. (VII-6a) is:

$$\frac{d}{dt}\rho_{21} = -i\omega_{21}\rho_{21} - 2i\Big\{\gamma_e B_e \cos(\omega_e t)\,\{(S_x)_{21}\rho_{11} - \rho_{22}(S_x)_{21} + (S_x)_{2,1'}\rho_{1',1} - \rho_{2,2'}(S_x)_{2',1}\}$$

$$+ \gamma_n B_n \cos(\omega_n t)\,\{(S_x)_{21}\rho_{11} - \rho_{22}(S_x)_{21} + (S_x)_{2,1'}\rho_{1',1} - \rho_{2,2'}(S_x)_{2',1}\}$$

$$+ \gamma_n B_n \cos(\omega_n t)\,\{(I_x)_{2,2'}\rho_{2',1} - \rho_{2,1'}(I_x)_{1',1}\}\Big\} + R_{21,21}\rho_{2,1} \qquad \text{(VII-68)}$$

and similar equations can be written for $\rho_{2',1'}$, $\rho_{1',1}$, $\rho_{2',2}$, $\rho_{2',1}$, and $\rho_{2,1'}$.

As in the cases treated in the previous two subsections, the time dependent fields are assumed to be weak relative to the separation between hyperfine lines. This assumption is equivalent to the condition:

$$\gamma_e B_e\,,\ \gamma_n B_n \ll |\gamma_e a_0| \qquad \text{(VII-69)}$$

Thus if one chooses, for example, $\omega_e \approx \omega_{21}$ and $\omega_n \approx \omega_{2'2}$ then transitions $1 \to 2$ and $2 \to 2'$ are excited, whereas transitions $1' \to 2'$ and $1 \to 1'$ are too far off resonance to be excited. Consequently, the relevant density matrix elements are $\rho_{2,1}$, $\rho_{2',2}$, and $\rho_{2,1'}$. Their main Fourier components are $(Z_1)_{2,1}\exp(i\omega_e t) \equiv Z_e \exp(i\omega_e t)$, $(Z_1)_{2',2}\exp(i\omega_n t) \equiv Z_n \exp(i\omega_n t)$ and $(Z_1)_{2',1}\exp(i(\omega_e+\omega_n)t) \equiv Z_x \exp(i(\omega_e+\omega_n)t)$, respectively (cf. text concerning Eqs. (VII-10) and (VII-33) above).

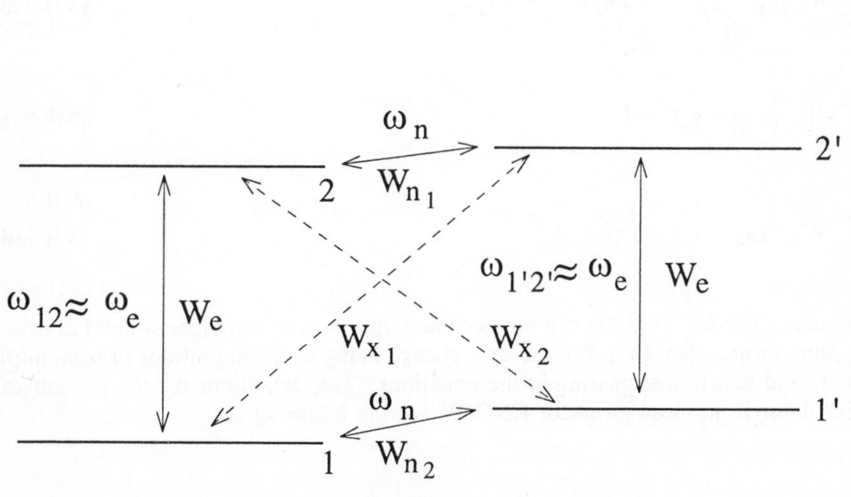

Fig. VII.2. A level scheme for ENDOR. Two electronic transitions and two nuclear transitions are coupled via relaxation mechanisms.

Performing the Fourier expansion and selecting the needed components under these conditions one obtains the following equations, analogous to Eq. (VII-20):

$$\left\{ i\Delta\omega_e + \frac{1}{(T_2)_e} \right\} Z_e + i d_e (\chi_{11} - \chi_{22}) + i d_n Z_x = i q \omega_e d_e \tag{VII-70a}$$

$$\left\{ i\Delta\omega_n + \frac{1}{(T_2)_n} \right\} Z_n + i d_n (\chi_{22} - \chi_{2'2'}) - i d_e Z_x = i q \omega_n d_n \tag{VII-70b}$$

$$\left\{ i(\Delta\omega_e + \Delta\omega_n) + \frac{1}{(T_2)_x} \right\} Z_x - i d_e Z_n + i d_n Z_e = 0 \tag{VII-70c}$$

where $\Delta\omega_e \equiv \omega_e - \omega_{21}$, $\Delta\omega_n \equiv \omega_n - \omega_{2'2}$, $d_e \equiv ½\, \gamma_e B_e$, $d_n \equiv ½\, \gamma_n B_n$, $1/(T_2)_e \equiv \mathbf{R}_{21,21}$, $1/(T_2)_n \equiv \mathbf{R}_{2'2,2'2}$ and $1/(T_2)_x \equiv \mathbf{R}_{2'1,2'1}$. The matrix elements will be expressed in terms of their real and imaginary components: $Z_\alpha \equiv Z_\alpha' + iZ_\alpha''$ (α = e, n or x), as in the equations for ELDOR. The set of equations for transition probabilities analogous to Eq. (VII-36) is:

$$\sum_{j\neq 1} W_{1j}(\chi_{11} - \chi_{jj}) = 2 d_e Z_e'' \tag{VII-71a}$$

$$\sum_{j\neq 2} W_{2j}(\chi_{22} - \chi_{jj}) = -2 d_e Z_e'' + 2 d_n Z_n'' \tag{VII-71b}$$

$$\sum_{j\neq 1'} W_{1'j}(\chi_{1'1'} - \chi_{jj}) = 0 \tag{VII-71c}$$

$$\sum_{j\neq 2'} W_{2'j}(\chi_{2'2'} - \chi_{jj}) = -2 d_n Z_n'' \tag{VII-71d}$$

Equations (VII-70), (VII-71) can be combined into a set of equations which has exactly the same form as Eq. (VII-37), the only change being in the definitions of some of the vectors and matrices appearing in the equations. The definitions for the present case which differ from those given for ENDOR are the following:

$$\boldsymbol{Z}' = \begin{pmatrix} Z_e' \\ Z_n' \\ Z_x' \end{pmatrix} \qquad \boldsymbol{Z}'' = \begin{pmatrix} Z_e'' \\ Z_n'' \\ Z_x'' \end{pmatrix} \qquad \boldsymbol{Q} = q \begin{pmatrix} \omega_e d_e \\ \omega_n d_n \\ 0 \end{pmatrix} \qquad \boldsymbol{0} = \begin{pmatrix} 0 \\ 0 \\ 0 \end{pmatrix} \tag{VII-72}$$

$$\boldsymbol{K} = \begin{pmatrix} \Delta\omega_e & 0 & d_n \\ 0 & \Delta\omega_n & -d_e \\ d_n & -d_e & \Delta\omega_e + \Delta\omega_n \end{pmatrix} \qquad \boldsymbol{R} = -\begin{pmatrix} \frac{1}{(T_2)_e} & 0 & 0 \\ 0 & \frac{1}{(T_2)_n} & 0 \\ 0 & 0 & \frac{1}{(T_2)_x} \end{pmatrix} \tag{VII-73}$$

$$\boldsymbol{d} = \begin{pmatrix} d_e & -d_e & 0 & 0 \\ 0 & d_n & 0 & -d_n \\ 0 & 0 & 0 & 0 \end{pmatrix} \qquad \boldsymbol{d}^{tr} = \begin{pmatrix} d_e & 0 & 0 \\ -d_e & d_n & 0 \\ 0 & 0 & 0 \\ 0 & -d_n & 0 \end{pmatrix} \tag{VII-74}$$

In the vector **Q** it is usually a good approximation to set the second element to zero, since $\omega_e/\omega_n = \gamma_e/\gamma_n \approx 660$ (for protons). However, at this stage the second element will be kept as it is. **W** and $\mathbf{d}^{tr}$ are modified as in Eq. (VII-37), with the obvious change that now the last row in $\mathbf{d}^{tr}$ has three 0's . The solutions are formally given by Eqs. (VII-46) - (VII-48), with the new definitions of the arrays. The most significant part of the solution is the value of Z_e'', since the (EPR) signal measured in ENDOR is proportional to this quantity. Its calculation requires the calculation of **S**, which is equal in the present case to:

$$\boldsymbol{S} = \begin{pmatrix} d_e^2\,\Omega_{e,e} & d_e d_n\,\Omega_{e,n} & 0 \\ d_e d_n\,\Omega_{n,e} & d_n^2\,\Omega_{n,n} & 0 \\ 0 & 0 & 0 \end{pmatrix} \tag{VII-75}$$

with the following definitions (omitting the prime from here on):

$$\Omega_{e,e} = 2\left\{W^{-1})_{1,1} - (W^{-1})_{1,2} - (W^{-1})_{2,1} + (W^{-1})_{2,2}\right\}$$

$$\Omega_{e,n} = 2\left\{W^{-1})_{1,2} - (W^{-1})_{1,2'} - (W^{-1})_{2,2} + (W^{-1})_{2,2'}\right\}$$

$$\Omega_{n,e} = 2\left\{W^{-1})_{2,1} - (W^{-1})_{2,2} - (W^{-1})_{2',1} + (W^{-1})_{2',2}\right\}$$

$$\Omega_{n,n} = 2\left\{W^{-1})_{2,2} - (W^{-1})_{2,2'} - (W^{-1})_{2',2} + (W^{-1})_{2',2'}\right\} \quad \text{(VII-76)}$$

The elements of the 3 × 3 matrix **F** are equal to:

$$F_{1,1} = 1 + ((T_2)_e \Delta\omega_e)^2 + (T_2)_e\left((T_2)_x d_n^2 + \Omega_{e,e} d_e^2\right)$$

$$F_{1,2} = d_e d_n (T_2)_e\left(\Omega_{e,n} - (T_2)_x\right)$$

$$F_{1,3} = d_n (T_2)_e\left((T_2)_e \Delta\omega_e + (T_2)_x (\Delta\omega_e + \Delta\omega_n)\right)$$

$$F_{2,1} = d_e d_n (T_2)_n\left(\Omega_{n,e} - (T_2)_x\right)$$

$$F_{2,2} = 1 + ((T_2)_n \Delta\omega_n)^2 + (T_2)_n\left((T_2)_x d_e^2 + \Omega_{n,n} d_n^2\right) \quad \text{(VII-77)}$$

$$F_{2,3} = -d_e (T_2)_n\left((T_2)_n \Delta\omega_n + (T_2)_x (\Delta\omega_e + \Delta\omega_n)\right)$$

$$F_{3,1} = d_n (T_2)_x\left((T_2)_e \Delta\omega_e + (T_2)_x (\Delta\omega_e + \Delta\omega_n)\right)$$

$$F_{3,2} = -d_e (T_2)_x\left((T_2)_n \Delta\omega_n + (T_2)_x (\Delta\omega_e + \Delta\omega_n)\right)$$

$$F_{3,3} = 1 + \left((T_2)_x(\Delta\omega_e + \Delta\omega_n)\right)^2 + (T_2)_x\left((T_2)_e d_n^2 + (T_2)_n d_e^2\right)$$

A particular case of special interest is that of exact resonance for both the electronic and nuclear transitions:

$$\Delta\omega_e = \Delta\omega_n = 0 \quad \text{(VII-78)}$$

Substituting this value of the off-resonance terms in the above equations, the form of **K** and of **F** is simplified. In Eq. (VII-77) for this case, all terms which depend on T_2

parameters but not on the $\Omega_{\alpha,\beta}$ parameters originate in the non-zero off-diagonal terms in **K** ("coherence terms"). Eq. (VII-48a) now results in:

$$Z_e' = (T_2)_e d_n Z_x''$$

$$Z_n' = -(T_2)_n d_e Z_x''$$

$$Z_x' = (T_2)_x \left(d_n Z_e'' - d_e Z_n''\right) \tag{VII-79}$$

and the inverse of **F** is given by:

$$F^{-1} = \frac{1}{r}\begin{pmatrix} F_{2,2} & -F_{1,2} & 0 \\ -F_{2,1} & F_{1,1} & 0 \\ 0 & 0 & \dfrac{r}{F_{3,3}} \end{pmatrix} \tag{VII-80}$$

with the definition

$$r = F_{1,1}F_{2,2} - F_{1,2}F_{2,1} \tag{VII-81}$$

It follows that

$$Z_e'' = \frac{q}{r}\left(\omega_e d_e (T_2)_e F_{2,2} - \omega_n d_n (T_2)_n F_{1,2}\right) \tag{VII-82a}$$

$$Z_n'' = \frac{q}{r}\left(\omega_n d_n (T_2)_n F_{1,1} - \omega_e d_e (T_2)_e F_{2,1}\right) \tag{VII-82b}$$

$$Z_x'' = 0 \tag{VII-82c}$$

which may be substituted into Eq. (VII-79) to give the final result for the Z_α'.

The value of Z_e'', which is essentially the ENDOR signal, may be calculated approximately as follows. First, assume $\omega_n d_n \ll \omega_e d_e$. From Eq. (VII-77) for the present case it is reasonable to assume that $F_{2,2} (T_2)_e$ and $F_{1,2} (T_2)_n$ are of the same order of magnitude, so the second term in Eq. (VII-82a) is negligible compared with the first term. Thus

$$Z_e'' = \frac{q}{F_{1,1} - F_{1,2}F_{2,1}/F_{2,2}}\left(\omega_e d_e (T_2)_e\right) \tag{VII-83}$$

The denominator in the last equation can be written as:

$$F_{1,1} - \frac{F_{1,2}\,F_{2,1}}{F_{2,2}} = 1 + (T_2)_e\left\{(T_2)_x d_n^2 + \Omega_{e,e} d_e^2\right\} - (T_2)_e d_e^2 \times$$

$$\times \frac{(T_2)_n d_n^2 \left(\Omega_{e,n} - (T_2)_x\right)\left(\Omega_{n,e} - (T_2)_x\right)}{1 + (T_2)_n\left\{(T_2)_x d_e^2 + \Omega_{n,n} d_n^2\right\}} =$$

$$\equiv 1 + (T_2)_e d_e^2\{\Omega_{e,e} - \xi\} + (T_2)_e (T_2)_x d_n^2 \tag{VII-84}$$

so that

$$Z_e'' = \frac{q\,(T_2)_e \omega_e d_e}{1 + (T_2)_e d_e^2\{\Omega_{e,e} - \xi\} + (T_2)_e (T_2)_x d_n^2} \tag{VII-85}$$

In this form it is easy to see the relation between the ENDOR signal for the present case of double resonance, and the result for a simple saturated line (Eqs. (VII-22), (VII-23) above). As $\Omega_{e,e}$ is a T_1-like parameter, the term $(T_2)_e\,(\Omega_{e,e} - \xi)\,d_e^2$ is the analog of the term $d^2\,T_1\,T_2$ there. The presence of ξ reduces the value of the denominator in Eq. (VII-85), so it enhances the ENDOR signal. The additional term (in the denominator) $(T_2)_e\,(T_2)_x\,d_n^2$ somewhat reduces the signal, but if $\Omega_{e,e}\,d_e^2 \gg (T_2)_x\,d_n^2$ the effect of this term may be neglected. Making also the assumption: $\Omega_{n,n}\,d_n^2 \gg (T_2)_x\,d_e^2$ it is possible to neglect the term $(T_2)_x\,d_e^2$ in the denominator of ξ. In a similar manner, we assume that $\Omega_{e,n}$, $\Omega_{n,e} \gg (T_2)_x$ so that ξ is given by

$$\xi \approx \frac{(T_2)_n d_n^2 (\Omega_{e,n})^2}{1 + (T_2)_n \Omega_{n,n} d_n^2} \tag{VII-86}$$

where $\Omega_{e,n} = \Omega_{n,e}$ as in the Example above. In this situation the measured signal is independent of the cross-relaxation parameter $(T_2)_x$ and is also independent of the **K** matrix. If $d_e^2 \approx d_n^2$, the conditions for this result can be summarized as:

$$\Omega_{e,e}\,,\ \Omega_{n,n}\,,\ |\,\Omega_{e,n}\,| \gg (T_2)_x \tag{VII-87}$$

(d). Saturation and relaxation in the slow motional regime

The case of a single line, treated in subsection (a) above, will now be re-examined for the case of slow tumbling. Eq. (VII-1) is now replaced by the stochastic Liouville equation:

$$\frac{\partial}{\partial t}\rho(\Omega,t) = \frac{1}{i\hbar}[H(\Omega,t),\rho(\Omega,t)] + \boldsymbol{R}\left(\rho(\Omega,t) - \rho_{(eq.)}\right) - \boldsymbol{\Gamma}_\Omega\left(\rho(\Omega,t) - \rho_{(eq.)}\right) \quad \text{(VII-88)}$$

where

$$H(\Omega,t) = H_0 + H_1(\Omega) + H_2(t) \quad \text{(VII-89)}$$

using the same notation as in the previous chapters. The superoperator **R** is included here in order to account for orientation independent relaxation mechanisms. The equation can be solved by the method of eigenfunctions, as shown in Sec. V.3. The irradiation term $\mathbf{H}_2(t)$ for a system with a single line spectrum is given by Eq. (VII-5) above, which is the same (up to a numerical factor) as Eq. (V-98). Thus the terms proportional to ω_1 in Eqs. (V-117), (V-118) already include the effect of $\mathbf{H}_2(t)$. In Chapter V it was mentioned that the ω_1 term in Eq. (V-116) is usually neglected, because $\omega_1 \ll \omega_0$. When saturation is present, however, this approximation does not hold any more. Eq. (VI-63) was obtained for such a situation, but with the additional assumption that $\mathbf{H}_1(\Omega)$ consists only of an axially symmetric g-tensor interaction. This case will now be examined with the formalism developed in the present Chapter for multiple resonance.

The two equations leading to Eq. (VI-63), viz. Eqs. (VI-59) and (VI-61), can be written as:

$$\left\{i\,\Delta\omega + \gamma_{LMO} + \frac{1}{T_2}\right\}\left(Z'_{LM} + iZ''_{LM}\right) - i\sum_{L'M'} F(L,M,L'M')\left(Z'_{L'M'} + iZ''_{L'M'}\right)$$

$$+\,i\,d\left((\chi_{LM})_{11} - (\chi_{LM})_{22}\right) = i\,q'\,\omega_0\,d\,\delta_{L,0}\,\delta_{M,0} \quad \text{(VII-90)}$$

$$\frac{1}{2}\left[\gamma_{LMO} + \frac{1}{T_1}\right]\left((\chi_{LM})_{11} - (\chi_{LM})_{22}\right) = 2d\,Z''_{LM} \quad \text{(VII-91)}$$

with $Z_{LM} \equiv (\mathbf{C}_{(1)LMO})_{ij}$, $q' \equiv q/(8\pi^2)^{1/2}$ and F(L,M,L'M') is defined in Eq. (VI-60). The relaxation terms $1/T_1$ and $1/T_2$ were included here, as in Eq. (VI-63), to account for the effect of relaxation processes other than rotational diffusion. If the g-tensor is axially symmetric, only the elements with M = M' = 0 are involved, and then the equations can be written in matrix form as:

$$\begin{bmatrix} X_{0,0} & X_{0,2} & 0 & 0 & \cdots & \cdots \\ X_{2,0} & X_{2,2} & X_{2,4} & 0 & \cdots & \cdots \\ 0 & X_{4,2} & X_{4,4} & X_{4,6} & 0 & \cdots \\ 0 & 0 & X_{6,4} & X_{6,6} & X_{6,8} & \cdots \\ \cdots & \cdots & \cdots & \cdots & \cdots & \cdots \\ \cdots & \cdots & \cdots & \cdots & \cdots & \cdots \end{bmatrix} \begin{bmatrix} C_0 \\ C_2 \\ C_4 \\ C_6 \\ \cdots \\ \cdots \end{bmatrix} = \begin{bmatrix} V_0 \\ V_2 \\ V_4 \\ V_6 \\ \cdots \\ \cdots \end{bmatrix} \tag{VII-92}$$

Each "matrix element" $X_{L,L'}$ is actually a 3 by 3 sub-matrix of **X** , and each "vector element" C_L , V_L is a 3-dimensional column sub-vector of **C** or **V**. The general sub-matrix is:

$$X_{L,L'} = \begin{bmatrix} (\gamma_{L00} + \frac{1}{T_2})\,\delta_{L,L'} & -\Delta\omega\,\delta_{L,L'} + F(L,0,L',0) & 0 \\ \Delta\omega\,\delta_{L,L'} - F(L,0,L',0) & (\gamma_{L00} + \frac{1}{T_2})\,\delta_{L,L'} & d\,\delta_{L,L'} \\ 0 & -2d\,\delta_{L,L'} & \tfrac{1}{2}(\gamma_{L00} + \frac{1}{T_1})\,\delta_{L,L'} \end{bmatrix} \tag{VII-93}$$

and the sub-vectors are:

$$C_L = \begin{bmatrix} Z'_{L,0} \\ Z''_{L,0} \\ \Delta\chi_{L,0} \end{bmatrix} \qquad V_L = \delta_{L,0} \begin{bmatrix} 0 \\ q'\omega_0 d \\ 0 \end{bmatrix} \tag{VII-94}$$

with γ_{L00} = RL(L+1) and , from Eq. (VI-60):

$$F(L,0,L',0) = \frac{F}{\hbar}\sqrt{\frac{8\pi^2}{5}}\,\sqrt{(2L+1)(2L'+1)}\begin{pmatrix} L & 2 & L' \\ 0 & 0 & 0 \end{pmatrix}^2 \tag{VII-95}$$

Here F is the interaction constant $F^{(2,0)}$ and explicit values for the 3-j coefficients are given in Eqs. (VI-11), (VI-12).

The set of equations in Eq. (VII-92) is the generalization of Eq. (VII-21) to the case in which rotational diffusion is present. As for non-saturated lines, the major change in the presence of such motion is that the rigid limit equations are essentially repeated for each set of values of the rotation matrix indices (L,M,K), and it is the relaxation T_2-like parameter which changes depending on these indices. The equations for different (L,M,K) values are coupled through magnetic interaction parameters. As there is an infinite number of values of such indices, the set of equations is infinite, and can only be solved by truncation. Such truncation is allowed when the off-diagonal coupling terms are small compared with the diagonal elements of the matrix. The approximate condition for such truncation is that the maximum value of L included in the calculation satisfies: $L^2 \gg F/R$. In general the dimension of the truncated equations may be quite large, and therefore the equations can only be solved numerically. Analytical solutions are possible in extreme situations, when the motion is either very fast or very slow.

If the motion is extremely slow the dimensions of the equations become prohibitively large, but then it is physically more reasonable to describe the effect of motion phenomenologically by means of T_2, T_1 parameters. These parameters express implicitly the effect of motion. One then has to solve just the equations described in subsection VII.1.(a) above.

If the motion is very fast, i.e., $R \gg F$ then the truncation condition mentioned above is satisfied even for L = 2. In that case one only needs in **X** the blocks $X_{0,0}$, $X_{0,2}$, $X_{2,0}$ and $X_{2,2}$, and in the vectors one only needs the blocks C_0, C_2 and V_0, V_2. In Eq. (VII-92) for this case the matrix and the vectors are:

$$\boldsymbol{X} = \begin{bmatrix} \frac{1}{T_2} & -\Delta\omega + f(0,0) & 0 & 0 & f(0,2) & 0 \\ \Delta\omega - f(0,0) & \frac{1}{T_2} & d & -f(0,2) & 0 & 0 \\ 0 & -2d & \frac{1}{2T_1} & 0 & 0 & 0 \\ 0 & f(0,2) & 0 & \frac{1}{T_2}+6R & -\Delta\omega + f(2,2) & 0 \\ -f(0,2) & 0 & 0 & \Delta\omega - f(2,2) & \frac{1}{T_2}+6R & d \\ 0 & 0 & 0 & 0 & -2d & \frac{1}{2T_1}+3R \end{bmatrix} \qquad \text{(VII-96)}$$

$$\boldsymbol{C} = \begin{bmatrix} Z'_0 \\ Z''_0 \\ \Delta\chi_0 \\ Z'_2 \\ Z''_2 \\ \Delta\chi_2 \end{bmatrix} \qquad \boldsymbol{V} = \begin{bmatrix} 0 \\ q'\omega_0 d \\ 0 \\ 0 \\ 0 \\ 0 \end{bmatrix} \qquad \text{(VII-97)}$$

In these equations the following definitions are used:

$$f(0,0) \equiv F(0,0,0,0) = 0 \qquad f(2,2) \equiv F(2,0,2,0) = \frac{2}{7}\left[\frac{F}{\hbar}\sqrt{8\pi^2/5}\right]$$

$$f(0,2) \equiv F(0,0,2,0) = \frac{1}{5}\left[\frac{F}{\hbar}\sqrt{8\pi^2/5}\right] \tag{VII-98}$$

From the simple form of **V** in Eq. (VII-97) it follows that the j'th element of the 6-dimensional vector **C** is equal to

$$C_j = \left(X^{-1}\right)_{j,2} V_2 \tag{VII-99}$$

Thus if the inverse of **X** is found only its second column is necessary. Using a standard method of matrix inversion (e.g., Cramer's determinant rule) one may invert **X** and solve for **C**. In order to express as simply as possible the results of this procedure, it is convenient to define:

$$D \equiv \det(X) \qquad r \equiv \frac{1}{T_2} \qquad W \equiv \frac{1}{2T_1} \qquad k \equiv 3R \qquad \Delta \equiv \Delta\omega$$

$$b \equiv f(0,2) \qquad c \equiv f(2,2) \qquad r' \equiv r + 2k \qquad W' \equiv W + k \tag{VII-100}$$

and then define further:

$$\alpha \equiv (r')^2 + (\Delta - c)^2 \qquad \beta \equiv W(r^2 + \Delta^2) + 2rd^2 \qquad \gamma \equiv r'd^2$$

$$\delta \equiv rr'W \qquad \epsilon \equiv r'b^2 \tag{VII-101}$$

The calculation then results in expressions for the determinant:

$$D = W'\left\{\alpha\beta + b^2\left\{2(\gamma+\delta) + W\left(b^2 - 2\Delta(\Delta - c)\right)\right\}\right\} + 2\beta\gamma + 2rWb^2d^2 \tag{VII-102}$$

and for the relevant elements of $\mathbf{X}^{-1}$:

$$(X^{-1})_{1,2} = \frac{W}{D}\left\{W'\{\alpha\Delta - b^2(\Delta - c)\} + 2\gamma\Delta\right\}$$

$$(X^{-1})_{2,2} = \frac{W}{D}\left\{W'(\alpha r + \epsilon) + 2\gamma r\right\}$$

$$(X^{-1})_{3,2} = \frac{2d}{D}\left\{W'(\alpha r + \epsilon) + 2\gamma r\right\}$$

$$(X^{-1})_{4,2} = \frac{bW}{D}\left\{W'\{\Delta(\Delta - c) - b^2 - rr'\} - 2rd^2\right\}$$

$$(X^{-1})_{5,2} = \frac{bW}{D} W'\{r(\Delta - c) + r'\Delta\}$$

$$(X^{-1})_{6,2} = \frac{2bdW}{D} W'\{r(\Delta - c) + r'\Delta\} \tag{VII-103}$$

The power absorption is then proportional to

$$Z_0'' = q'\omega_0 \frac{dW}{D}\left\{W'(\alpha r + \epsilon) + 2\gamma r\right\} \tag{VII-104}$$

For fast motion, the rate constant is large compared with either the phenomenological relaxation constants or the interaction constants:

$$k \gg r, W \qquad and \qquad k \gg \Delta, b, c \tag{VII-105}$$

so that

$$D \approx k\left\{4k^2\beta + b^2\left\{4k(d^2 + rW) + W\left(b^2 - 2\Delta(\Delta - c)\right)\right\}\right\}$$

$$+ 4kd^2\beta + 2rWb^2d^2 \tag{VII-106}$$

$$Z_0'' \approx q'\omega_0 \frac{dW}{D}\left\{k(4k^2 r + 2kb^2) + 4kd^2 r\right\} \tag{VII-107}$$

For appreciable saturation the constant d, which is similar in magnitude to ω_1, is of the same order of magnitude as b,c (and Δ, if $\Delta \neq 0$). Also, since rotational diffusion is assumed here to be the chief relaxation mechanism, T_2 and T_1 are quite long, giving rise to narrow lines in the absence of diffusion. In such a situation:

$$d \gg \frac{1}{T_2} > \frac{1}{T_1} \qquad and \qquad k \gg d \tag{VII-108}$$

and therefore

$$D \approx 4k^2d^2\left(2kr + b^2\right) \tag{VII-109}$$

$$Z_0'' \approx q'\omega_0 \frac{dW}{2d^2} \tag{VII-110}$$

which is the same as the limit of Eq. (VII-22b) for strong saturation. Therefore the result for very fast motion in the presence of strong saturation is not sensitive to the type of motional mechanism causing relaxation.

VII.2 Two Dimensional Magnetic Resonance

Numerous multiple pulse methods have been developed in NMR. These methods have greatly enhanced the capabilities and the versatility of NMR. For some time this progress did not reach EPR, due to the difference of three orders of magnitude in the time scales involved. However, in recent years the technical barrier has been overcome, and several pulsed techniques have been employed to a variety of applications. Some of these will be treated here, in order to show how the SLE formalism can be used in this context.

(a) Calculation of the density matrix in a multiple pulse experiment

The time span of a multiple pulse experiment (Fig. VII.3) can be divided into sections, each of which is either an interval of free evolution of the spins (no irradiation) or a period in which intense irradiation (a pulse) is applied to the system. The signal is measured for a set of time values during the last interval t_n, starting either "immediately" (after the instrumental "dead-time") after the last pulse or from a certain instant later on. By varying the length of time intervals t_1, ..., t_{n-1} from one signal measurement to the next one obtains a signal which is a function of two or more time variables, as well as variables characterizing the pulses.

Fourier transforming one or more time variables to the frequency domain, the result is a spectrum which depends on one or more frequency variables, and possibly also on one or more time variables. For example, for a three pulse sequence one may keep the length t_2 of the second interval fixed, varying t_1 from one signal acquisition to the next. Fourier transforming with respect to t_3 (the time during which the measurement is made) and t_1 one obtains a spectrum which is a function of two frequency variables ν_1, ν_2 and also depends parametrically on the value of t_2.

The key to calculating the effect of such an experiment is to calculate separately the outcome of each section of the experiment, since the end result for one section is the initial condition for the next section. One thus has to deal at each stage either with the effect of free evolution for a given time interval, or with the result of operating with a single pulse. Calculating the evolution of the density matrix section by section one can get the final form of the density matrix, from which the final magnetization can be found.

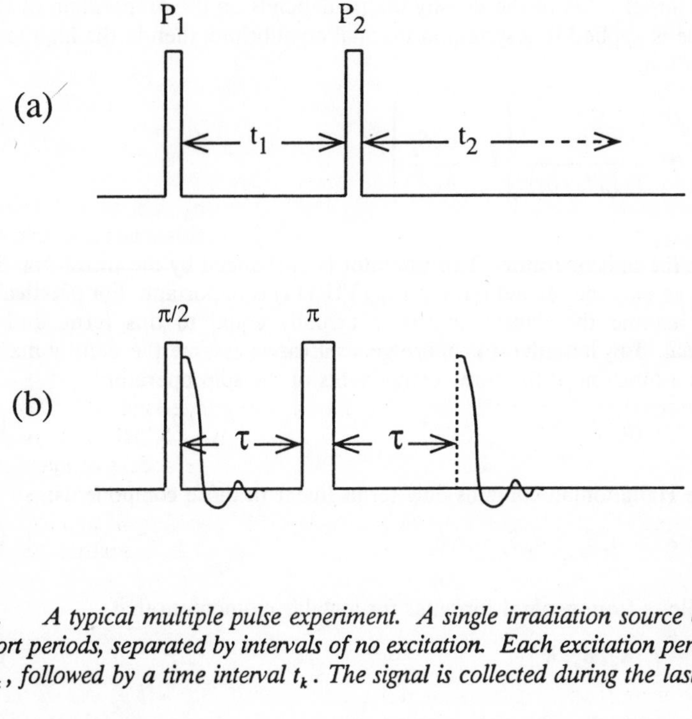

Fig. VII.3. A typical multiple pulse experiment. A single irradiation source operates at several short periods, separated by intervals of no excitation. Each excitation period defines a pulse P_k, followed by a time interval t_k. The signal is collected during the last interval.

Calculation during the pulses can be simplified if they operate for very short times, because then relaxation during the pulses can be neglected. For example, suppose an oscillating magnetic field is applied along **x** during a period τ, with a constant amplitude B_1, to a system of isolated spins **S** which is originally at equilibrium. Its interaction with the spins is given by

$$H_1 = \gamma B_1 \cos(\omega t)\, S_x \tag{VII-111}$$

This is added to the effect of the constant magnetic field, which is parallel to the **z** axis. Thus, for the duration of the pulse the Hamiltonian is equal in the rotating frame to:

$$H = H_0 + H_1 = \Delta\omega\, S_z + \omega_1 S_x \tag{VII-112}$$

where $\Delta\omega \equiv \omega - \omega_0$, $\omega_0 \equiv \gamma B_0$ and $\omega_1 \equiv \gamma B_1$. Neglecting relaxation, the density matrix is changed by the pulse through a unitary transformation:

$$\rho(t=\tau) = \exp\left[-\frac{i}{\hbar}\int_0^\tau H\,dt\right] \rho(t=0) \exp\left[\frac{i}{\hbar}\int_0^\tau H\,dt\right] \qquad \text{(VII-113)}$$

In general the amplitude ω_1 of the pulse may be a function of time, but here it is assumed for simplicity to be constant. Then the integral in the exponent is equal to: $\mathbf{H}\tau$. The initial value of the density matrix depends on the preparation of the system. If the pulse is applied to a system in thermal equilibrium, then in the high temperature approximation:

$$\rho(t=0) \approx \frac{1}{Tr\{\rho(t=0)\}}\left\{I - \frac{\omega_0 S_z}{k_B T}\right\} \qquad \text{(VII-114)}$$

where **I** is the unit operator. This operator is unchanged by the transformation in Eq. (VII-113), so only the second term in Eq. (VII-114) is important. For practical purposes one may assume the density matrix is actually equal to this term, and therefore proportional to $\mathbf{S}_z$ initially. For more general cases, assume the density matrix can be written as a function of the three components of the spin operator:

$$\rho(t=0) = F(S_x, S_y, S_z) \qquad \text{(VII-115)}$$

If also the Hamiltonian contains only terms linear in those components:

$$H = a \cdot S = a_x S_x + a_y S_y + a_z S_z \qquad \text{(VII-116)}$$

then it follows from general formulas for angular momentum that

$$\rho(t=\tau) = F(K_x, K_y, K_z) \qquad \text{(VII-117)}$$

where the vectorial operator **K** is equal to:

$$K = u(u \cdot S)(1 - \cos(v)) + S\cos(v) + S \times u(\sin(v)) \qquad \text{(VII-118)}$$

with $\mathbf{v} \equiv \hbar^{-1}\,{}_0\int^\tau \mathbf{a}\,dt$ (= $\mathbf{a}\tau/\hbar$ for a pulse of constant amplitude), $v \equiv |\mathbf{v}| = \sqrt{(\mathbf{v}\cdot\mathbf{v})}$ and $\mathbf{u} \equiv \mathbf{v}/v$. In the present case, $\mathbf{a} = (\hbar\omega_1, 0, \hbar\Delta\omega)$ so that $\mathbf{v} = (\tau\omega_1, 0, \tau\Delta\omega)$ and $v = \tau\{(\omega_1)^2 + (\Delta\omega)^2\}^{1/2} \equiv \tau\omega_{\text{eff}}$. Then $\mathbf{u} = (\sin(\phi), 0, \cos(\phi))$ where ϕ is defined by: $\cos(\phi) = (\Delta\omega)/\omega_{\text{eff}}$ and $\sin(\phi) = \omega_1/\omega_{\text{eff}}$. It is also convenient to define $v_x = \cos(v)$ and $v_y = \sin(v)$. The last equation can be written with this notation as:

$$\begin{bmatrix} K_x \\ K_y \\ K_z \end{bmatrix} = \begin{bmatrix} \sin^2(\phi) + v_x\cos^2(\phi) & v_y\cos(\phi) & (1-v_x)\sin(\phi)\cos(\phi) \\ -v_y\cos(\phi) & v_y & v_y\sin(\phi) \\ (1-v_x)\sin(\phi)\cos(\phi) & -v_y\sin(\phi) & \cos^2(\phi)+v_x\sin^2(\phi) \end{bmatrix} \begin{bmatrix} S_x \\ S_y \\ S_z \end{bmatrix} \qquad \text{(VII-119)}$$

This is a rotation of the spin vector, and the density matrix is related (through Eq. (VII-117)) to this rotated vector exactly as the initial density matrix was related to the spin vector itself, which is independent of this rotation.

The magnetization is calculated by taking the trace of the density matrix multiplied by some component of the spin operator. The rotation described here results in a corresponding rotation of the magnetization. This is the reason for the well-known fact, that pulses act by rotating the magnetization vector. Pulses are usually designed to rotate the magnetization by specific angles, in most cases either $\pi/2$ or π. This means that, given a certain off-resonance value $\Delta\omega$, the rotation angle $v = \tau\omega_{eff}$ is equal to $\pi/2$ or π. For these angles the matrix in Eq. (VII-119) has a very simple form. For example, if a $\pi/2$ pulse is applied exactly on resonance ($\Delta\omega = 0$) then

$$K_x = S_x \qquad K_y = S_y + S_z \qquad K_z = -S_y \qquad \text{(VII-120)}$$

and then the equilibrium initial condition leads to

$$\rho\,(t=\tau) = q\,S_y \qquad \text{(VII-121)}$$

where $q = \omega_0/(k_B T)$. For more general cases the result is less simple, but may still be calculated directly from Eq. (VII-119).

The situation is more complicated in the intervals between pulses. In those intervals $\mathbf{H_1}$ does not operate, but the time is usually sufficiently long that relaxation effects must be considered. Instead of the unitary transformation in Eq. (VII-113) which solves an ordinary Liouville-von Neumann equation, one has to solve the more general problem of Eq. (V-4). The solution must lead to a decay in time, and is formally described by a non-unitary transformation (see beginning of Chap. V). The convenient way to combine the two types of transformations, for the pulses and for the separating intervals, depends mainly on the type of relaxation process involved. Two different types of relaxation will be discussed in the next two subsections.

(b) Multiple pulse experiments on chemically exchanging systems

The mathematical treatment of relaxation by chemical exchange is relatively simple, so it is convenient to describe it by an exponential transformation (cf. Eq. (V-3)). On the basis of Eq. (III-64) one may write:

$$\rho\,(t) = \exp(Lt)\,\rho\,(0) \qquad \text{(VII-122)}$$

where $\boldsymbol{\rho}$ is regarded as a vector. The elements of this vector may be arranged in a single column according to their quantum order (see subsection I.8.(d) and the Examples in

Sec. III.3). However, since from an algebraic viewpoint also matrices form a vector space, one may write ρ in the conventional form of matrices, i.e., as a two dimensional array, even though it is regarded as a vector for Eq. (VII-122). The exponential is a four-index Liouville space operator, defined by the operator in the exponent:

$$L = L_0 + L_1 + R \tag{VII-123}$$

where

$$L_0 = \frac{i}{\hbar}\{H_0 \otimes I - I \otimes H_0\} \tag{VII-124a}$$

$$L_1 = \frac{1}{\tau}\left\{\frac{1}{2}\left(X \otimes X^{-1} + X^{-1} \otimes X\right) - I \otimes I\right\} \tag{VII-124b}$$

$$R = -\frac{1}{T_2}\{I \otimes I\} - \left[\frac{1}{T_1} - \frac{1}{T_2}\right] R_D \tag{VII-124c}$$

In the last equation $\mathbf{R_D}$ is a Liouville space operator, operating only on the diagonal of ρ:

$$\left(R_D\right)_{ij,mn} = \delta_{i,m}\,\delta_{j,n}\,\delta_{i,j} \tag{VII-125}$$

$\mathbf{H_0}$ is the static Hamiltonian, which may include intra-molecular interaction terms. With these expressions it is possible to follow the free evolution of the system, i.e., its evolution in the intervals between pulses.

For the pulses one has to find the corresponding unitary transformations. In a multiple pulse experiment, different pulses may not be identical, so the j'th pulse is characterized by its Hamiltonian:

$$H^{(j)} = H_0 + H_1^{(j)} \tag{VII-126}$$

which acts during a time period τ_j. The related propagator is:

$$U^{(j)} = \exp\left[-\frac{i}{\hbar}\int_0^{\tau_j} H^{(j)} dt\right] = \exp\left[-\frac{i}{\hbar} H^{(j)} \tau_j\right] \tag{VII-127}$$

The second equality in this equation follows from the assumption that $\mathbf{H}^{(j)}$ is time independent. With each of these propagators the density matrix is transformed as:

$$\rho\,(t = t_0 + \tau_j) = U^{(j)}\,\rho\,(t = t_0)\,U^{(j)\dagger} \tag{VII-128}$$

For example, suppose one performs a two pulse experiment, in which the time interval between the two pulses P_1 and P_2 is t_1, and the time from the second pulse (to measurement) is t_2. A short notation for such a sequence is: P_1 - t_1 - P_2 - t_2. The density matrix at the time of measurement is equal to:

$$\rho_{final} = \exp(Lt_2)\left\{U^{(2)}\left\{\exp(Lt_1)\{U^{(1)}\rho(t=0)\,U^{(1)\dagger}\}\right\}U^{(2)\dagger}\right\} \tag{VII-129}$$

Here $\rho(t = 0)$ is the initial density matrix, which is normally an equilibrium one. Eq. (VII-129) is calculated step by step, from the inner brackets out. The first stage is operating on $\rho(t = 0)$ with $U^{(1)}$, then the result is the initial density matrix for the application of Eq. (VII-122) with $t = t_1$, and so on.

Each pulse transformation is convenient to calculate in a basis in which $H^{(i)}$ has a simple form. In practice this occurs in a basis in which the total spin operators S^2, S_z are diagonal. For the free evolution periods the calculation is easiest in a basis which diagonalizes H_0. In many relevant cases in magnetic resonance H_0 is diagonal in the same basis in which S^2 and S_z are diagonal, so all stages of the calculation can be performed in the same basis.

Having chosen a convenient basis, one has to diagonalize the total Liouville superoperator **L** by a transformation of the form

$$\Theta^{-1} L\,\Theta = \Lambda \tag{VII-130}$$

A general element of the diagonal supermatrix Λ is:

$$\Lambda_{ij,mn} = \delta_{i,m}\,\delta_{j,n}\,\lambda_{ij} \tag{VII-131}$$

so a general element of the Liouvillian is equal to:

$$L_{ij,mn} = \sum_{k,l} \Theta_{ij,kl}\,\lambda_{kl}\,\Theta^{-1}{}_{kl,mn} \tag{VII-132}$$

It then follows (see Chap. I) that

$$\left\{\exp(Lt)\right\}_{ij,mn} = \sum_{k,l} \Theta_{ij,kl}\exp(\lambda_{kl}t)\,\Theta^{-1}{}_{kl,mn} \tag{VII-133}$$

Applying this, for example, to the two pulse experiment mentioned above, a density matrix element is equal to:

$$\{\rho_{final}\}_{i,j} = \sum \Theta_{ij,kl}\exp(\lambda_{kl}t_2)\,\Theta^{-1}{}_{kl,mn}\left\{U^{(2)}{}_{mp}\left\{\Theta_{pq,rs}\exp(\lambda_{rs}t_1)\,\Theta^{-1}{}_{rs,tu}\,\times\right.\right.$$

$$\left.\left.\times\left\{U^{(1)}{}_{tw}\,\rho(0)_{wx}\left(U^{(1)}{}_{ux}\right)^*\right\}\right\}\left(U^{(2)}{}_{nq}\right)^*\right\} \tag{VII-134}$$

The summation is over all repeated indices i.e., all indices except for i,j. The most difficult part in the calculation is the diagonalization of **L**. However, as mentioned above, this task is simplified using the basis of H_0, because in that basis:

$$\left(L_0\right)_{ij,mn} = -\frac{i}{\hbar}\left(E_i\,\delta_{i,m}\,\delta_{j,n} - E_j\,\delta_{i,m}\,\delta_{j,n}\right) = -i\,\omega_{ij}\,\delta_{i,m}\,\delta_{j,n} \tag{VII-135}$$

Thus all off-diagonal elements in **L** originate in the exchange term $\mathbf{L_1}$. The actual diagonalization is carried out numerically, unless the dimensions of the system are very small.

An important experimental method, developed originally in NMR specifically for studying chemical exchange, is known as "2D (two dimensional) exchange NMR". In this method one performs a three pulse experiment: P_1 - t_1 - P_2 - T - P_3 - t_2 (see Fig. VII.3). The technique, sometimes referred to by the acronym NOESY (for Nuclear Overhauser Effect Spectroscopy), has also been used in EPR under the name of 2D ELDOR (due to its similarity to ELDOR) for studying Heisenberg exchange. In this method one measures the signal for many values of t_2 , and the whole sequence is repeated for many values of t_1 . The interval T is kept constant in these repetitions of the sequence. The resulting values of the magnetization are arranged in a two dimensional array $M(t_1, t_2)$ which is Fourier transformed in both variables, yielding the frequency domain signal $M(\omega_1, \omega_2)$. The signal depends implicitly on the occurrence of exchange during the intermediate interval T. The two dimensional frequency domain spectrum has peaks only along the diagonal of the (ω_1, ω_2) plane if no exchange occurs. However, if exchange does occur then off-diagonal peaks appear as well, and their location and shape contain information on the exchange process.

Example VII.3: A two spin system in a liquid crystalline solution

Assume the 2D exchange NMR sequence is applied, with on resonance pulses, to a system of two S= ½ spins undergoing chemical exchange. The three pulses are identical (all with the same phase), so

$$H^{(1)} = H^{(2)} = H^{(3)} = H_0 + \omega_1 S_x \tag{E-1}$$

and

$$U^{(1)} = U^{(2)} = U^{(3)} = \exp\left[-\frac{i}{\hbar}\tau\left(H_0 + \omega_1 S_x\right)\right] \tag{E-2}$$

The internal Hamiltonian is

$$H_0 = \delta S_{z1} - \delta S_{z2} + \frac{1}{2}D\left(3\,S_{z1}S_{z2} - S_1\cdot S_2\right) \tag{E-3}$$

Here δ is the chemical shift and D is a dipolar interaction, which is not averaged to zero due to the ordering effect of the liquid crystal.

In the spin basis $\{|++\rangle, |+-\rangle, |-+\rangle, |--\rangle\}$ the internal Hamiltonian has the matrix

$$H_0 = \begin{pmatrix} D & 0 & 0 & 0 \\ 0 & -D+2\delta & -\tfrac{1}{4}D & 0 \\ 0 & -\tfrac{1}{4}D & -D-2\delta & 0 \\ 0 & 0 & 0 & D \end{pmatrix} \tag{E-4}$$

This can be diagonalized by transforming to a new basis: $\{|++\rangle, |2\rangle, |3\rangle, |--\rangle\}$ where

$$|2\rangle = a_2 \left\{ \frac{1}{4} D\,|+-\rangle + (2\delta - r)\,|-+\rangle \right\} \tag{E-5a}$$

$$|3\rangle = a_3 \left\{ \frac{1}{4} D\,|+-\rangle + (2\delta + r)\,|-+\rangle \right\} \tag{E-5b}$$

with $r = ((\tfrac{1}{4}D)^2 + (2\delta)^2)^{1/2}$ and

$$a_2 = \frac{1}{\sqrt{2r^2 - 4\delta r}} \qquad a_3 = \frac{1}{\sqrt{2r^2 + 4\delta r}} \tag{E-6}$$

The eigenvectors of $\mathbf{H_0}$ are the vectors comprising the new basis, and the corresponding eigenvalues are:

$$\lambda_{++} = D \qquad \lambda_{+-} = -D + r \qquad \lambda_{-+} = -D - r \qquad \lambda_{--} = D \tag{E-7}$$

In this new basis one may use Eq. (VII-136) with

$$\hbar\omega_{12} = 2D - r \qquad \hbar\omega_{13} = 2D + r \qquad \hbar\omega_{14} = 0$$

$$\hbar\omega_{23} = 2r \qquad \hbar\omega_{24} = -2D + r \qquad \hbar\omega_{34} = -2D - r \tag{E-8}$$

The exchange process is asssumed to cause effectively an exchange of spin 1 with spin 2, so the non-zero elements of **X** (in the original basis) are:

$$\langle ++|X|++\rangle = \langle --|X|--\rangle = 1 \quad ; \quad \langle +-|X|-+\rangle = \langle -+|X|+-\rangle = 1 \tag{E-9}$$

The matrix of **X** in the original basis is thus

$$X = \begin{pmatrix} 1 & 0 & 0 & 0 \\ 0 & 0 & 1 & 0 \\ 0 & 1 & 0 & 0 \\ 0 & 0 & 0 & 1 \end{pmatrix} \tag{E-10}$$

When $\mathbf{H_0}$ is diagonalized, also **X** is transformed to the same basis. According to Eqs. (E-5),(E-6) the transformation of any matrix **A** is:

$$A \rightarrow A' \equiv T^{-1} A T \tag{E-11}$$

with the transformation

$$T = \begin{pmatrix} 1 & 0 & 0 & 0 \\ 0 & \frac{D}{4} a_2 & \frac{D}{4} a_3 & 0 \\ 0 & (2\delta - r)\, a_2 & (2\delta + r)\, a_3 & 0 \\ 0 & 0 & 0 & 1 \end{pmatrix} \tag{E-12}$$

Since the matrix of $\mathbf{H_0}$ is real symmetric, $\mathbf{T}^{-1}$ is $\mathbf{T}^{t}$ (the transpose of $\mathbf{T}$). Therefore

$$X' = \begin{pmatrix} 1 & 0 & 0 & 0 \\ 0 & -\frac{1}{4}\frac{D}{r} & 2\frac{\delta}{r} & 0 \\ 0 & 2\frac{\delta}{r} & \frac{1}{4}\frac{D}{r} & 0 \\ 0 & 0 & 0 & 1 \end{pmatrix} \tag{E-13}$$

The 4 × 4 Hilbert space matrices can now be used to construct the 16 × 16 Liouville matrix **L** according to Eqs. (VII-124)-(VII-126). After diagonalizing it one can derive the complete solution for $\boldsymbol{\rho}$, and this leads also to the value of the measured magnetization.

****

(c) Multiple pulse experiments on systems with rotational diffusion

When rotational diffusion is present, the evolution of the density matrix between pulses is described by:

$$\frac{\partial}{\partial t}\rho(\Omega,t) = \frac{1}{i\hbar}[H_0 + H_1(\Omega), \rho(\Omega,t)] - \Gamma_\Omega\left(\rho(\Omega,t) - \rho_{(eq.)}\right) \tag{VII-136}$$

Since rotational diffusion is usually fast on the NMR time scale, the relaxation treatment discussed here is usually applicable to EPR. In common applications, the internal Hamiltonian $\mathbf{H_0} + \mathbf{H_1}(\Omega)$ refers to a single electron, which may or may not interact with some nucleus or nuclei. Here $\mathbf{H_0}$ may include, in addition to the isotropic part of Zeeman interactions, also the isotropic part of hyperfine interactions. $\mathbf{H_1}(\Omega)$ includes the anisotropic part of the Zeeman interactions and of intra-molecular interactions. It is of the general form of Eq. (V-97). During a pulse the density matrix evolves as:

$$\frac{\partial}{\partial t}\rho(\Omega,t) = \frac{1}{i\hbar}[H_0 + H_1(\Omega) + H_2(t), \rho(\Omega,t)] - \Gamma_\Omega\left(\rho(\Omega,t) - \rho_{(eq.)}\right) \qquad \text{(VII-137)}$$

$\mathbf{H}_2(t)$ is the irradiation field, which may be taken as:

$$H_2(t) = 2\,\omega_1\, S_x \cos(\omega t) \qquad \text{(VII-138)}$$

An EPR pulse is applied at a microwave frequency, close to the electron's Larmor frequency, and therefore very far from the rf nuclear resonances. Thus $\mathbf{H}_2(t)$ depends only on the electronic spin.

For the intervals between pulses one could use directly the solutions given in Chap. V (Eqs. (V-116) - (V-118)) but for the pulses one would need to solve with a different Hamiltonian. It is therefore convenient to follow here the general procedure developed above: first use exponential operators as in Eq. (VII-123) for the evolution of the density matrix, and only then apply the eigenfunctions method of Chapter V. For the first stage define:

$$L_f = \frac{i}{\hbar}\left\{\left(H_0 + H_1(\Omega)\right) \otimes I - I \otimes \left(H_0 + H_1(\Omega)\right)\right\} - \Gamma_\Omega \qquad \text{(VII-139a)}$$

$$L_p = L_f - \frac{i}{\hbar}\left\{H_2(t) \otimes I - I \otimes H_2(t)\right\} \qquad \text{(VII-139b)}$$

These are the Liouville space operators for free evolution and during pulses, respectively. In the general case Eq. (VII-123) is applied with $\mathbf{L} = \mathbf{L}_f$ for each free evolution period, and with $\mathbf{L} = \mathbf{L}_p$ for each pulse. If the rate constant $1/\tau$ for rotational diffusion is small relative to the duration of the pulse τ_P, i.e., $\tau_P/\tau \ll 1$, then the relaxation operator $\boldsymbol{\Gamma}_\Omega$ can be neglected in the expression for the pulses, $\mathbf{L}_p$ (but not in $\mathbf{L}_f$, which applies to free evolution !). Moreover, if the intensity of the pulse ω_1 is large compared with the width of the spectrum (which is not so common in EPR !), then all time independent interactions can be neglected during the pulse, so that

$$L_p \approx -\frac{i}{\hbar}\left\{H_2(t) \otimes I - I \otimes H_2(t)\right\} \qquad \text{(VII-140)}$$

As long as these approximations hold, the density matrix undergoes a unitary transformation with each pulse, as in Eq. (VII-128). Then Eqs. (VII-129) and (VII-134) apply to a two pulse sequence, for example. It must be remembered, however, that $\mathbf{L}_f$ for rotational diffusion has a very different structure from $\mathbf{L}_f$ of chemical exchange. If one is interested in saturated lines, one has to use Fourier decomposition of the time dependence of $\mathbf{H}_2(t)$, leading to the solutions outlined in Chap. V and discussed in subsections VI.1.(d) and VII.1.(d) above.

Example VII.4: Two dimensional electron spin echo spectroscopy

One of the simplest yet most useful experiments in pulsed magnetic resonance is Hahn's spin echo technique, employing the two pulse sequence $\pi/2$ - τ - π - t . The magnetization of non-interacting spins is refocused at time t = τ after the second pulse, where τ is the time interval between the two pulses. If the spins have a distribution of resonance frequencies, leading to a fast decay of the FID after the first pulse, they are forced to contribute coherently to the signal at the time of refocusing. If the signal is Fourier transformed with respect to time (after the last pulse), a signal which would be inhomogeneously broadened following a single pulse is turned into a narrower signal, with only the homogeneous line width, after the refocusing is achieved.

Slow rotational diffusion has been studied in EPR by a related technique, called two dimensional electron spin echo (2D ESE) spectroscopy. In this method one repeats the basic spin echo experiment, measuring the signal only at the point t = τ, as a function of two variables. First the spin echo experiment is conducted with a fixed value of τ for many values of the Larmor resonance frequency. This is done by sweeping slowly the "constant" magnetic field throughout a range of values. Second, the whole sweep is repeated for many values of τ. The experiment is thus CW with respect to the off-resonance values and pulsed with respect to τ. A one dimensional Fourier transform of the τ dependence yields an unconventional two dimensional frequency domain spectrum. The alternative of measuring the signal as a function of t (for many values of t) as well, and transforming over both time variables, is utilized in ordinary two dimensional NMR spectroscopy, but is not always practical in EPR. This is because such a 2D FT experiment is useful only if the irradiation strength ω_1 is large compared with the total width of the spectrum. EPR spectra of doublets may be about 100 gauss wide, whereas typical EPR pulses have an ω_1 which is smaller by an order of magnitude.

The resulting two dimensional spectrum of 2D ESE looks like an ordinary one dimensional spectrum along the "magnetic field" axis. This is an inhomogeneously broadened line shape. Parallel to the other axis, representing the Fourier transform with respect to τ, the signal has only homogeneous broadening. The two dimensional array thus correlates with each frequency value in the ordinary (one dimensional) spectrum the corresponding homogeneous width(see Fig. VII.4).

In order to study systematically the homogeneous width as a function of frequency it is convenient to use a contour plot, which is a "topographical map" of the spectrum (see Fig. VII.5). The contour plots are normalized for each frequency value separately, so as to be "flat" along the magnetic field axis. Thus all height variations in this plot result from homogeneous broadening only.

Using the theory outlined in the present subsection such spectra were simulated with various models of random motion, in order to fit the experimental spectra. The fitting procedure consisted of two stages. First the fitting was done for the contour plots. Only then was fitting performed also for the line shape in the magnetic field direction. The model of Brownian rotational motion gave the best fit to the experimental results.

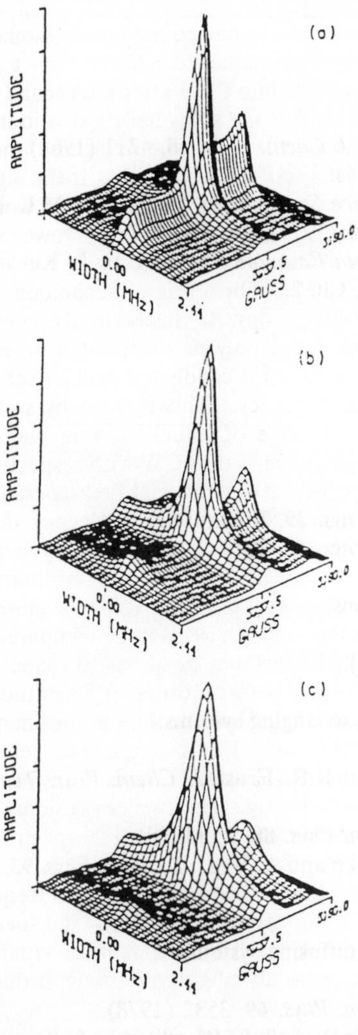

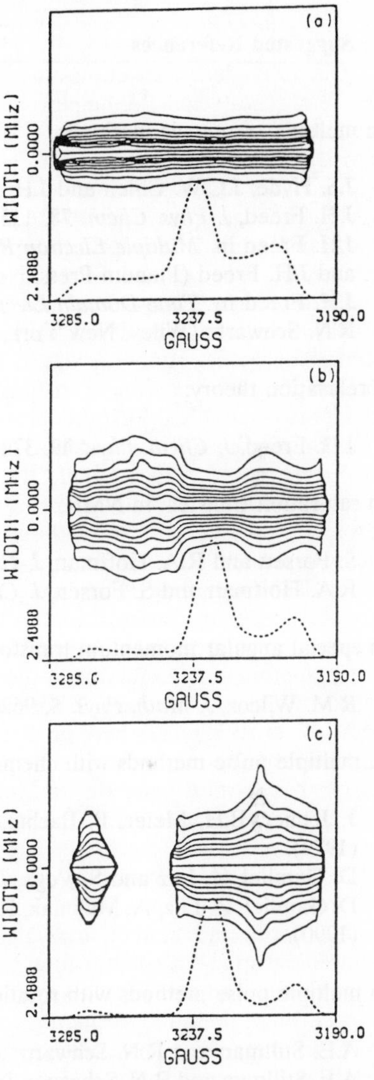

Left: Fig. VII.4. 2D ESE spectra of Tempone dissolved in 85% glycerol/water at (a) -100° C, (b) -75° C and (c) -65° C.
Right: Fig. VII.5. Normalized contour plots for the spectra of Fig. VII.4 (see text for normalization procedure). Each successive contour line represents a 10% change relative to the normalized maximum. At the bottom of each plot, the line shape for 0 MHz (i.e., along the magnetic field axis) is shown. Regions with very low signal amplitudes have been deleted because of their low signal to noise ratio.

****

Suggested References

* On multiple resonance methods:

1. J.S. Hyde, J.C.W. Chien and J.H. Freed, *J. Chem. Phys.* **48**, 4211 (1968).
2. J.H. Freed, *J. Phys. Chem.* **78**, 1155 (1974).
3. J.H. Freed in *"Multiple Electron Resonance Spectroscopy"*, edited by M.L. Dorio and J.H. Freed (Plenum Press, New York, 1979), Ch. 3.
4. J.H. Freed in *"Time Domain Electron Spin Resonance"*, edited by L. Kevan and R.N. Schwartz (Wiley, New York, 1979), Ch. 2.

* On relaxation theory:

5. J.H. Freed, *J. Chem. Phys.* **49**, 376 (1968).

* On saturation transfer in NMR:

6. S. Forsen and R.A. Hoffman, *J. Chem. Phys.* **39**, 2892 (1963).
7. R.A. Hoffman and S. Forsen, *J. Chem. Phys.* **45**, 2049 (1966).

* On special angular momentum transformations:

8. R.M. Wilcox, *J. Math. Phys.* **8**, 962 (1967).

* On multiple pulse methods with chemically exchanging systems:

9. J. Jeener, B.H. Meier, P. Bachmann and R.R. Ernst, *J. Chem. Phys.* **71**, 4546 (1979).
10. D. Gamliel, Z. Luz and S. Vega, *J. Chem. Phys.* **88**, 25 (1988).
11. D. Gamliel, Z. Luz, A. Maliniak, R. Poupko and A. Vega, *J. Chem. Phys.* **93**, 5379 (1990).

* On multiple pulse methods with rotationally diffusing systems:

12. A.E. Stillman and R.N. Schwartz, *J. Chem. Phys.* **69**, 3532 (1978).
13. A.E. Stillman and R.N. Schwartz, in *"Time Domain ESR"*, edited by L. Kevan and R.N. Schwartz (Wiley-Interscience, New York, 1979), Chap. 5.
14. L.J. Schwartz, A.E. Stillman and J.H. Freed, *J. Chem. Phys.* **77**, 5410 (1982).
15. G.L. Millhauser and J.H. Freed, *J. Chem. Phys.* **81**, 37 (1984).
16. J. Gorcester and J.H. Freed, *J. Chem. Phys.* **88**, 4678 (1988).
17. J. Gorcester, G.L. Millhauser and J.H. Freed in *"Modern Pulsed and Continuous-Wave Electron Spin Resonance"*, edited by L. Kevan and M.K. Bowman (Wiley, New York, 1988), Ch. 3.
18. J. Gorcester, G.L. Millhauser and J.H. Freed in *"Advanced EPR: Applications in Biology and Biochemistry"*, edited by A. Hoff (Elsevier, Amsterdam, 1990), Ch. 5.

19. B.R. Patyal, R.H. Crepeau, D. Gamliel and J.H. Freed, *Chem. Phys. Lett.* **175**, 445 (1990).
20. B.R. Patyal, R.H. Crepeau, D. Gamliel and J.H. Freed, *Chem. Phys. Lett.* **175**, 453 (1990).
21. S. Lee, B.R. Patyal and J.H. Freed, *J. Chem. Phys.* **98**, 3665 (1993).
22. D.E. Budil, K.A. Earle and J.H. Freed, *J. Phys. Chem.* **97**, 1294 (1993).
23. K.A. Earle, D.E. Budil and J.H. Freed, *J. Phys. Chem.* **97**, 13289 (1993).
24. S. Lee, D.E. Budil and J.H. Freed, *J. Chem. Phys.* **101**, 5529 (1994).

CHAPTER VIII

EXPERIMENTAL METHODS

VIII.1 Magnetic Resonance Spectroscopy: General

Magnetic resonance spectroscopy is traditionally associated with the absorption of electromagnetic radiation by nuclear or electron spin ensembles embedded in an external magnetic field. The discovery of microwaves resonance absorption by electrons was first published by Zavoisky in 1945, leading the way to electron paramagnetic resonance (EPR) spectroscopy. Closely related to EPR spectroscopy is nuclear magnetic resonance (NMR), where the resonance absorption is of radiowaves by nuclei. This method was discovered independently by Purcell and Bloch and published in 1946. Since its discovery half a century ago, magnetic resonance has gained an immense momentum, in particular with its second generation spectroscopies, such as electron-nuclear (or electron-electron) double resonance (ENDOR and ELDOR), pulsed NMR and EPR, and time resolved magnetic resonance. Magnetic resonance, at large, is one of the most sensitive and informative spectroscopies at the molecular level. It was, and still is, implemented in numerous basic and applied disciplines, in all branches of science. It is not surprising that several prizes were awarded in this field; Nobel Prizes to F. Bloch and E. Purcell (1952, physics), and R. Ernst (1991, chemistry) and Wolf Prizes to H. Gutowsky, H. McConnel and J. Waugh (1983/4, chemistry) to E. Hahn (1983/4, physics), to R. Ernst and A. Pines (1991, chemistry).

Understandably, no attempt is made in this book to cover thoroughly the experimental aspects of magnetic resonance spectroscopy. However, since the main theme of this book is based on magnetic resonance experiments, and for a coherent presentation, we give only a general outline of these experimental methods that are not covered in textbooks.

The reaction mechanisms and spin dynamics in numerous chemical and biological reactions, involve transient paramagnetic states and/or species as intermediates. To establish a structure-function relationship in such reactions, the identification of the paramagnetic intermediates and analyzing their dynamics, is of prime importance.

In general, increased time resolution is accompanied by low spectral resolution due to the uncertainty principle. Thus, ultrafast optical spectroscopy, which is frequently used in studying fast processes, suffers from relatively poor spectral resolution, implying a lack of details on vibrational, fine and hyperfine interactions. On the other hand, magnetic resonance has the capacity of much higher spectral resolution (down into the kHz region in EPR, and even a fraction of an Hz in NMR). Fortunately, it is the hyperfine structure combined with the unique features of spin polarization, that carry detailed information about the electronic and spatial properties of the paramagnetic transients. Obviously, such high spectral resolution, in magnetic resonance, is compensated by relatively low time resolution, which presently approaches a few nanoseconds. In principle, this can be accomplished by means of high-sensitive EPR methods, when probing the formation kinetics of paramagnetic species.

NMR is the most known and extensively described spectroscopy in magnetic resonance. The experimental modes associated with NMR spectroscopy are extremely versatile, reaching a very high degree of sophistication experimentally and theoretically. Currently, NMR spectroscopy is associated with extremely high magnetic fields combined with the detection of rarely abundant magnetic nuclei in the liquid phase, as well as in the solid state, by using complex pulse techniques. Advances in NMR spectroscopy have reached levels of sophistication, where structure determination of high molecular-weight macromolecules and proteins are of common (and complicated) practice. In addition to its vast functions in basic and analytical applications, NMR spectroscopy is routinely employed in medical diagnostics in terms of magnetic resonance imaging (MRI), which is of extreme value.

VIII.2 Special Experimental Techniques

(a) *Time-resolved EPR*

The experimental techniques that unveil the problems of electron spin dynamics comprise steady state EPR, ENDOR, time-resolved continuous wave (CW)- and FT-EPR methods. Most magnetic resonance studies, where the Stochastic Liouville Equation (SLE) has been applied, involve conventional EPR and NMR spectroscopies. The recent advancements in time-resolved EPR (TREPR) spectroscopy are associated with improving the time resolution in both CW and pulsed EPR methods, as well as the recent accomplishments in high field EPR. We emphasize this point, since time-resolved optical spectroscopic methods have traditionally been used to study dynamic processes, even though different paramagnetic states were involved in many of these processes.

In general terms, TREPR is normally associated with light-induced processes, which generate paramagnetic states such as doublets and/or triplets. The celebrated process of the primary charge separation and subsequent electron transfer in photosynthesis is a main impetus in the recent advances of TREPR spectroscopy.

The application of the SLE, described in chapters VI and VII, is associated with several experimental techniques in magnetic resonance. Among these methods, TREPR spectroscopy is a leading technique in studying reaction and electron spin dynamics in conjunction with photoexcitation. Many photophysical and photochemical reactions in solids and liquids, which involve paramagnetic intermediates, exhibit electron spin polarization (ESP) effects, indicating deviation from thermal spin equilibrium.

The present fields where EPR spectroscopy is applied are related to: transient radicals in photochemical reactions, relaxation processes in large paramagnetic assemblies, utilizing of spin labels and photoexcited triplets as a probes in studying molecular dynamics of large assemblies. Thus, sufficiently high time-resolution, compared to spin relaxation rates, allows for the observation of ESP effects in three different modes:

1. Detection of spin-polarized triplets of chromophores (D), described by Eq. (VIII-1). The system is photoexcited into its singlet state from which, by spin-orbit intersystem crossing (SO-ISC), the paramagnetic triplet state is formed. In chapter VI, the SLE method was applied in the analysis of the triplet dynamics of photoexcited C_{60} and C_{70}. These carbon clusters were found to be unique chromophores, with respect to their triplet state detection by EPR, not only in the solid phase, but also **in the liquid**

phase. This point is emphasized, since these molecules are the only examples, where triplet EPR spectra are feasible in the liquid phase.

$$D \xrightarrow{h\nu} {}^{*1}D \xrightarrow{ISC} {}^{*3}D \qquad \text{(VIII-1)}$$

2. In reactions (VIII-2) and (VIII-3), the chromophore, D, is photoexcited, but in addition to the process described by Eq. (VIII-1), an electron transfer reaction occurs to produce the two doublets. The doublets of the donor (D) and acceptor (A) pairs are uncorrelated (diffusing radicals) and in most case are spin polarized. The electron transfer can proceed by either of two different routes, i.e., singlet and/or triplet precursors, described by Eqs. (VIII-2) and (VIII-3). TREPR spectroscopy allows to determine from the signal phases and shapes the exact route of formation and spin polarization mechanism.

$$D + A \xrightarrow{h\nu} {}^{*1}D + A \rightarrow {}^{*1}[D^{\cdot+} ... A^{\cdot-}] \xrightarrow{diff} D^{\cdot+} + A^{\cdot-} \qquad \text{(VIII-2)}$$

or

$$D + A \xrightarrow{h\nu} {}^{*1}D + A \xrightarrow{ISC} {}^{*3}D + A \rightarrow {}^{3}[D^{\cdot+} ... A^{\cdot-}] \xrightarrow{diff} D^{\cdot+} + A^{\cdot-} \qquad \text{(VIII-3)}$$

3. Detection of correlated radical-ion pairs held together in the photosynthetic reaction center, in micelles, by Coulombic interactions ($D^{\cdot+}A^{\cdot-}$), or by a spacer *s* at a fixed distance, (D*s*A). Similar to the case described by Eqs. (VIII-2) and (VIII-3), the radical pairs are spin polarized, but by other polarization mechanisms, and can also be formed either via a singlet or a triplet precursor. The reaction routes are described by Eq. (VIII-4) and (VIII-5).

$$DsA \xrightarrow{h\nu} {}^{*1}DsA \xrightarrow{ET} {}^{1}[D^{\cdot+}sA^{\cdot-}] \leftrightarrow {}^{3}[D^{\cdot+}sA^{\cdot-}] \qquad \text{(VIII-4)}$$

or

$$DsA \xrightarrow{h\nu} {}^{*1}DsA \xrightarrow{ISC} {}^{*1}DsA \xrightarrow{ET} {}^{1}[D^{\cdot+}sA^{\cdot-}] \leftrightarrow {}^{3}[D^{\cdot+}sA^{\cdot-}] \qquad \text{(VIII-5)}$$

Reactions (VIII-1)-(VIII-5) can be studied by different modes of TREPR, i.e., (CW) continuous wave- and/or pulsed- TREPR. In addition, electron spin polarization (ESP), i.e., the formation of spin states in non-Boltzmann equilibrium, increases the sensitivity and provides important information on the spin dynamics. A common feature in all the reactions outlined schematically above, is their direct relevance to the primary photochemical processes in photosynthesis, carried out with a variety of model compounds.

(b) Direct-detection (CW) and Fourier transform (pulsed) EPR

Three types of TREPR detection are commonly used (Fig. VIII.1):

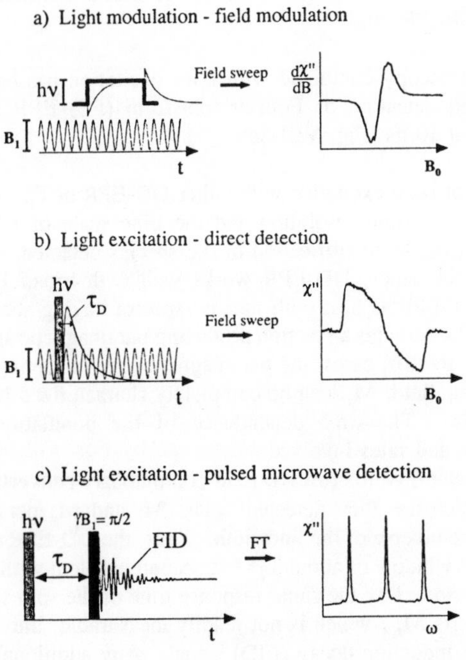

Fig. VIII.1. *Schematic representation of three TREPR experiments. On the left-hand-side are shown both the exciting light pulses and the microwave (EPR) probe together with the time domain signal. On the right-hand-side typical EPR spectra are shown: top: a derivative spectrum (in emission) taken at high modulation frequency; middle: a triplet spectrum exhibiting low-field emission and high-field absorption, taken at τ_D after the laser pulse; bottom: an FT-spectrum (absorption), taken at τ_D after the laser pulse.*

(i) the earliest mode of signal detection employs magnetic field modulation combined with photoexcitation such as light modulation. This method suffers from relatively poor time resolution of longer than about 100 μs and will not be discussed further here;

(ii) the most commonly used method is based on diode (direct) EPR detection (field modulation disconnected) combined with laser excitation. The time resolution of CW-TREPR is limited to $\sim$50 ns. The principles of this method of detection are depicted in Fig. VIII-1b. Only one point in the EPR spectrum is followed as a function of time after each laser pulse, but a series of complete spectra are built up by slowly sweeping the magnetic field. It should be pointed out, however, that at very short times after the laser pulse, spectra can be severely distorted. Nevertheless, it is the most useful method for detection of broad transient signals, presently encountered by TREPR; and

(iii) pulsed EPR spectroscopy combined with laser excitation has been used with either electron spin echo detection or Fourier transform (FT) EPR detection with time resolution of about 10 ns (Fig. VIII-3c).

The combination of laser excitation with either DD-EPR or FT-EPR provides a good match between experimental time resolution and the time scale of electron- and energy-transfer reactions, permitting an identification of the species detected. The two techniques are complementary to each other; DD-EPR works well with broad EPR lines and large spectral widths, while FT-EPR is best with narrow spectra having sharp lines. In typical experiments, the laser pulse initiates a reaction producing paramagnetic species, free radicals, radical pairs, or triplets. In most cases, the paramagnetic species having only magnetization along the applied magnetic field, $\mathbf{M}_z$, can be completely characterized by the populations in the different spin levels. The time dependence of the populations is sufficient for determining the reactions and rates involved.

In DD-EPR, a weak CW microwave field continuously converts a small portion of $\mathbf{M}_z$ into $\mathbf{M}_x$ and $\mathbf{M}_y$ which are then detected. The $\mathbf{M}_x$ and $\mathbf{M}_y$ persist for a time T_2^*, approximately equal to the inverse of the linewidth. Thus, the DD-EPR signal represents the integrated $\mathbf{M}_z$ over a time window of about T_2^*. A second problem with DD-EPR is that the signal at early times is distorted by the finite response time of the spin system. In FT-EPR, a microwave pulse converts $\mathbf{M}_z$, which is not readily measurable, into $\mathbf{M}_x$ and $\mathbf{M}_y$, which form the observable free induction decay (FID) signal. Any additional triplets or radicals born after the microwave pulse generate new $\mathbf{M}_z$, and do not contribute to the FID. Any chemical process destroying radicals or triplets produces line broadening and affects line shapes, but will not change the integrated intensity of the lines in the spectrum. Thus, FT-EPR gives a spectrum of the $\mathbf{M}_z$ originally present at the time of the microwave pulse with little complication from reactions occurring after the pulse.

Under favorable conditions, the FID in the FT-EPR experiment is closely related to the signal in DD-EPR. Namely, the DD-EPR signal at time t_d after the laser pulse is the Fourier transform of the FID between times 0 and t_d. However, for $t_d < T_2^*$, severe spectral distortions may occur. Fortunately, for species with broad lines, such as triplet states, T_2^* is a few nanoseconds and the two problems just discussed are of no practical importance. It is precisely these samples with short T_2^*, in which their FIDs decay rapidly, are most difficult to measure by FT-EPR.

(c) Multiple resonance methods: electron-nuclear and electron-electron double resonance (ENDOR and ELDOR)

EPR spectroscopy is ideal for elucidating hyperfine structure (hfs) of small molecules or molecules of high symmetry. In many reactions, however, the radicals involved are large and asymmetric in their structure, so that hfs can no longer be resolved due to inhomogeneous broadening of the EPR lines. Double resonance spectroscopies provide the experimentalist with techniques that put together the advantages of specific spectroscopies. Within this context we should mention electron-nuclear double resonance (ENDOR), electron-electron double resonance (ELDOR), optical detection of magnetic resonance (ODMR), fluorescence-detected magnetic resonance (FDMR) and reaction yield detected magnetic resonance (RYDMR) spectroscopies. We mention some methods that are related to the main theme of this book.

ENDOR is extremely useful in obtaining precise measurements of hyperfine coupling constants when these are not readily obtained from the ESR spectrum. Experimentally, this method is based on the simultaneous application of two resonant electromagnetic radiation fields (i.e., the two selection rules of NMR and EPR) to the system. In other words, while pumping the EPR transitions, the NMR transitions are driven and detected, thus improving the sensitivity by several orders of magnitude as compared to NMR spectroscopy. The main advantage of ENDOR becomes apparent when one considers coupling to a group of equivalent nuclei. For example, the EPR spectrum of triphenylmethyl radical contains (7x7x4) = 196 lines, while the ENDOR spectrum contains six lines. Moreover, as discussed briefly above, ENDOR spectroscopy is characterized by resolution enhancement, which is reflected by the narrow lines as typified by NMR spectra.

An ELDOR experiment is characterized by simultaneous microwave irradiation at two EPR frequencies. Thus, saturating one hyperfine component should affect the intensity of a second one. The magnitude of the change in EPR line intensity is governed by the relaxation mechanism(s) that couple the spin systems, that being pumped and the observed one, i.e., the more effective is the coupling, the greater is the ELDOR response. Thus, ELDOR spectroscopy is suitable for studying spin dynamic processes such as relaxation, exchange, molecular tumbling (reorientation), and solid state dynamics. Some examples, that are treated by the SLE method, are presented in Chapter VII.

Suggested References

1. *Modern Pulsed and Continuous-Wave Electron Spin Resonance*, L. Kevan L and M. K. Bowman MK (Eds.) Wiley, New York (1990).

2. A. Schweiger, *Angew. Chem. Int. Ed. Engl.*, **30**, 265 (1991).

3. M. Bixon, J. Fajer, G. Feher, J. H. Freed, D. Gamliel, A. J. Hoff, H. Levanon, K. Möbius, R. Nechushtai, J. R. Norris, A. Scherz, J. L. Sessler, and D. Stehlik (Special Issue H. Levanon Ed.) *Israel. J. Chem.* **32**, 369 (1992).

4. G. L. Closs, P. Gautam, D. Zhang, P. J. Krusic, S. A. Hill and E. Wasserman, *J. Phys. Chem.*, **96**, 5228 (1992).

5. H. Levanon, and M. K. Bowman, in: *Photosynthetic Reaction Centers*, J. R. Norris and J. Deisenhofer (Eds.) (Academic Press, New York, 1993) Vol. II, Ch. 13, pp. 387.

6. K. Hasharoni, H. Levanon, J. von Gersdorff, H. Kurreck, and K. Möbius *J. Chem. Phys.* **98** 2916 (1993).

7. H. Levanon, S. Michaeli and V. Meiklyar, in: *Recent Advances in the Chemistry and Physics of Fullerenes and Related Materials*, K. Kadish and R. S. Ruoff (Eds.). The Electrochemical Society, Pennington N.J. 894 (1994) .

APPENDIX A

ANGULAR MOMENTUM IN QUANTUM MECHANICS

A.1 The Angular Momentum Operators

Angular momentum is the variable which corresponds to rotational motion. Its importance derives from the fact that classically it is one of the "conserved quantities" under certain conditions, and quantum mechanically the eigenstates of a related operator are eigenstates of the Hamiltonian in the equivalent cases. In classical physics, the angular momentum vector is defined by the vector product (cross product):

$$\boldsymbol{L} = \boldsymbol{r} \times \boldsymbol{p} \tag{A-1}$$

where **r** = (x,y,z) is the position vector and **p** = mv = m(d**r**/dt) is the linear momentum vector. Its components are

$$L_x = y p_z - z p_y \qquad L_y = z p_x - x p_z \qquad L_z = x p_y - y p_x \tag{A-2}$$

In one dimensional motion (y = z = 0) the angular momentum is zero. In two dimensional motion in the **x-y** plane (z = 0), the only non-zero component of angular momentum is

$$L_z = m\left[x\frac{dy}{dt} - y\frac{dx}{dt}\right] \tag{A-3}$$

For example, for motion along a circle of radius r with a uniform angular velocity:

$$\boldsymbol{r} = \left(r\cos(\omega t), r\sin(\omega t), 0\right) \quad \Rightarrow \quad L_z = m r^2 \omega \tag{A-4}$$

The angular momentum is conserved (i.e., it is time independent) if there is no torque, namely if: **r** × (d**p**/dt) = 0 . The quantum mechanical analog to such a conservation rule can be found from the Heisenberg picture equation of motion, Eq. (I-38). That equation implies that an operator which commutes with the Hamiltonian and has no explicit time dependence will have no time dependence at all. It will then be diagonalized in the same basis as **H**.

In quantum mechanics the linear momentum operator is defined as:

$$\boldsymbol{p} \rightarrow -i\hbar\nabla_r = -i\hbar\frac{\partial}{\partial \boldsymbol{r}} \tag{A-5}$$

so the quantum mechanical operators representing the three components of the angular momentum vector are:

$$L_x = -i\hbar\left[y\frac{\partial}{\partial z} - z\frac{\partial}{\partial y}\right] \qquad L_y = -i\hbar\left[z\frac{\partial}{\partial x} - x\frac{\partial}{\partial z}\right]$$

$$L_z = -i\hbar\left[x\frac{\partial}{\partial y} - y\frac{\partial}{\partial x}\right] \tag{A-6}$$

The natural coordinates for describing three dimensional rotational motion are polar (spherical) coordinates r,θ,ϕ to which the Cartesian coordinates are related by:

$$x = r\sin(\theta)\cos(\phi) \qquad y = r\sin(\theta)\sin(\phi) \qquad z = r\cos(\theta) \tag{A-7}$$

In terms of these coordinates:

$$L_x = i\hbar\left[\sin(\phi)\frac{\partial}{\partial\theta} + \cot(\theta)\cos(\phi)\frac{\partial}{\partial\phi}\right] \tag{A-8a}$$

$$L_y = i\hbar\left[-\cos(\phi)\frac{\partial}{\partial\theta} + \cot(\theta)\sin(\phi)\frac{\partial}{\partial\phi}\right] \tag{A-8b}$$

$$L_z = -i\hbar\frac{\partial}{\partial\phi} \tag{A-8c}$$

from which it follows that

$$L^2 \equiv L_x^2 + L_y^2 + L_z^2 = -\hbar^2\left[\frac{1}{\sin(\theta)}\frac{\partial}{\partial\theta}\left[\sin(\theta)\frac{\partial}{\partial\theta}\right] + \frac{1}{\sin^2(\theta)}\frac{\partial^2}{\partial\phi^2}\right] \tag{A-9}$$

The quantum mechanical operator which corresponds to the conserved angular momentum is *not* **L** but $\mathbf{L}^2$. The operator $\mathbf{L}^2$ represents, both in classical and in quantum mechanics, the magnitude of the angular momentum vector. In classical mechanics all components of **L** can be conserved simultaneously, but in quantum mechanics only the magnitude of the vector and the value of one of its three components, for example $\mathbf{L}_z$ can be well defined simultaneously. The spherical harmonics $Y^l_m(\theta,\phi)$ are eigenfunctions of $\mathbf{L}^2$ and $\mathbf{L}_z$:

$$L^2 Y^l_m(\theta,\phi) = \hbar^2 l(l+1) Y^l_m(\theta,\phi) \tag{A-10a}$$

$$L_z Y^l_m(\theta,\phi) = \hbar m Y^l_m(\theta,\phi) \tag{A-10b}$$

These functions are defined for any non-negative integer l = 0, 1, 2, ... and for m = -l, -l+1, ..., l-1, l. The numbers l and m are known as the quantum numbers of angular momentum. The functions are orthogonal, with the scalar product

$$\int_0^{2\pi} d\phi \int_0^{\pi} d\theta \sin(\theta) \left(Y^l{}_m(\theta,\phi)\right)^* Y^{l'}{}_{m'}(\theta,\phi) = \delta_{l,l'}\,\delta_{m,m'} \qquad \text{(A-11)}$$

However, the functions are not eigenfunctions of $\mathbf{L}_x$ and $\mathbf{L}_y$, because the fundamental quantum commutation rules

$$[r_a, p_b] = i\hbar\,\delta_{a,b} \qquad (a,b = x,y \text{ or } z) \qquad \text{(A-12)}$$

imply

$$[L_x, L_y] = i\hbar L_z \qquad [L_y, L_z] = i\hbar L_x \qquad [L_z, L_x] = i\hbar L_y \qquad \text{(A-13)}$$

Since the different components of **L** do not commute, they cannot be diagonalized in the same basis. Nevertheless, $\mathbf{L}_x$ and $\mathbf{L}_y$ will only mix spherical harmonics of the same rank l. It is convenient to define the non-hermitian combinations:

$$L_+ \equiv L_x + iL_y = \hbar e^{i\phi}\left[\frac{\partial}{\partial\theta} + i\cot(\theta)\frac{\partial}{\partial\phi}\right] \qquad \text{(A-14a)}$$

$$L_- \equiv L_x - iL_y = \hbar e^{-i\phi}\left[-\frac{\partial}{\partial\theta} + i\cot(\theta)\frac{\partial}{\partial\phi}\right] \qquad \text{(A-14b)}$$

These operators also mix different spherical harmonics of the same rank, but in a simpler manner, changing the azimuthal quantum number m by one unit:

$$L_+ Y^l{}_m(\theta,\phi) = \hbar\left(l(l+1) - m(m+1)\right)^{1/2} Y^l{}_{m+1}(\theta,\phi) \qquad \text{(A-15a)}$$

$$L_- Y^l{}_m(\theta,\phi) = \hbar\left(l(l+1) - m(m-1)\right)^{1/2} Y^l{}_{m-1}(\theta,\phi) \qquad \text{(A-15b)}$$

Because of this property they are called "raising" and "lowering" operators, respectively. Their commutation relations can be derived from Eq. (A-13):

$$[L_z, L_+] = \hbar L_+ \qquad [L_z, L_-] = -\hbar L_- \qquad [L_+, L_-] = 2\hbar L_z \qquad \text{(A-16)}$$

In quantum mechanics there is also, in addition to the angular momentum **L**, also another angular momentum **S** which has no classical analog. **L** is called the orbital angular momentum, and it operates in real space on functions of θ,ϕ; **S** is called spin angular momentum, and it operates in abstract space on "spin functions". Just as $\mathbf{L}^2$, $\mathbf{L}_z$ operate on the (2l+1) spherical harmonics of rank l, mixing them among themselves, the components of **S** operate on the (2s+1) components of a vector wave function, in which the spatial part is the same for all components. This can be written as an ordinary spatial wave function multiplied by a vector in that abstract space. The components of **S** commute with those of **L**, and satisfy the same commutation rules as (A-13) among themselves. There is, however, one important difference between **L** and **S**: **S** can have

a half-integer quantum number s = 1/2, 3/2, ... and not only integer values like l. The corresponding "azimuthal" quantum number m_s still follows the rule:
m_s = -s, -s+1, ..., s-1, s with respect to the main quantum number s.

A.2 Angular Momentum as the Generator of Rotations

Suppose a wave function is given in terms of the position vector in a known frame of reference: $\psi = \psi(\mathbf{r})$. Now the system is rotated (or: the system of axes is rotated in the opposite sense), so that **r** is transformed to **r'** = **R r** by a transformation **R**. In Cartesian coordinates, the components of the new position vector can be calculated by

$$\boldsymbol{r}' = \begin{pmatrix} x' \\ y' \\ z' \end{pmatrix} = \begin{pmatrix} R_{xx} & R_{xy} & R_{xz} \\ R_{yx} & R_{yy} & R_{yz} \\ R_{zx} & R_{zy} & R_{zz} \end{pmatrix} \begin{pmatrix} x \\ y \\ z \end{pmatrix} = \boldsymbol{R}\boldsymbol{r} \tag{A-17}$$

Fig. A.1 shows a special case of such a rotation. This is a rotation in the **x-y** plane by an angle α. From elementary trigonometry:

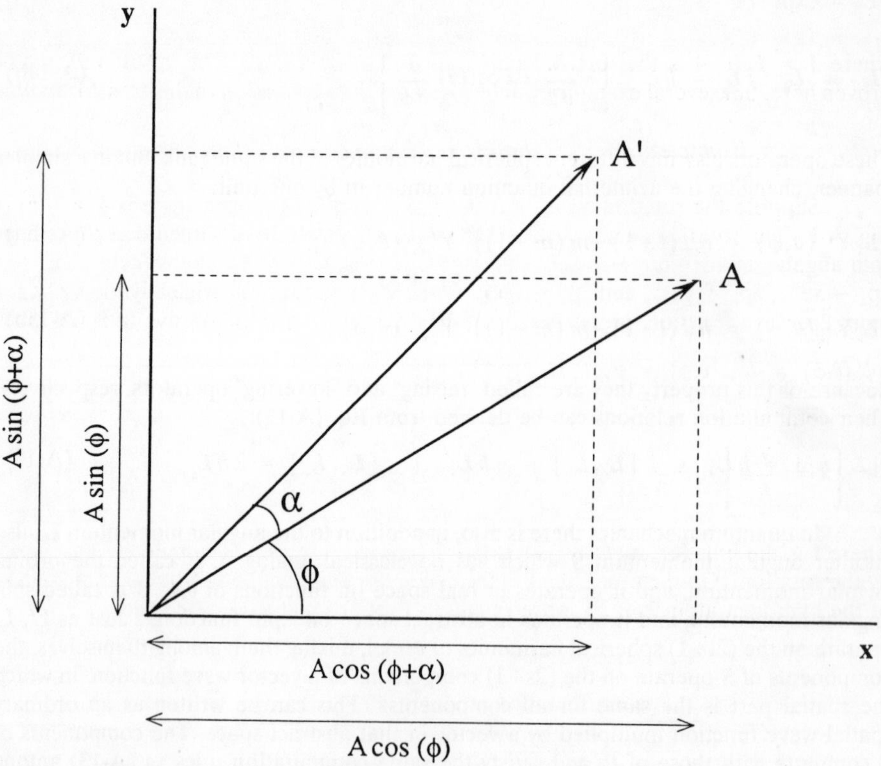

Fig. A.1. *Rotation of a vector in the x-y plane by an angle α.*

$$A_x' = A_x \cos(\alpha) - A_y \sin(\alpha) \qquad\qquad A_y' = A_x \sin(\alpha) + A_y \cos(\alpha) \tag{A-18}$$

The rotation matrix for this special case is:

$$\boldsymbol{R} = \begin{pmatrix} \cos(\alpha) & -\sin(\alpha) & 0 \\ \sin(\alpha) & \cos(\alpha) & 0 \\ 0 & 0 & 1 \end{pmatrix} \tag{A-19}$$

Since the rotation in this case is about the **z** axis, one may define the **z** axis to be the rotation axis.

In the general case, the rotation axis is the vector $\boldsymbol{\Phi} = (\phi_x, \phi_y, \phi_z)$, and $\Phi = |\boldsymbol{\Phi}|$ is the magnitude of the rotation angle. If the system is symmetrical with respect to rotations, then the wave function should be transformed so that the dependence of the new wave function on **r'** will be the same as the depenedence of the original wave function on **r**. Mathematically, this means that if the transformed function is: $\mathbf{U}\,\psi(\mathbf{r}) = \psi'(\mathbf{r})$ then: $\psi'(\mathbf{r}') = \psi(\mathbf{r})$. It can be shown that this condition is fulfilled for rotations if the transformation operator is

$$\boldsymbol{U} = \exp(-i\boldsymbol{\Phi}\cdot\boldsymbol{J}/\hbar) \tag{A-20}$$

where $\mathbf{J} = \mathbf{L} + \mathbf{S}$ is the total angular momentum operator. This result will not be proven here, but several examples will be given for the application of this transformation.

Example A.1: Rotation about the z axis - effect on wave functions

Suppose the rotation axis is just the **z** axis, and the rotation angle is $\Phi = \alpha$ as in Fig. A.1. The rotation vector is then $\boldsymbol{\Phi} = (0, 0, \alpha)$. It will be assumed that there is no spin angular momentum. Consider the hydrogen atom wave functions (orbitals): $2p_x \sim xe^{-cr}$, $2p_y \sim ye^{-cr}$, and $2p_z \sim ze^{-cr}$. If the wave function is originally the hydrogen atom $2p_z$ wave function, proportional to $Y^1{}_0(\theta,\phi)$, then the rotated wave function is

$$\psi'(\theta,\phi) = \boldsymbol{U}\psi(\theta,\phi) = \exp(-i\alpha J_z/\hbar)Y^1{}_0(\theta,\phi) =$$

$$= \left\{ \boldsymbol{I} + \frac{-i\alpha}{\hbar}\left[-i\hbar\frac{\partial}{\partial\phi}\right] + \ldots \right\}\cos(\theta) = \cos(\theta)$$

where **I** is the unit (or: identity) operator. The wave function is completely unaffected by the rotation. On the other hand, if the original wave function is a linear combination of the hydrogen atom $2p_x$, $2p_y$ wave functions proportional to $Y^1{}_1(\theta,\phi)$ then

$$\psi'(\theta,\phi) = U\psi(\theta,\phi) = \left\{ I + \frac{-i\alpha}{\hbar}\left[-i\hbar\frac{\partial}{\partial\phi}\right] + \frac{1}{2!}\frac{-\alpha^2}{\hbar^2}\left[-\hbar^2\frac{\partial^2}{\partial\phi^2}\right] + \dots \right\} \sin(\theta)\, e^{i\phi}$$

$$= \left\{ I + \left[-\alpha\frac{\partial}{\partial\phi}\right] + \frac{1}{2!}\left[\alpha^2\frac{\partial^2}{\partial\phi^2}\right] + \dots \right\} \sin(\theta)\, e^{i\phi} =$$

$$= \left\{ 1 + (-i\alpha) + \frac{1}{2!}(-\alpha^2) + \dots \right\} \sin(\theta)\, e^{i\phi} = e^{-i\alpha}\psi(\theta,\phi)$$

In this case the exponential operator of $\mathbf{L}_z$ modifies the wave function. Similarly, if the original wave function is proportional to $Y^1_{-1}(\theta,\phi)$ it is multiplied by $e^{i\alpha}$ when **U** operates on it. These two spherical harmonics are therefore eigenfunctions of the given transformation operator **U**, with different eigenvalues. The $2p_x$, $2p_y$ orbitals are linear combinations of them, so they are mixed by the operator. This can be seen directly as follows:

$$\exp(-i\alpha J_z/\hbar)\left(x e^{-cr}\right) = \left\{ I + \left[-\alpha\frac{\partial}{\partial\phi}\right] + \frac{1}{2!}\left[\alpha^2\frac{\partial^2}{\partial\phi^2}\right] + \dots \right\} r\sin(\theta)\cos(\phi)\, e^{-cr} =$$

$$= r\sin(\theta)\, e^{-cr}\left\{ \cos(\phi) + \alpha\sin(\phi) - \frac{1}{2!}\alpha^2\cos(\phi) + \dots \right\} =$$

$$= r\sin(\theta)\, e^{-cr}\{\cos(\phi)\cos(\alpha) + \sin(\phi)\sin(\alpha)\} = x e^{-cr}\cos(\alpha) + y e^{-cr}\sin(\alpha)$$

The rotation about **z** simply rotates the $2p_x$ orbital in the **x-y** plane, and the same would happen to the $2p_y$ orbital.

****

Example A.2: Effect of rotation on operators

Again, assume the system is rotated about the **z** axis, and the rotation vector is $\mathbf{\Phi} = (0, 0, \alpha)$. How are operators transformed by this rotation ? Take, for example, the angular momentum operator $\mathbf{J}_x$. It is transformed as:

$$J_x' = UJ_xU^{-1} = \exp(-i\alpha J_z/\hbar)\, J_x \exp(i\alpha J_z/\hbar) =$$

$$= \left\{ I + \frac{-i\alpha}{\hbar} J_z + \frac{1}{2!}\frac{-\alpha^2}{\hbar^2} J_z^{\,2} + \ldots \right\} J_x \left\{ I + \frac{-i\alpha}{\hbar} J_z + \frac{1}{2!}\frac{-\alpha^2}{\hbar^2} J_z^{\,2} + \ldots \right\} =$$

$$= J_x + \frac{-i\alpha}{\hbar}[J_z, J_x] + \frac{1}{2!}\frac{-\alpha^2}{\hbar^2}\left(J_z^{\,2}J_x + J_xJ_z^{\,2}\right) + \frac{\alpha^2}{\hbar^2} J_zJ_xJ_z + \ldots =$$

$$= J_x + \alpha J_y - \frac{1}{2}\frac{\alpha^2}{\hbar^2}\left(J_z^{\,2}J_x + (J_zJ_x - i\hbar J_y)J_z - 2J_z(J_zJ_x - i\hbar J_y)\right) + \ldots =$$

$$= J_x\left[1 - \frac{1}{2}\alpha^2 + \ldots\right] + J_y\left[\alpha - \frac{1}{3!}\alpha^3 + \ldots\right] = J_x\cos(\alpha) + J_y\sin(\alpha)$$

The expansion in the second line over all powers in each exponent leads to the expansion (in the third line) over products of operators with all possible degrees (degree = sum of powers of operators appearing in a product). Repeated application of the commutation relations (A-13), to transform all operators of the same degree to the same form, thus leads to a simple result: the transformation simply rotates the operator, in exactly the same manner as it would rotate a function.

****

A.3 Matrix Representations for Angular Momentum

It was mentioned above that on the spherical harmonics $Y^l_m(\theta,\phi)$ the operation of the orbital angular momentum operators is very simple. Thus if a spherical harmonic is written as a Hilbert space vector $|l,m\rangle$, the matrices of the angular momentum operators would be very simple. In problems in which the Hamiltonian is written in terms of angular momentum operators, this has clear implications for the structure of the Hamiltonian matrix in that basis. The basis itself is infinite, so the matrix of any operator in it is infinite dimensional. However, if the basis is broken into subsets, each with its own l value, these operators would be represented in each such subset by a finite matrix with a simple structure. The dimensions of these finite matrices are $2l+1=1, 3, 5, \ldots$ for integer spin, which is the only possibility with orbital angular momentum. The full matrix would consist of blocks (with a simple internal structure) on the diagonal and zeroes outside these blocks:

$$\begin{pmatrix} x & 0 & 0 & 0 & 0 & 0 & 0 & 0 & 0 & \cdots \\ 0 & x & x & x & 0 & 0 & 0 & 0 & 0 & \cdots \\ 0 & x & x & x & 0 & 0 & 0 & 0 & 0 & \cdots \\ 0 & x & x & x & 0 & 0 & 0 & 0 & 0 & \cdots \\ 0 & 0 & 0 & 0 & x & x & x & x & x & \cdots \\ 0 & 0 & 0 & 0 & x & x & x & x & x & \cdots \\ 0 & 0 & 0 & 0 & x & x & x & x & x & \cdots \\ 0 & 0 & 0 & 0 & x & x & x & x & x & \cdots \\ 0 & 0 & 0 & 0 & x & x & x & x & x & \cdots \\ . & . & . & . & . & . & . & . & . & \cdots \end{pmatrix} \tag{A-21}$$

Spin angular momentum operates on a different kind of functions, which may be denoted as $|s,m_s\rangle$, but due to its identical commutation relations it operates there in exactly the same manner as orbital angular momentum operates in the $|l,m\rangle$ basis. Thus for any integer s value the submatrix of **S** in the $|s,m_s\rangle$ basis is identical to the submatrix of **L** in the $|l,m\rangle$ basis. For half-integer s values the submatrix of **S** has no counterpart in **L**. It is therefore possible to construct the matrix of the angular momentum operator **J** (where now **J** is either **L** or **S**) by constructing the relevant submatrices. The basis is denoted as $|j,m\rangle$ and the matrices of $\mathbf{J}_x$, $\mathbf{J}_y$ and $\mathbf{J}_z$ are calculated from:

$$\langle j',m'|J_z|j,m\rangle = \hbar m\, \delta_{j',j}\,\delta_{m',m} \tag{A-22}$$

$$\langle j',m'|J_+|j,m\rangle = \hbar\big(j(j+1) - m(m+1)\big)^{1/2}\, \delta_{j',j}\,\delta_{m',m+1} \tag{A-23}$$

$$\langle j',m'|J_-|j,m\rangle = \hbar\big(j(j+1) - m(m-1)\big)^{1/2}\, \delta_{j',j}\,\delta_{m',m-1} \tag{A-24}$$

and from the relations $\mathbf{J}_x = \frac{1}{2}(\mathbf{J}_+ + \mathbf{J}_-)$, $\mathbf{J}_y = -\frac{1}{2}i\,(\mathbf{J}_+ - \mathbf{J}_-)$.

The full matrix of **J** in the infinite basis of functions $|j,m\rangle$ is a reducible representation of the group of three dimensional rotations. For each j value, the (2j+1) × (2j+1) block constitutes an irreducible representation for the group of three dimensional rotations. In the case of orbital angular momentum, the rotations are in real three-dimensional space. In the case of spin angular momentum, the rotations are in an abstract spin space. The mathematical descriptions, however, are the same for both. When both **L** and **S** are present, the relevant space is a direct product of the two relevant spaces, and the wave functions are a direct product of the orbital and spin wave functions.

In addition to their use for representing angular momentum, such matrices can also be used as a basis for the space of matrices of the appropriate dimension. This requires having a sufficient number of independent matrices. For any given j, the dimension of the matrices is (2j+1) × (2j+1), and one therefore needs $(2j+1)^2$ operators, including the unit operator. The required operators can be constructed, for example, by products of powers of the elementary angular momentum operators $\mathbf{J}_x$, $\mathbf{J}_y$, $\mathbf{J}_z$. In general this may be complicated, but for j = ½ and even for j = 1 it is simple and may be useful in problems involving angular momentum.

Example A.3: *The j = ½ case*

The situation is especially simple in the j = ½ case, where one needs just four independent matrices. These can be chosen as the unit 2 × 2 matrix **I** and the j = ½ matrices of $\mathbf{J}_x$, $\mathbf{J}_y$, and $\mathbf{J}_z$. It is conventional, however, to work with the Pauli matrices, which are proportional to those of the angular momentum components, because they have a simpler normalization condition. In this subspace, the angular momentum operators are represented by the matrices:

$$J_x = \frac{1}{2}\hbar\begin{bmatrix}0 & 1\\ 1 & 0\end{bmatrix} \qquad J_y = \frac{1}{2}\hbar\begin{bmatrix}0 & -i\\ i & 0\end{bmatrix} \qquad J_z = \frac{1}{2}\hbar\begin{bmatrix}1 & 0\\ 0 & -1\end{bmatrix} \tag{A-25}$$

These operators are related to the Pauli matrices σ_i by

$$J_i = \frac{1}{2}\hbar\,\sigma_i \qquad (i = x,y,z) \tag{A-26}$$

In this case **J** must be a spin angular momentum operator, so the notation **S** could be used instead. The Pauli matrices clearly have the same commutation relations as angular momentum components (Eq. (A-13)), and in addition to this they satisfy the normalization condition

$$(\sigma_i)^2 = I \tag{A-27}$$

for each i. Any 2 x 2 matrix can be written as a linear combination of I, σ_x, σ_y, σ_z. In particular, any 2 × 2 hermitian matrix can be expressed as:

$$\begin{bmatrix}a_0+a_z & a_x+ia_y\\ a_x-ia_y & a_0-a_z\end{bmatrix} = a_0\begin{bmatrix}1 & 0\\ 0 & 1\end{bmatrix} + a_x\begin{bmatrix}0 & 1\\ 1 & 0\end{bmatrix} + a_y\begin{bmatrix}0 & -i\\ i & 0\end{bmatrix} + a_z\begin{bmatrix}1 & 0\\ 0 & -1\end{bmatrix} \tag{A-28}$$

Using the relation between the spin operators and the Pauli matrices, the general hermitian matrix in this space is equal to

$$\begin{bmatrix}a_0+a_z & a_x+ia_y\\ a_x-ia_y & a_0-a_z\end{bmatrix} = a_0 I + \frac{2}{\hbar}\,\mathbf{a}\cdot\mathbf{S} \tag{A-29}$$

with $\mathbf{a} = (a_x, a_y, a_z)$.

Suggested References

1. D.M. Brink and G.R. Satchler, *"Angular Momentum"*, 2nd edition, corrected printing (Clarendon Press, Oxford, 1971).
2. R.N. Zare, *"Angular Momentum"* (Wiley, New York, 1988).

APPENDIX B

ROTATION MATRICES

In Appendix A it was mentioned that the most general rotation can be regarded formally as being generated by a rotation vector (ϕ_x, ϕ_y, ϕ_z). In practice, however, the rotation is usually expressed as a sequence of three rotation operations performed in a known order about certain axes. First one rotates by an angle α about the **z** axis, then by β about the new axis $\mathbf{y}_1$, and finally by an angle γ about the **z'** axis (see Fig. B.1).

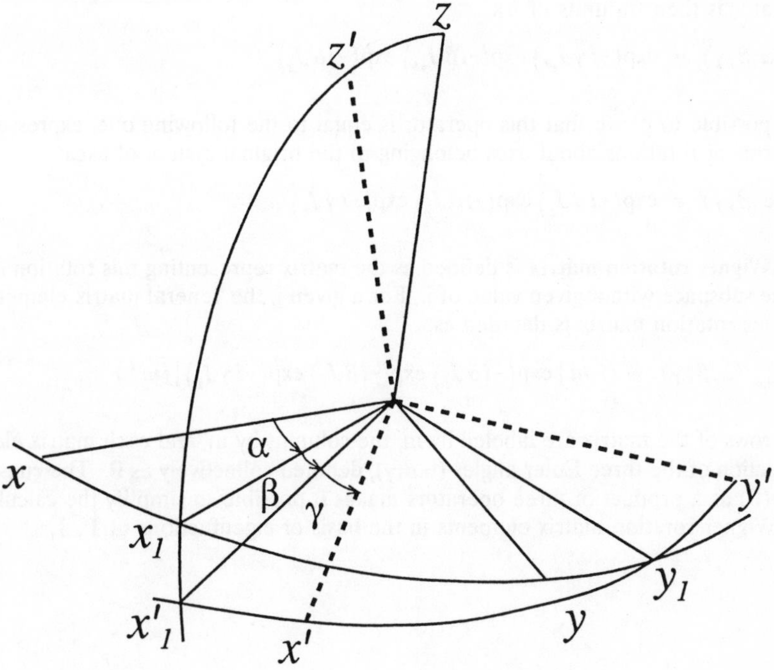

Fig. B.1. *The rotation defined by the Euler angles* (α,β,γ).

The angles of these rotations are known as the Euler angles, and the order is significant because rotations around different axes do not commute in general. This conventional description of rotations has its quantum mechanical counterpart in the standard description of rotation operators, to be presented here.

It was seen in Appendix A that the most general rotation in quantum mechanics is formally given by the operator

$$U = \exp\left(-i\left(\phi_x J_x + \phi_y J_y + \phi_z J_z\right)/\hbar\right) \tag{B-1}$$

This operator is not convenient to use in practice, because it contains in the exponent a sum of three non-commuting operators. For non-commuting operators A and B: $e^{A+B} \neq e^A e^B$ so if one has wave functions which are eigenfunctions of $\mathbf{J}_z$, for example, it is not straightforward to operate on them with such a combined operator. Consequently it is very difficult to simplify calculations with an operator such as that appearing in Eq. (B-1). One therefore uses an alternative description of rotations, equivalent to the classical description of rotations with three Euler angles. The rotation operator is then (in units of $\hbar$):

$$D(\alpha,\beta,\gamma) = \exp\left(-i\gamma J_{z'}\right)\exp\left(-i\beta J_{y1}\right)\exp\left(-i\alpha J_z\right) \tag{B-2}$$

It is possible to prove that this operator is equal to the following one, expressed solely in terms of rotations about axes belonging to the original system of axes:

$$D(\alpha,\beta,\gamma) = \exp\left(-i\alpha J_z\right)\exp\left(-i\beta J_y\right)\exp\left(-i\gamma J_z\right) \tag{B-3}$$

The **Wigner rotation matrix** is defined as the matrix representing this rotation operator in the subspace with a given value of j. For a given j, the general matrix element of the Wigner rotation matrix is denoted as:

$$D^j_{m,m'}(\alpha,\beta,\gamma) = \langle j,m \,|\, \exp\left(-i\alpha J_z\right)\exp\left(-i\beta J_y\right)\exp\left(-i\gamma J_z\right) \,|\, jm' \rangle \tag{B-4}$$

The rows of the matrix are labeled by m, the columns by m' and each matrix element is a function of the three Euler angles (α,β,γ), denoted collectively as Ω. The construction of $D(\Omega)$ as a product of three operators makes it possible to simplify the calculation of the Wigner rotation matrix elements in the basis of eigenfuctions of $\mathbf{J}^2$, $\mathbf{J}_z$:

$$D^j_{m,m'}(\alpha,\beta,\gamma) =$$

$$= \sum_{n,n'} \langle j,m\,|\exp(-i\alpha J_z)|\,j,n\rangle\langle j,n\,|\exp(-i\beta J_y)|\,j,n'\rangle\langle j,n'\,|\exp(-i\gamma J_z)|\,jm'\rangle =$$

$$= \sum_{n,n'} \langle j,m\,|\exp(-i\alpha n)|\,j,n\rangle\langle j,n\,|\exp(-i\beta J_y)|\,j,n'\rangle\langle j,n'\,|\exp(-i\gamma m')|\,jm'\rangle =$$

$$= e^{-im\alpha}\langle j,m\,|\exp(-i\beta J_y)|\,j,m'\rangle e^{-im'\gamma} \equiv e^{-im\alpha} d^j_{m,m'}(\beta)\, e^{-im'\gamma} \tag{B-5}$$

The **reduced Wigner functions** $d^j_{m,m'}(\beta)$ are the only non-trivial part of the full Wigner functions. They are closely related to the θ-dependent part of the spherical harmonics, and in general $D^j_{m,m'}(\alpha,\beta,\gamma)$ is proportional to $Y^j_m(\beta,\alpha)$.

It follows from the definition (B-2) of the transformation operators that they are unitary, and so are the Wigner matrices:

$$\left(D^j_{m,m'}(\alpha,\beta,\gamma)\right)^{-1} = \left(D^j_{m,m'}(\alpha,\beta,\gamma)\right)^{\dagger} = \left(D^j_{m',m}(\alpha,\beta,\gamma)\right)^{*} \tag{B-6}$$

so that

$$\sum_{m'}\left(D^j_{m',m}(\alpha,\beta,\gamma)\right)^{*} D^j_{m',m''}(\alpha,\beta,\gamma) = \delta_{m,m''} \tag{B-7}$$

Example B.1: Construction of the rotation matrix for j = 1

The eigenfunctions of $\mathbf{J}^2$ and $\mathbf{J}_z$ in the j = 1 subspace are $|1,-1\rangle$, $|1,0\rangle$ and $|1,1\rangle$. These may either be spin functions, or orbital angular momentum functions. In the latter case they are simply $Y^1_m(\theta,\phi)$ with m = -1,0 and 1. However, the rotation matrices are the same for both cases. The angular momentum matrices are:

$$J_x = \frac{\hbar}{\sqrt{2}}\begin{pmatrix}0&1&0\\1&0&1\\0&1&0\end{pmatrix} \qquad J_y = \frac{\hbar}{\sqrt{2}}\begin{pmatrix}0&-i&0\\i&0&-i\\0&i&0\end{pmatrix} \qquad J_z = \hbar\begin{pmatrix}1&0&0\\0&0&0\\0&0&-1\end{pmatrix}$$

and $\mathbf{J}^2 = 2\hbar^2\,\mathbf{I}$ where $\mathbf{I}$ is the 3 × 3 unit matrix. From Eq. (B-5), the Wigner rotation matrix is a product of three matrices, depending on α, β and γ respectively. In a system of units with $\hbar=1$, the first and third matrices are:

$$e^{-i\alpha J_z} = \begin{pmatrix} e^{-i\alpha} & 0 & 0 \\ 0 & 1 & 0 \\ 0 & 0 & e^{i\alpha} \end{pmatrix} \qquad e^{-i\gamma J_z} = \begin{pmatrix} e^{-i\gamma} & 0 & 0 \\ 0 & 1 & 0 \\ 0 & 0 & e^{i\gamma} \end{pmatrix}$$

For the second matrix one needs to calculate all powers of $\mathbf{J}_y$. It turns out that

$$J_y^2 = \frac{1}{2}\begin{pmatrix} 1 & 0 & -1 \\ 0 & 2 & 0 \\ -1 & 0 & 1 \end{pmatrix} \qquad J_y^3 = \frac{1}{2\sqrt{2}}\begin{pmatrix} 0 & -2i & 0 \\ 2i & 0 & -2i \\ 0 & 2i & 0 \end{pmatrix} = J_y$$

so that $\mathbf{J}_y^4 = \mathbf{J}_y^2$, $\mathbf{J}_y^5 = \mathbf{J}_y^3 = \mathbf{J}_y$, ...
The expansion in powers of $\mathbf{J}_y$ is then very simple:

$$\exp(-i\beta J_y) = I - i\beta J_y + \frac{1}{2!}(-i\beta)^2 J_y^2 + \frac{1}{3!}(-i\beta)^3 J_y^3 + \ldots =$$

$$= I - J_y^2 + J_y^2\left[1 - \frac{1}{2!}\beta^2 + \ldots\right] - iJ_y\left[\beta - \frac{1}{3!}\beta^3 + \ldots\right] =$$

$$= \left(I - J_y^2\right) + J_y^2\cos(\beta) - iJ_y\sin(\beta)$$

Therefore the reduced Wigner's matrix $d^j_{m,m'}(\beta)$ for this case is:

$$\begin{array}{c|ccc} & m'=1 & m'=0 & m'=-1 \\ m=1 & \frac{1}{2}(1+\cos(\beta)) & -\frac{1}{\sqrt{2}}\sin(\beta) & \frac{1}{2}(1-\cos(\beta)) \\ m=0 & \frac{1}{\sqrt{2}}\sin(\beta) & \cos(\beta) & -\frac{1}{\sqrt{2}}\sin(\beta) \\ m=-1 & \frac{1}{2}(1-\cos(\beta)) & \frac{1}{\sqrt{2}}\sin(\beta) & \frac{1}{2}(1+\cos(\beta)) \end{array}$$

The trigonometric identities $\frac{1}{2}(1 + \cos(\beta)) = \cos^2(\beta/2)$ and $\frac{1}{2}(1 - \cos(\beta)) = \sin^2(\beta/2)$ can be used in order to rewrite all elements of the form $1 \pm \cos(\beta)$. The full Wigner's rotation matrix is:

$$D(\alpha,\beta,\gamma) = \begin{pmatrix} e^{-i(\alpha+\gamma)}\cos^2\left(\frac{\beta}{2}\right) & -e^{-i\alpha}\frac{\sin(\beta)}{\sqrt{2}} & e^{i(\alpha-\gamma)}\sin^2\left(\frac{\beta}{2}\right) \\ e^{-i\gamma}\frac{\sin(\beta)}{\sqrt{2}} & \cos(\beta) & -e^{i\gamma}\frac{\sin(\beta)}{\sqrt{2}} \\ e^{i(\alpha-\gamma)}\sin^2\left(\frac{\beta}{2}\right) & e^{i\alpha}\frac{\sin(\beta)}{\sqrt{2}} & e^{i(\alpha+\gamma)}\cos^2\left(\frac{\beta}{2}\right) \end{pmatrix}$$

Suggested References

Same as for Appendix A.

APPENDIX C

ANGULAR MOMENTUM COUPLING COEFFICIENTS

For a single particle there is only one angular momentum operator **J**. For composite systems one needs to deal with more than one such operator. The combination of such operators, each of which is a vector, leads to the definition of total angular momentum and to the more general concept of a tensor operator. The practical way to construct useful combinations is through a systematic treatment of operator products. We shall thus start by treating such products, and then tensor operators will be defined.

Suppose a system consists of two elementary parts, with angular momentum operators $\mathbf{J}_1$, $\mathbf{J}_2$ respectively. The two parts may be two electrons, or two protons, or a proton and an electron etc. One basis for the space of wave functions for the whole system is that of "angular momentum product functions", which are direct products of eigenfunctions of the two angular momentum operators. If $|j_k, m_k\rangle$ are defined (for k=1,2) by:

$$J_k^2\,|j_k,m_k\rangle = j_k(j_k+1)\,|j_k,m_k\rangle \qquad and \qquad J_{kz}\,|j_k,m_k\rangle = m_k\,|j_k,m_k\rangle \tag{C-1}$$

(taking $\hbar=1$), the product function is $|j_1, j_2, m_1, m_2\rangle \equiv |j_1, m_1\rangle \otimes |j_2, m_2\rangle$. The direct product sign is usually omitted, so $|j_1, j_2, m_1, m_2\rangle = |j_1, m_1\rangle|j_2, m_2\rangle$ is the conventional way of writing this function. The product function is an eigenfunction of the direct product operators $\mathbf{J}_1^2 \otimes \mathbf{I}_2$, $\mathbf{J}_{1z} \otimes \mathbf{I}_2$, $\mathbf{I}_1 \otimes \mathbf{J}_2^2$ and $\mathbf{I}_1 \otimes \mathbf{J}_{2z}$, where $\mathbf{I}_k$ is the unit operator in the k'th subspace. These operators are usually abbreviated as $\mathbf{J}_1^2$, $\mathbf{J}_{1z}$, $\mathbf{J}_2^2$ and $\mathbf{J}_{2z}$. For example,

$$J_1^2 \otimes I_2\,|j_1,m_1\rangle|j_2,m_2\rangle = \left(J_1^2\,|j_1,m_1\rangle\right) \otimes \left(I_2\,|j_2,m_2\rangle\right) = j_1(j_1+1)\,|j_1,m_1\rangle|j_2,m_2\rangle \tag{C-2}$$

Another important basis for the same space is that of simultaneous eigenfunctions of $\mathbf{J}_1^2$, $\mathbf{J}_2^2$, $\mathbf{J}^2$ and $\mathbf{J}_z$, where $\mathbf{J} = \mathbf{J}_1 + \mathbf{J}_2$ is the total angular momentum operator. Thus

$$J^2 = J_1^2 + J_2^2 + 2\left(J_{1x}J_{2x} + J_{1y}J_{2y} + J_{1z}J_{2z}\right) =$$

$$= J_1^2 + J_2^2 + J_{1+}J_{2-} + J_{1-}J_{2+} + 2J_{1z}J_{2z} \tag{C-3a}$$

$$J_z = J_{1z} + J_{2z} \tag{C-3b}$$

These functions are written as $|j_1, j_2, J, M\rangle$, and they satisfy the conditions:

$$J_k^2\,|j_1,j_2,J,M\rangle = j_k(j_k+1)\,|j_1,j_2,J,M\rangle \qquad (k=1,2) \tag{C-4a}$$

$$J^2\,|j_1,j_2,J,M\rangle = J(J+1)\,|j_1,j_2,J,M\rangle \tag{C-4b}$$

$$J_z\,|j_1,j_2,J,M\rangle = M\,|j_1,j_2,J,M\rangle \tag{C-4c}$$

The transformation between these two basis sets makes it possible to expand a member of one basis in terms of the members of the other basis. Using the projection operators $|j_1,j_2,m_1,m_2\rangle\langle j_1,j_2,m_1,m_2|$ (see Ch. 1) one may write:

$$|j_1,j_2,J,M\rangle = \sum_{m_1,m_2} |j_1,j_2,m_1,m_2\rangle\langle j_1,j_2,m_1,m_2|j_1,j_2,J,M\rangle \tag{C-5}$$

The transformation element is $\langle j_1,j_2,m_1,m_2|j_1,j_2,J,M\rangle$. The elements of the transformation matrix are known as **Clebsch-Gordan coefficients**, and there are many different abbreviated notations for them. We shall use here the notation $\langle j_1,j_2,m_1,m_2|J,M\rangle \equiv \langle j_1,j_2,m_1,m_2|j_1,j_2,J,M\rangle$. Since $\mathbf{J_1}$, $\mathbf{J_2}$ operate independently, it is possible to show that this coefficient is equal to zero unless:

$$M = m_1 + m_2 \qquad and \qquad |j_1-j_2| \le J \le |j_1+j_2| \tag{C-6}$$

The inequalities for J are known as the "triangle condition", since they are similar to the inequalities relating the lengths of three sides of a triangle. Therefore the allowed values for J are: $J = -|j_1 - j_2|, -|j_1 - j_2| + 1, \ldots, |j_1 + j_2| - 1, |j_1 + j_2|$ and the allowed values for M are $M = -(m_1 + m_2), -(m_1 + m_2) + 1, \ldots, (m_1 + m_2) - 1, (m_1 + m_2)$. A more symmetric form for the transformation coefficients is the **Wigner 3-j symbol**:

$$\begin{pmatrix} j_1 & j_2 & J \\ m_1 & m_2 & M \end{pmatrix} = \frac{1}{\sqrt{2J+1}}\,(-1)^{j_1-j_2+M}\,\langle j_1,j_2,m_1,m_2|J,-M\rangle \tag{C-7}$$

This symbol vanishes unless $m_1 + m_2 + M = 0$, and J satisfies the triangle condition as above.

Example C.1: *Coupling two electronic spins*

Suppose the system consists of two electrons, with no orbital angular momentum. Thus $\mathbf{J_k}$ are the spin operators of the two electrons, for which $j_1 = j_2 = ½$. Omitting the j_k values, the product functions are $|+,+\rangle \equiv |m_1=½, m_2=½\rangle$, $|+,-\rangle \equiv |½,-½\rangle$, $|-,+\rangle \equiv |-½,½\rangle$ and $|-,-\rangle \equiv |-½,-½\rangle$. According to the triangle rule, $0 \le J \le 1$, so the allowed values of J are 0 and 1. Therefore the basis $\{|j_1,j_2,J,M\rangle\}$ in which $\mathbf{J}^2$ and $\mathbf{J_z}$ are diagonal consists of the following wave functions (omitting the j_k values): $|0,0\rangle \equiv |J=0, M=0\rangle$, $|1,1\rangle$, $|1,0\rangle$ and $|1,-1\rangle$. The first function is called a singlet state, and the other three constitute the triplet state for the two electrons. The Clebsch-Gordan coefficients can be calculated from general expressions, but in this simple case it is easy to calculate them directly by evaluating the result of operating with $\mathbf{J}^2$ and $\mathbf{J}_z$ on the product functions. The condition $M = m_1 + m_2$ implies that $|1,1\rangle = |+,+\rangle$ and $|1,-1\rangle = |-,-\rangle$. For the other two states, it is clear that $M = 0$, and:

$$J^2\,|+,-\rangle = \left\{J_1^{\,2} + J_2^{\,2} + J_{1+}J_{2-} + J_{1-}J_{2+} + 2J_{1z}J_{2z}\right\}|+,-\rangle =$$

$$= \left\{\frac{1}{2}\left[\frac{1}{2}+1\right] + \frac{1}{2}\left[\frac{1}{2}+1\right] + 0\cdot 0 + 2\cdot\frac{1}{2}\cdot\left[-\frac{1}{2}\right]\right\}|+,-\rangle + \{1\cdot 1\}\,|-,+\rangle =$$

$$= |+,-\rangle + |-,+\rangle$$

Similarly,

$$J^2\,|-,+\rangle = |+,-\rangle + |-,+\rangle$$

Taking the normalized symmetric and anti-symmetric combinations of these states:

$$J^2\psi_S \equiv J^2\left\{\frac{1}{\sqrt{2}}(|+,-\rangle + |-,+\rangle)\right\} = 2\cdot\frac{1}{\sqrt{2}}(|+,-\rangle + |-,+\rangle) = 1(1+1)\,\psi_S$$

$$J^2\psi_A \equiv J^2\left\{\frac{1}{\sqrt{2}}(|+,-\rangle - |-,+\rangle)\right\} = 0\cdot\frac{1}{\sqrt{2}}(|+,-\rangle - |-,+\rangle) = 0(0+1)\,\psi_A$$

These results are exactly the defining relations of eigenfunctions of $\mathbf{J}^2$ with eigenvalues of 1 and 0 respectively. The conclusion is that $\psi_S = |1,0\rangle$ and $\psi_A = |0,0\rangle$. Summarizing, the Clebsch-Gordan coefficients for this case are:

$$\langle \tfrac{1}{2},\tfrac{1}{2},m_1,m_2\,|\,1,1\rangle = \begin{cases} 1 & \textit{for}\ \ m_1 = m_2 = \tfrac{1}{2} \\ 0 & \textit{otherwise} \end{cases}$$

$$\langle \tfrac{1}{2},\tfrac{1}{2},m_1,m_2\,|\,1,-1\rangle = \begin{cases} 1 & \textit{for}\ \ m_1 = m_2 = -\tfrac{1}{2} \\ 0 & \textit{otherwise} \end{cases}$$

$$\langle \tfrac{1}{2},\tfrac{1}{2},m_1,m_2\,|\,1,0\rangle = \begin{cases} 1/\sqrt{2} & \textit{for}\ \ m_1 = -m_2 = \pm\tfrac{1}{2} \\ 0 & \textit{otherwise} \end{cases}$$

$$\langle \tfrac{1}{2},\tfrac{1}{2},m_1,m_2\,|\,0,0\rangle = \begin{cases} \pm 1/\sqrt{2} & \textit{for}\ \ m_1 = -m_2 = \pm\tfrac{1}{2} \\ 0 & \textit{otherwise} \end{cases}$$

The 3-j coefficients for this case are

$$\begin{pmatrix} \frac{1}{2} & \frac{1}{2} & 1 \\ m_1 & m_2 & -1 \end{pmatrix} = \begin{cases} -1/\sqrt{3} & m_1 = m_2 = \frac{1}{2} \\ 0 & otherwise \end{cases}$$

$$\begin{pmatrix} \frac{1}{2} & \frac{1}{2} & 1 \\ m_1 & m_2 & 1 \end{pmatrix} = \begin{cases} -1/\sqrt{3} & m_1 = m_2 = -\frac{1}{2} \\ 0 & otherwise \end{cases}$$

$$\begin{pmatrix} \frac{1}{2} & \frac{1}{2} & 1 \\ m_1 & m_2 & 0 \end{pmatrix} = \begin{cases} 1/\sqrt{6} & m_1 = -m_2 = \pm\frac{1}{2} \\ 0 & otherwise \end{cases}$$

$$\begin{pmatrix} \frac{1}{2} & \frac{1}{2} & 0 \\ m_1 & m_2 & 0 \end{pmatrix} = \begin{cases} \pm 1/\sqrt{2} & m_1 = -m_2 = \pm\frac{1}{2} \\ 0 & otherwise \end{cases}$$

****

Suggested References

1,2. Same as for Appendix A.

3. L.C. Biedenharn and J.D. Louck, *"Angular Momentum in Quantum Physics"* (Addison-Wesley, Reading, MA, 1981).

APPENDIX D

IRREDUCIBLE SPHERICAL TENSORS

The concept of a vector can be generalized to the so called "Cartesian tensors", in which the single index of the vector is replaced by a set of several indices. For example, if A, B, C,... are vectors, with components A_i, B_j, C_k,... respectively, one may define the two dimensional tensor AB as a quantity having two indices, with components $(AB)_{ij} = A_iB_j$. The term Cartesian refers to the fact that the indices are related to the three spatial dimensions defined by Cartesian coordinates. In this example, each component of a two-dimensional tensor is defined as a product of components of vectors, which are regarded as one-dimensional tensors. More generally, a two dimensional tensor may be defined with no reference to any vectors. For example, a matrix M with elements $M_{i,j}$ is a two-dimensional Cartesian tensor. Similarly one can define a three dimensional tensor, for example using vectors A, B, C to define ABC by: $(ABC)_{ijk} = A_iB_jC_k$, and so on. A Liouville space matrix can thus be regarded as a four-dimensional tensor, etc.

However, in problems involving angular momentum it is not convenient to work with such tensors. It is much more natural to form linear combinations of Cartesian tensors which have simple transformation properties under rotations. Such combinations do exist, and they are known as **irreducible spherical tensors.** There are two equivalent definitions for such tensors. Proving the equivalence both ways is not trivial (see Ref. 3), so the definitions will be given here without proving the relationship between them.

The **algebraic definition** is:
An irreducible spherical tensor of rank k, $\mathbf{T_k}$, is a set of (2k+1) operators indexed by q = -k, -k+1, ..., k possessing the following commutation relations with the angular momentum operators:

$$[J_z, T_k^q] = \hbar q T_k^q \tag{D-1a}$$

$$[J_\pm, T_k^q] = \hbar\left(k(k+1) - q(q\pm1)\right)^{1/2} T_k^{q\pm1} \tag{D-1b}$$

The **geometrical definition** is:
An irreducible spherical tensor of rank k, $\mathbf{T_k}$, is a set of (2k+1) operators which transform under rotation as

$$T_k^{q\prime} = \sum_p T_k^p D^k{}_{p,q}(\alpha,\beta,\gamma) \tag{D-2}$$

Here $\Omega=(\alpha,\beta,\gamma)$ is the set of Euler angles specifying the rotation, $D^k{}_{p,q}$ are Wigner's rotation matrices, and the prime on the left hand side of the equation indicates that it refers to the new (transformed) system of axes, whereas the unprimed operator on the right hand side refers to the old system of axes (prior to the transformation).

An important example is the set of operators T_l^m (with m = -l, -l+1, ..., l) defined as multiplying the operand by the spherical harmonic wave function $Y^l_m(\theta,\phi)$: $T_l^m\psi = Y^l_m(\theta,\phi)\psi$. In fact, since the set of spherical harmonics is the most natural set of functions for which Eqs. (D-1),(D-2) are obtained, one may regard the concept of irreducible spherical tensors as a generalization of the definition of spherical harmonics.

It is instructive to look at a simple example of the class of spherical tensors related to spherical harmonics. The first rank tensor T_1 is defined as the set of operators $\{ T_1^{-1}, T_1^0, T_1^1 \}$ which act on a function by multiplying it with the spherical harmonics $Y_1^{-1}(\theta,\phi)$, $Y_1^0(\theta,\phi)$ and $Y_1^1(\theta,\phi)$ respectively:

$$T_1^{-1}\psi = Y_1^{-1}(\theta,\phi)\,\psi = \left[\frac{3}{8\pi}\right]^{1/2} \sin(\theta)\, e^{-i\phi}\,\psi \qquad \text{(D-3a)}$$

$$T_1^{0}\psi = Y_1^{0}(\theta,\phi)\,\psi = \left[\frac{3}{4\pi}\right]^{1/2} \cos(\theta)\,\psi \qquad \text{(D-3b)}$$

$$T_1^{1}\psi = Y_1^{1}(\theta,\phi)\,\psi = -\left[\frac{3}{8\pi}\right]^{1/2} \sin(\theta)\, e^{i\phi}\,\psi \qquad \text{(D-3c)}$$

How is this first rank tensor related to first rank Cartesian tensors, i.e., ordinary vectors? The answer can be found easily if the vector is described in terms of the spherical coordinates θ and ϕ. For an ordinary vector **A** in three-dimensional space it is known that

$$A_x = A\sin(\theta)\cos(\phi) \qquad A_y = A\sin(\theta)\sin(\phi) \qquad A_z = A\cos(\theta) \qquad \text{(D-4)}$$

where A is the magnitude (length) of **A**, θ is the polar angle - the angle **A** makes with the **z** axis, and ϕ is the azimuthal angle - the angle the projection of **A** on the **x-y** plane makes with the **x** axis. Choosing a unit vector, namely A = 1, it follows that

$$Y_1^{-1}(\theta,\phi) = \left[\frac{3}{8\pi}\right]^{1/2} (A_x - iA_y) = \left[\frac{3}{4\pi}\right]^{1/2} \frac{1}{\sqrt{2}}(A_x - iA_y) \qquad \text{(D-5a)}$$

$$Y_1^{0}(\theta,\phi) = \left[\frac{3}{4\pi}\right]^{1/2} A_z \qquad \text{(D-5b)}$$

$$Y_1^{1}(\theta,\phi) = -\left[\frac{3}{8\pi}\right]^{1/2} (A_x + iA_y) = \left[\frac{3}{4\pi}\right]^{1/2} \frac{-1}{\sqrt{2}}(A_x + iA_y) \qquad \text{(D-5c)}$$

As expected, each component of the first rank irreducible spherical tensor can be expressed as a linear combination of the components of a corresponding Cartesian vector. The overall factor of $(3/4\pi)^{1/2}$ in this particular case is needed only for normalization of the integrals over the spherical harmonics. Like the magnitude A, it is irrelevant to the transformations.

Following the definitions given above one may define a zero rank tensor $\mathbf{T}_0^0$ as a single operator which commutes with all angular momentum operators (from the algebraic definition) and is invariant under rotations (from the geometrical definition). Thus $\mathbf{T}_0^0$ must be the operation of multiplying with a scalar. Since it has only one component, it is also a zero rank Cartesian tensor - no linear combinations are needed.

Another important class of irreducible spherical tensors is the class of tensors formed from angular momentum operators. For example, the first rank tensor of this type must be constructed from the angular momentum vector **J**. From Eqs. (A-16),(D-1) it is clear that $\mathbf{T}_1^{\pm1}$ are proportional to $\mathbf{J}_\pm$ and $\mathbf{T}_1^0$ is proportional to $\mathbf{J}_z$. Assuming with no loss of generality that $\mathbf{T}_1^0 = \mathbf{J}_z$ one has from Eq. (D-1b):

$$[J_+, T_1^{-1}] = \hbar\sqrt{2}\, T_1^0 = \frac{1}{\sqrt{2}}[J_+, J_-] \tag{D-6a}$$

$$[J_-, T_1^{1}] = \hbar\sqrt{2}\, T_1^0 = -\frac{1}{\sqrt{2}}[J_-, J_+] \tag{D-6b}$$

These equations imply: $\mathbf{T}_1^{-1} = (1/\sqrt{2})\,\mathbf{J}_-$ and $\mathbf{T}_1^{1} = -(1/\sqrt{2})\,\mathbf{J}_+$.

Let us now return to the general case with spherical tensors. If one multiplies components of two irreducible tensors and sums over the products, the result is always a tensor, i.e. a quantity defined by several indices, but is generally not an irreducible tensor. For example, suppose one defines an operator F which multiplies functions by

$$F = Y_1^{-1}(\theta,\phi)\,Y_1^0(\theta,\phi) + Y_1^{-1}(\theta,\phi)\,Y_1^1(\theta,\phi) \tag{D-7}$$

This function F is a single function, but is is not a scalar - it is not a zero rank tensor. Or suppose one defines $F_{i,j}$ as a set of operators which multiply functions by

$$F_{i,j} = Y_1^i(\theta,\phi)\,Y_1^j(\theta,\phi) \tag{D-8}$$

This is a set of six functions (and not nine, because $F_{i,j} = F_{j,i}$) which is not an irreducible spherical tensor. Nevertheless, linear combinations of its components can be constructed to form the scalar Y_0^0 and the five-component second rank tensor $Y_2^m(\theta,\phi)$. There are two ways of looking at such products. On the one hand, each product (each $F_{i,j}$) is a linear combination of some irreducible spherical tensor components. On the other hand, specific components of irreducible tensors can be formed from linear combinations of the $F_{i,j}$. More generally, it is possible to decompose products of tensor components into sums of irreducible tensors, or to combine them so as to generate new irreducible tensors. In order to get a specific irreducible tensor as a result of such a linear combination, the component products must be added with appropriate coefficients. It turns out that these are exactly the Clebsch-Gordan coefficients encountered above. The formula for generating the irreducible tensor $\mathbf{V}_K$ out of the irreducible tensors $\mathbf{T}_L$, $\mathbf{U}_M$ is:

$$V_K^q = \sum_{r,s} T_L^r \, U_M^s \, \langle L, M, r, s \,|\, K, q \rangle \tag{D-9}$$

An important special case is that of forming a scalar, i.e., a tensor of rank 0, out of two tensors of non-zero rank. Using Eq. (D-9) it is apparent that in such a case r+s = 0, and also that |L-M| ≤ 0 ≤ L+M . Therefore L = M, and so the relevant 3-j symbol is

$$\begin{pmatrix} L & L & 0 \\ r & -r & 0 \end{pmatrix} = \frac{1}{\sqrt{2{\cdot}0+1}} (-1)^{L-L+0} \langle L, L, r, -r \,|\, 0, 0 \rangle = (-1)^{L-r} \frac{1}{\sqrt{2{\cdot}L+1}} \tag{D-10}$$

where the last equality on the right hand side is based on standard formulas for 3-j coefficients. Therefore if $\mathbf{W_0}$ is a scalar formed from two irreducible tensors in this manner, it is equal to:

$$W_0^0 = \sum_m T_L^m \, U_L^{-m} \, (-1)^{L-m} \sqrt{2L+1} \tag{D-11}$$

This applies, for example, to Hamiltonians, because a Hamiltonian is always invariant under rotations, so it is a scalar, but it may depend on various tensors, which do have a non-trivial orientation dependence.

Another important special case is that of forming a second rank tensor out of a first rank tensor, by "squaring" it. In this case Eq. (D-9) is used with: $\mathbf{T_L^r} = \mathbf{T_1^r}$, $\mathbf{U_M^s} = \mathbf{T_1^s}$ and K = 2 . The relevant coefficients are

$$\langle 1, 1, r, s \,|\, 2, q \rangle = \sqrt{5} \, (-1)^q \begin{pmatrix} 1 & 1 & 2 \\ r & s & -q \end{pmatrix} \tag{D-12}$$

The values of these coefficients can be found in standard tables to be:

$$\begin{pmatrix} 1 & 1 & 2 \\ r & s & -q \end{pmatrix} = (-1)^q \left(\frac{2!\,2!\,(2-q)!\,(2+q)!}{5!\,(1+r)!\,(1-r)!\,(1+s)!\,(1-s)!} \right)^{1/2} \tag{D-13}$$

Therefore:

$$V_2^0 = \frac{1}{\sqrt{6}} \left(T_1^1 \, T_1^{-1} + T_1^{-1} \, T_1^1 + 2 \, T_1^0 \, T_1^0 \right) \tag{D-14a}$$

$$V_2^{\pm 1} = \frac{1}{\sqrt{2}} \left(T_1^{\pm 1} \, T_1^0 + T_1^0 \, T_1^{\pm 1} \right) \tag{D-14b}$$

$$V_2^{\pm 2} = T_1^{\pm 1} \, T_1^{\pm 1} \tag{D-14c}$$

For example, for the tensor defined as: $\mathbf{T_1^{-1}} = (1/\sqrt{2})\,\mathbf{J_-}$, $\mathbf{T_1^0} = \mathbf{J_z}$ and $\mathbf{T_1^1} = -(1/\sqrt{2})\,\mathbf{J_+}$

one obtains:

$$V_2^0 = \frac{1}{\sqrt{6}}\left(-\frac{1}{2}(J_+J_- + J_-J_+) + 2(J_z)^2\right) =$$

$$= \frac{1}{\sqrt{6}}\left(3(J_z)^2 - J^2\right) \tag{D-15a}$$

$$V_2^{\pm 1} = \mp\frac{1}{2}(J_\pm J_z + J_z J_\pm) \tag{D-15b}$$

$$V_2^{\pm 2} = \frac{1}{2}(J_\pm)^2 \tag{D-15c}$$

Example D.1: *An orientation dependent Hamiltonian*

In magnetic resonance one is often interested in quantities which depend on orientation in real space. At the same time, spin operators which depend on orientation in spin space are always present in calculations for magnetic resonance. The Hamiltonian is a sum of scalars, each of which is formed by tensor products as indicated above. Using **A** for tensors of spin operators and **F** for tensors of orientation dependent interaction parameters, the Hamiltonian is of the general form

$$H = \sum_\lambda \sum_L (-1)^L \frac{1}{\sqrt{2L+1}} \sum_m (-1)^m A_\lambda^{(L,m)} F_\lambda^{(L,-m)}$$

The labeling of the individual components by λ implies that there may be several physically different interactions, all of which have the same tensor rank. Because of this additional label, written as a subscript, the two labels which specify rank and component (L and m) are both written here as superscriprts, to avoid confusion. In practice, the magnetic resonance Hamiltonian is constructed in the general case from several zero rank and second rank tensors. This includes chemical shielding, dipolar interactions and the hyperfine interaction. The spin operators are expressed in the laboratory frame of reference, because in this way the result of the measurement can be expressed simply as a transverse magnetization component. The tensor of interaction constants must also be written in the same frame of reference, but this is not convenient. It has a simple form in the molecular frame of reference, and the latter is generally tilted in an arbitrary manner with respect to the laboratory frame of reference. The tilt angles are functions of time, and change rapidly with molecular tumbling. It is therefore necessary to write $F_\lambda^{(L,-m)}$ as the result of a transformation from the molecular frame. If $\Omega=(\alpha,\beta,\gamma)$ is the orientation of the laboratory axes with respect to molecular axes, then

$$H = \sum_{\lambda} \sum_{L} (-1)^L \frac{1}{\sqrt{2L+1}} \sum_{m,m'} (-1)^m A_{\lambda}^{(L,m)} F_{\lambda}^{(L,-m')'} D^L_{-m',-m}(\Omega)$$

It may be convenient to use the orientation of the molecule with respect to the laboratory, rather than the other way round. This orientation is $-\Omega=(-\gamma,-\beta,-\alpha)$ and one can use the relation

$$D^L_{-m',-m}(\Omega) = (-1)^{m-m'} D^L_{m,m'}(-\Omega)$$

to obtain finally:

$$H = \sum_{\lambda} \sum_{L} (-1)^L \frac{1}{\sqrt{2L+1}} \sum_{m,m'} (-1)^{m'} A_{\lambda}^{(L,m)} F_{\lambda}^{(L,-m')'} D^L_{m,m'}(-\Omega)$$

Suggested References

Same as for Appendix C.

APPENDIX E

REPRESENTATIONS OF FINITE GROUPS

The systematic study of symmetry in physics and chemistry is based on the application of mathematical group theory. There are many textbooks in which the detailed theory of groups or its numerous applications are presented. Here we shall only review some definitions and theorems from that theory, together with some simple examples, as a background to the application of symmetry considerations in Chapter III. A **group** is defined in algebra as a set of elements $G=\{A_1, A_2, ...\}$ in which a combination rule is defined, associating with any ordered pair of elements (A_i, A_j) an element of the group A_r. This combination rule is often called "the group multiplication", but it can be either ordinary addition, or ordinary multiplication, or some more sophisticated operation. For convenience in notation we shall regard the operation as multiplication, writing it as, in this example: $A_i A_j = A_r$. Thus the result A_r of the operation will be called the "product" of A_i and A_j. However, defining a set G with a combination rule is not sufficient. G is a group only if the following requirements are satisfied:

(a) The combination rule is associative: $A_i (A_j A_k) = (A_i A_j) A_k$

(b) There is a unit (or identity) element E, such that for any element A: $AE = EA = A$.

(c) For each element A in the group there is an inverse A^{-1} such that: $AA^{-1} = A^{-1}A = E$.

For example, if the group elements are ordinary numbers (integers or real numbers) and the combination rule is addition, the "unit element" is the number zero, and the "inverse" of A is -A. If the group elements are ordinary non-zero numbers (integers or real numbers) and the combination rule is multiplication, the "unit element" is the number one, and the "inverse" of A is $1/A$.

Commutativity is not assumed in the general definition of a group. If the group multiplication is commutative, namely $AB = BA$ for each pair of elements in G, then the group is called **abelian**. A subset $H \subseteq G$, containing some (or all) of the elements of a group G, is called a **subgroup** of G if for each pair of elements A,B belonging to H, also the product AB belongs to H ($AB \in H$). Any group G is trivially a subgroup of itself. If H is a subgroup of G and is different from G ($H \subset G$), then H is called a **proper subgroup** of G. For any given group G, the group $G_E \equiv \{E\}$ containing only the unit element is trivially a subgroup of G.

The usefulness of group theory results from the fact that symmetry operations, i.e., operations which change some aspect of the system without changing some of its important characteristics, form groups. This is true for translation operations, for rotations, and for other geometrical operations, as well as for some more abstract operations in physics. We shall be concerned with groups of geometrical symmetry operations, as well as with groups of more abstract operations, related to Hilbert space operators.

The number of elements in a group is called the **order** of the group. This order may be finite or infinite. Infinite order groups occur in the study of some systems, which are symmetric under the continuous change of a variable. This is the case with the group of three dimensional rotations, which is relevant to angular momentum theory. Molecules which are axially symmetric, i.e., are not changed by rotation at any angle about a certain axis, also have a symmetry group of infinite order. Here we shall deal with finite groups, i.e., groups of finite order. Such groups occur naturally for molecules which are not physically changed by some discrete set of "geometrical" transformations. Such transformations may exchange the locations of some atoms without changing the configuration of the molecule or its Hamiltonian. If all atoms are labeled, such an operation exchanges the labels but does not change any physical characteristic of the system. It is possible to show that for any subgroup H of an arbitrary group G, the order of the subgroup divides the order of the group:

$$\frac{g}{h} = k \qquad (k \textit{ is an integer}) \tag{E-1}$$

where g is the order of G, h the order of H, and the integer k is called the **index** of H in G.

If A is an element belonging to a finite group G of order N, then the set of all powers of A (calculated with the group multiplication) is necessarily finite, because there cannot be more than N different values of these powers. Thus there is a minimal power n such that $A^n = E$, where obviously $n \leq N$. The set $\{A, A^2, ..., A^n = E\}$ with the group multiplication rule is a subgroup of G, and it is called a **cyclic** group. Cyclic groups are necessarily abelian. Rotations by an angle of $2\pi/n$ about a given axis, where n is an integer, are a cyclic group. For arbitrary n this is a subgroup of the group of all three dimensional rotations. For finite symmetry groups, a subgroup of this type can be formed by choosing a value of n appropriate to the group. For example, the group C_{3v} contains as symmetry operations the rotations by $2\pi/3$ about a certain axis. Thus the group of rotations $\{R(2\pi/3), R(4\pi/3), R(2\pi)\}$ where $R(\alpha)$ is rotation by α, is a cyclic subgroup of C_{3v}.

Given a set of *n* objects, the set of all permutations of these objects forms a group. If these objects are physical objects differing only in label, such as identical atoms occupying different locations in a molecule, such a permutation is a physical symmetry operation. Writing the labels of the objects as 1,2,...n (in this order!), the permutation rearranges this set of labels as $p_1, p_2, ..., p_n$ where each p_i is one of the original labels, and none of the p_i is repeated twice. The standard notation for such a permutation is:

$$\begin{bmatrix} 1 & 2 & 3 & \cdots & n \\ p_1 & p_2 & p_3 & \cdots & p_n \end{bmatrix} \tag{E-2}$$

The group multiplication is, in this case, the succesive operation of permutations. The order of the group is n! .

For any order n, the set of all square n × n matrices with the standard matrix multiplication forms a group. The unit element is the standard unit matrix, in which $(E)_{ij} = \delta_{ij}$. Since matrices can be used to represent physical operators such as rotation of vectors (in real or abstract space), this type of group is very useful. It can be used to represent geometrical operations in three dimensional space, or abstract rotations in Hilbert space, generated by the level-shift operators of Chapter I.

Two groups G, H are **isomorphic** if they have the same order, and there is a one-to-one rule relating each element $A \in G$ with some element $B \in H$, such that the multiplication is preserved by this relation. That is, different elements in G are related to different elements in H, and if $B_1 = f(A_1)$ is the element related to A_1 in this manner, and $B_2 = f(A_2)$ is the element related to A_2, then $f(A_1 A_2) = f(A_1)\, f(A_2)$. In this equation the product of the A_i is the product defined in G, and the product of $f(A_i)$ is the product defined in H. If two groups are related in a similar manner, except that several elements $A_1, A_2, \ldots$ in one group are related to the same element B_1 in the other group (the relation is not one-to-one), the two groups are **homomorphic**. Isomorphism is thus a special kind of homomorphism. If a homomorphism is established between some group of geometrical operations and a group of matrices, the matrices can be regarded as representing those operations. If the two groups are isomorphic, the representation is said to be **faithful**. Note that two groups of matrices, for example, can be isomorphic even if their dimension is different. For example, the cyclic group {(-1), (1)} consisting of two one dimensional matrices is isomorphic to the group consisting of the two two dimensional matrices

$$\begin{bmatrix} 0 & 1 \\ 1 & 0 \end{bmatrix} \quad and \quad \begin{bmatrix} 1 & 0 \\ 0 & 1 \end{bmatrix}$$

In an arbitrary group two elements A,B are said to be conjugate if there is some element X in the same group such that

$$B = XAX^{-1} \quad or \quad A = X^{-1}BX \tag{E-3}$$

A set of mutually conjugate elements is called a **class**. The class of an element A may be found by applying to it all group elements in the following way:

$$A_1 A A_1^{-1} \;,\; A_2 A A_2^{-1} \;\cdots \tag{E-4}$$

In general there will be repetitions in the sequence. It is easy to show that each group element belongs to one, and only one, class. In any group, the unit element is in a class by itself. In any abelian group, each element is in a class by itself.

It is possible to show that any finite group has at least one matrix representation. If the elements of a finite group are represented by matrices, then the matrices representing conjugate elements have the same trace (sum of diagonal elements). This is because conjugate elements are represented by conjugate matrices, which are realted by a similarity transformation (Eq. (E-3) for matrices). The trace of a matrix is invariant under a similarity transformation.

Suppose one has a matrix representation of dimension k of a group G. Then each element in G is represented by a $k \times k$ matrix, and these matrices form a group H. The representation is called **reducible** if there is a similarity transformation which will transform all elements of H to block-diagonal matrices, with equal dimension of the blocks. For example, it will transform each element of H to a matrix B in which $B_{ij} = 0$ for all elements in which $i > m$, $j \leq m$ or in which $i \leq m$, $j > m$, where $m < k$ is the dimension of the first block, and k-m is the dimension of the second block, as shown in Eq. (E-5).

$$\begin{pmatrix} A_{1,1} & A_{1,2} & \cdots & A_{1,m} & 0 & \cdots & 0 \\ A_{2,1} & A_{2,2} & \cdots & A_{2,m} & 0 & \cdots & 0 \\ \cdots & \cdots & \cdots & \cdots & \cdots & \cdots & \cdots \\ A_{m,1} & A_{m,2} & \cdots & A_{m,m} & 0 & \cdots & 0 \\ 0 & 0 & \cdots & 0 & A_{m+1,m+1} & \cdots & A_{m+1,m+k} \\ \cdots & \cdots & \cdots & \cdots & \cdots & \cdots & \cdots \\ 0 & 0 & \cdots & 0 & A_{m+k,m+1} & \cdots & A_{m+k,m+k} \end{pmatrix} \tag{E-5}$$

If the representation is not reducible, it is called an **irreducible representation.** The irreducible representations of a group characterize it completely. Equivalent representations can be considered, for this purpose, as identical. The number of different (non-equivalent) irreducible representations of a group is equal to the number of its classes. Within each class the elements are mutually conjugate, so the matrices which represent them in a given representation have the same trace. This trace is also called the **character** of the matrix.

The matrices of a representation can be regarded as transformations operating on some set of vectors, or basis functions. A general function, on which the group operations can be applied, can be decomposed into distinct parts, each of which can act as a basis function for one of the irreducible representations. Each such part is said to belong to that representation. It is possible to project out of the function the sum of those components which belong to a particular irreducible representation, by using the projection operator

$$P^{\Gamma} = \frac{n_{\Gamma}}{g} \sum_{R} \left(\chi^{\Gamma}(R)\right)^{*} R \tag{E-6}$$

where n_Γ is the dimension of the representation Γ. Here R stands for the actual operation of an element of G (in our applications, usually a geometrical symmetry operation), and $\chi(R)$ is the character of the matrix representing R in the irreducible representation Γ. If Γ is multi-dimensional, one can define separate projection operators for each of the basis functions, which are said to belong to different rows of the representation:

$$P^{\Gamma}{}_{i} = \frac{n_{\Gamma}}{g} \sum_{R} \left(D^{\Gamma}(R)\right)_{i,i}{}^{*} R \tag{E-7}$$

Here $D^\Gamma(R)$ is the full matrix representing R in Γ, and obviously $\chi^\Gamma(R) = \sum_i D_{ii}{}^\Gamma(R)$.

In an abelian group all irreducible representations are one dimensional. In this case the character of a matrix is simply equal to the single matrix element. The character of any element in the group is some complex root of the number 1, i.e., of the form $\exp(2\pi iM/N)$ where M,N are integers. N is the order of the element, namely the power of the element which is equal to E, the unit element. M depends on the representation. In a cyclic group $\{A, A^2, ..., A^n = E\}$ $\chi(A) = \exp(2\pi iM/n)$, where M can be used as an index of the irreducible representation.

A **direct product** of two groups G, G' is defined if the two groups have no element in common except for the unit element, they have the same multiplication rule, and their elements commute. Then $G \times G' \equiv \{g_i g'_j | \, g_i \in G, g'_j \in G'\}$ is a group, and G,G' are subgroups of this direct product. The irreducible representations of a direct product can be constructed by taking a direct product of the matrix representations of g_i and g'_j in the irreducible representations of G, G'.

In general, if there is an irreducible representation Γ for a group G, it is trivial to **subduce** it to a representation Γ_H of the subgroup H. One only has to use for Γ_H those matrices of Γ representing the elements of H. However, Γ_H will not be irreducible in the general case. It is also possible, starting from some irreducible reperesentation Δ_H of H, to **induce** a representation Δ of G. Also in this case, the induced representation is not necessarily irreducible. The study of induced and subduced representations makes it possible to relate irreducible representations of G to irreducible representations of H.

Suggested References

References 8-11 of Chapter III.

Subject Index